WORKSHEETS
FOR CLASSROOM OR LAB PRACTICE

with contributions from

Steve Ouellette

James J. Ball

Indiana State University

BEGINNING AND INTERMEDIATE ALGEBRA

FOURTH EDITION

Margaret L. Lial

American River College

John Hornsby

University of New Orleans

Terry McGinnis

Boston San Francisco New York
London Toronto Sydney Tokyo Singapore Madrid
Mexico City Munich Paris Cape Town Hong Kong Montreal

Reproduced by Pearson Addison-Wesley from electronic files supplied by the author.

Publishing as Pearson Addison-Wesley, 75 Arlington Street, Boston, MA 02116.

ISBN-13: 978-0-321-51687-9
ISBN-10: 0-321-51687-7

1 2 3 4 5 6 BRR 10 09 08 07

Name: Date:
Instructor: Section:

Chapter 1 THE REAL NUMBER SYSTEM

1.1 Fractions

Learning Objectives
1 Learn the definition of *factor*.
2 Write fractions in lowest terms.
3 Multiply and divide fractions.
4 Add and subtract fractions.
5 Solve applied problems that involve fractions.
6 Interpret data in a circle graph.

Key Terms
Use the vocabulary terms listed below to complete each statement in exercises 1-15.

proper fraction **improper fraction** **mixed number**

factor **product** **prime** **composite** **lowest terms**

basic principal of fractions **greatest common factor** **reciprocals**

quotient **least common denominator (LCD)** **difference**

circle graph

1. A fraction in which the numerator is less than the denominator is called a(n) ____________________.

2. A(n) ____________________of a given number is any number that divides evenly (without remainder) into the given number.

3. Given several denominators, the smallest number that is divisible by all the denominators is called the ____________________.

4. The ____________________ of a list of integers is the largest common factor of those integers.

5. A fraction in which the numerator is greater than the denominator is called a(n) ____________________.

6. The ____________________ states that, if the numerator and denominator of a fraction are multiplied or divided by the same nonzero number, the value of the fraction is not changed.

7. Pairs of numbers whose product is 1 are called ____________________ of each other.

Name: Date:
Instructor: Section:

8. A(n) __________________ includes a whole number and a fraction written together and is understood to be the sum of the whole number and the fraction.

9. A(n) __________________ number has at least one factor other than itself and 1.

10. A(n) __________________ is a circle divided into sectors, or wedges, whose sizes show the relative magnitudes of the categories of data being represented.

11. The answer to a subtraction problem is called the __________________.

12. A fraction is in __________________ when there are no common factors in the numerator and denominator.

13. The answer to a division problem is called the __________________.

14. The answer to a multiplication problem is called the __________________.

15. A natural number (except 1) is __________________ if it has only 1 and itself as factors.

Objective 1 Learn the definition of *factor*.

Identify the number as prime, composite, *or* neither.

1. 35 — 1.______________

2. 1 — 2.______________

3. 127 — 3.______________

Write the number as the product of prime factors.

4. 48 — 4.______________

5. 104 — 5.______________

6. 196 — 6.______________

Name: Date:
Instructor: Section:

Objective 2 Write fractions in lowest terms.

Write the fraction in lowest terms.

7. $\frac{30}{36}$ **7.**________________

8. $\frac{42}{15}$ **8.**________________

9. $\frac{48}{150}$ **9.**________________

10. $\frac{28}{336}$ **10.**________________

11. $\frac{144}{324}$ **11.**________________

Objective 3 Multiply and divide fractions.

Find the product and write it in lowest terms.

12. $\frac{7}{3} \cdot \frac{6}{14}$ **12.**________________

13. $\frac{12}{15} \cdot \frac{10}{15}$ **13.**________________

14. $4\frac{3}{8} \cdot 5\frac{3}{7}$ **14.**________________

Name: Date:
Instructor: Section:

Find the quotient and write it in lowest terms.

15. $\frac{5}{4} \div \frac{15}{24}$

15.________________

16. $\frac{12}{13} \div 6$

16.________________

17. $9\frac{5}{8} \div 3\frac{1}{2}$

17.________________

Objective 4 Add and subtract fractions.

Find the sum and write it in lowest terms.

18. $\frac{3}{8} + \frac{7}{8}$

18.________________

19. $\frac{5}{12} + \frac{5}{18}$

19.________________

20. $6\frac{3}{5} + 5\frac{1}{2}$

20.________________

Find the difference and write it in lowest terms.

21. $\frac{11}{12} - \frac{5}{12}$

21.________________

22. $\frac{7}{15} - \frac{3}{10}$

22.________________

23. $11\frac{1}{6} - 2\frac{2}{3}$

23.________________

Name: Date:
Instructor: Section:

Objective 5 Solve applied problems that involve fractions.

Solve the problem

24. A triangle has sides of length $\frac{1}{2}$ foot, $1\frac{1}{4}$ feet, and $1\frac{1}{8}$ feet. What is the distance around this triangle?

24.________________

25. Saul sold $\frac{3}{5}$ bushel of potatoes, $\frac{2}{5}$ bushel of apples, $\frac{3}{4}$ bushel of pears, $\frac{1}{4}$ bushel of peppers, and $1\frac{1}{4}$ bushels of tomatoes. How many bushels of fruits and vegetables did he sell?

25.________________

26. Pete has $12\frac{2}{3}$ cords of firewood for sale. If he sells the firewood in face cord lots (a face cord equals $\frac{1}{3}$ of a cord), how many face cords does he have for sale?

26.________________

27. A cake recipe calls for $1\frac{2}{3}$ cups of sugar. A caterer has 20 cups of sugar on hand. How many cakes can she make?

27.________________

28. Marissa bought 8 yards of fabric. She used $1\frac{3}{4}$ yards for a blouse and $3\frac{5}{8}$ yards for a skirt. How many yards were left?

28.________________

Name: Date:
Instructor: Section:

Objective 6 Interpret data in a circle graph.

Use the circle graph to answer the questions.

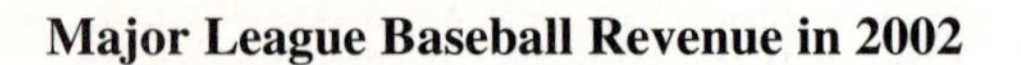

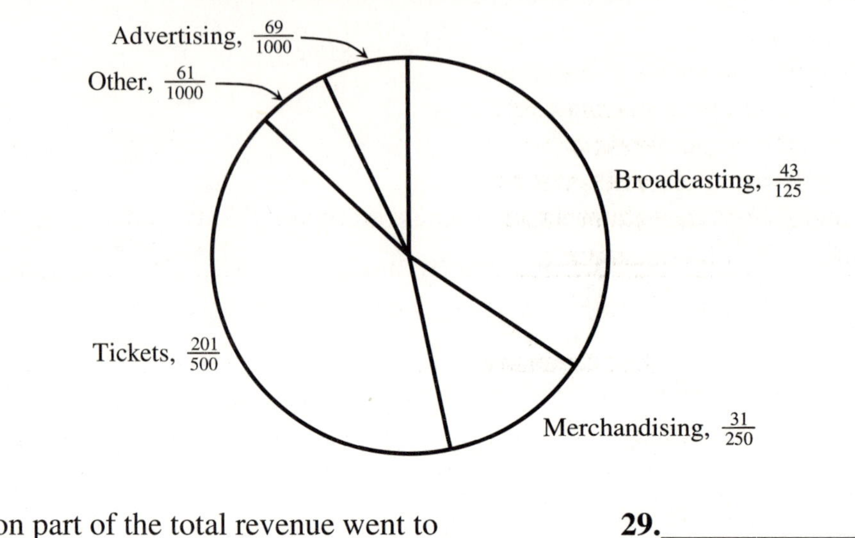

29. What fraction part of the total revenue went to broadcasting?

29.________________

30. If the annual revenue was $2.7 billion:

(a) How much more was spent on tickets than on broadcasting?

(b) How much was spent on areas other than advertising?

(c) How much more was spent on advertising than on "other" expenses?

30.________________

Name: Date:
Instructor: Section:

Chapter 1 THE REAL NUMBER SYSTEM

1.2 Exponents, Order of Operations, and Inequality

Learning Objectives

1. Use exponents.
2. Use the rules for order of operations.
3. Use more than one grouping symbol.
4. Know the meanings of $\neq$, $<$, $>$, $\leq$, and $\geq$.
5. Translate word statements to symbols.
6. Write statements that change the direction of inequality symbols.
7. Interpret data in a bar graph.

Key Terms

Use the vocabulary terms listed below to complete each statement in exercises 1-7.

exponent **base** **exponential expression** **grouping symbols**

order of operations **inequality** **bar graph**

1. A number or letter (variable) written with an exponent is a(n) ____________________.

2. The ____________________ is the number that is a repeated factor when written with an exponent.

3. A(n) ____________________ is a series of bars (or simulations of bars) arranged either vertically or horizontally to show comparisons of data.

4. The ____________________ is used to evaluate expressions containing more than one operation.

5. A(n) ____________________ is a number that indicates how many times a factor is repeated.

6. A(n) ____________________ is a statement that two expressions are not equal.

7. ____________________ are parentheses ( ), brackets [ ], or fraction bars.

Objective 1 Use exponents.

Find the value of each exponential expression.

1. 2^5 **1.**________________

Name: Date:
Instructor: Section:

2. 3^4 2.__________

3. $\left(\frac{2}{5}\right)^3$ 3.__________

4. $\left(\frac{2}{3}\right)^4$ 4.__________

5. $(.4)^2$ 5.__________

Objective 2 Use the rules for order of operations.

Find the value of each expression.

6. $7+2\cdot 4$ 6.__________

7. $3\cdot 15+10^2$ 7.__________

8. $20\div 5-3\cdot 1$ 8.__________

9. $\left(\frac{5}{6}\right)\left(\frac{3}{2}\right)-\left(\frac{1}{3}\right)^2$ 9.__________

10. $\frac{4\cdot 10+9\cdot 2}{2(4-2)}$ 10.__________

Objective 3 Use more than one grouping symbol.

Find the value of each expression.

11. $6[5+3(4)]$ 11.__________

Name: Date:
Instructor: Section:

12. $10+4[2(6)-3]$ **12.** ________

13. $4[3+2(9-2)]$ **13.** ________

14. $19+3[8(5-2)+6]$ **14.** ________

15. $3^3[(6+5)-2^2]$ **15.** ________

Objective 4 Know the meanings of ≠, <, >, ≤, and ≥.

Tell whether each statement is true *or* false.

16. $95>97$ **16.** ________

17. $3\cdot 4\div 2^2 \neq 3$ **17.** ________

18. $3.25 \geq 3.52$ **18.** ________

19. $2[7(4)-3(5)] \geq 45$ **19.** ________

20. $\frac{5+4\cdot 5}{14-2\cdot 3} \geq 2$ **20.** ________

Name: Date:
Instructor: Section:

Objective 5 Translate word statements to symbols.

Write each word statement in symbols.

21. Seven equals thirteen minus five. **21.**________________

22. The sum of nine and thirteen is greater than twenty-one. **22.**________________

23. Six is less than or equal to six. **23.**________________

24. Seven is greater than the quotient of fifteen and five. **24.**________________

Objective 6 Write statements that change the direction of inequality symbols.

Write each statement with the inequality symbol reversed.

25. $9 < 12$ **25.**________________

26. $12 \geq 8$ **26.**________________

27. $\frac{2}{3} < \frac{3}{4}$ **27.**________________

28. $.921 \leq .922$ **28.**________________

Name: Date:
Instructor: Section:

Objective 7 Interpret data in a bar graph.

Use the bar graph to answer the questions. Write your answer using signed numbers.

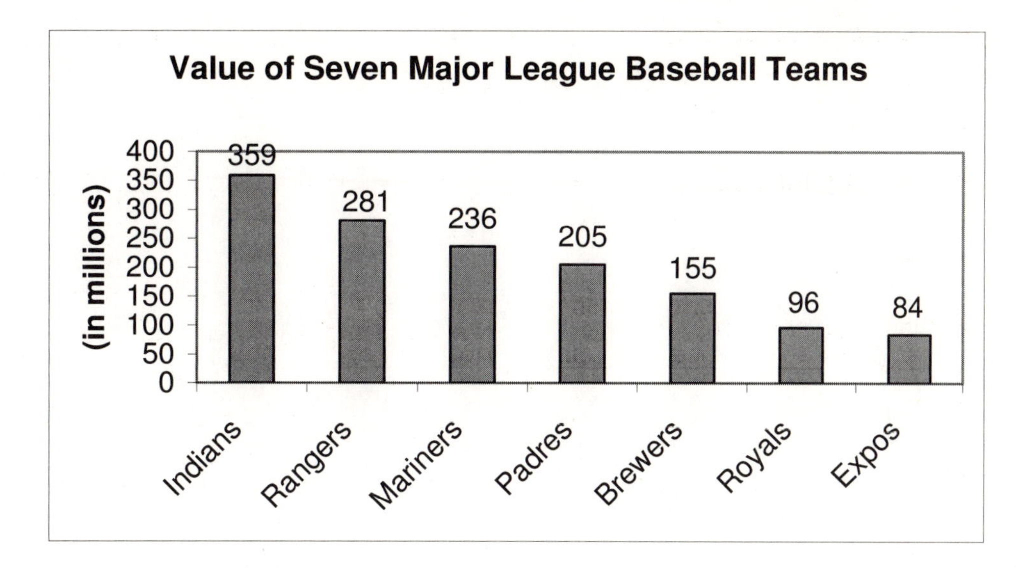

29. What is the difference in value between the Brewers and the Rangers?

29.________________

30. The Indians are worth \$60 million less than the Yankees, who are not on this list. What is the difference in value between the Yankees and the Expos?

30.________________

Name: Date:
Instructor: Section:

Chapter 1 THE REAL NUMBER SYSTEM

1.3 Variables, Expressions, and Equations

Learning Objectives

1 Evaluate algebraic expressions, given values for the variables.
2 Translate word phrases to algebraic expressions.
3 Identify solutions of equations.
4 Identify solutions of equations from a set of numbers.
5 Distinguish between *expressions* and *equations.*

Key Terms

Use the vocabulary terms listed below to complete each statement in exercises 1-6.

variable **algebraic expression** **equation** **solution**

set **elements**

1. A(n) ____________________ is a symbol, usually a letter, used to represent an unknown number.

2. A(n) ____________________ is a statement that two algebraic expressions are equal.

3. A(n) ____________________ is a collection of objects.

4. A(n) ____________________ of an equation is any replacement for the variable that makes the equation true.

5. ____________________ are the objects that belong to a set.

6. A(n) ____________________ is a sequence of numbers, variables, operation symbols, and/or grouping symbols (such as parentheses) formed according to the rules of algebra.

Objective 1 Evaluate algebraic expressions, given values for the variables.

Find the value of each expression if $x = 2$ *and* $y = -4$.

1. $x + 3$ **1.**________________

2. $8x^2 - 6x$ **2.**________________

Name: Date:
Instructor: Section:

3. $2(7x-3y)$ **3.**________________

4. $\dfrac{x^2+y}{x+1}$ **4.**________________

5. $\dfrac{2x+3y}{3x-y+2}$ **5.**________________

6. $\dfrac{3y^2+2x^2}{5x+y^2}$ **6.**________________

Objective 2 Translate word phrases to algebraic expressions.

Change the word phrases to algebraic expressions. Use x as the variable.

7. Four added to a number **7.**________________

8. The product of four less than a number and two **8.**________________

9. The difference between twice a number and 7 **9.**________________

10. Ten times a number, added to 21 **10.**________________

11. The sum of a number and 4 is divided by twice the number **11.**________________

12. Half a number is subtracted from two-thirds of the number **12.**________________

Name: Date:
Instructor: Section:

Objective 3 Identify solutions of equations.

Determine whether the given number is a solution of the equation.

13. $x+4=15$; 11 **13.**________________

14. $6b+2(b+3)=14$; 2 **14.**________________

15. $5+3x^2=19$; 2 **15.**________________

16. $\dfrac{x^2-7}{x}=6$; 2 **16.**________________

17. $3y+5(y-5)=22$; 4 **17.**________________

18. $x^2+2x+1=9$; 3 **18.**________________

Objective 4 Identify solutions of equations from a set of numbers.

Change the word statement to an equation. Use x *as the variable. Then find the solution of the equation from the set* $\{0, 2, 4, 6, 8, 10\}$.

19. The sum of a number and four is ten. **19.**________________

20. The sum of three times a number and five is 23. **20.**________________

21. The product of a number and five is 40. **21.**________________

22. Three times a number is equal to two more than twice the number. **22.**________________

Name: Date:
Instructor: Section:

23. 10 divided by a number is three more than the number. **23.**_______________

24. The quotient of a number and ten is one. **24.**_______________

Objective 5 Distinguish between *expressions* and *equations*.

Decide whether each of the following is an equation or an expression.

25. $3x+2y$ **25.**_______________

26. $8x=2y$ **26.**_______________

27. $9x+2y=2$ **27.**_______________

28. y^2-4y-3 **28.**_______________

29. $\dfrac{x+3}{15}=2x$ **29.**_______________

30. $5x^2+3$ **30.**_______________

Name: Date:
Instructor: Section:

Chapter 1 THE REAL NUMBER SYSTEM

1.4 Real Numbers and the Number Line

Learning Objectives
1 Classify numbers and graph them on number lines.
2 Tell which of two real numbers is less than the other.
3 Find additive inverses and absolute values of real numbers.
4 Interpret the meanings of real numbers from a table of data.

Key Terms
Use the vocabulary terms listed below to complete each statement in exercises 1-7.

number line **signed numbers** **set-builder notation**

coordinate **ordering numbers** **additive inverses** **absolute value**

1. Each number on a number line is called the ____________________ of the point that it labels.

2. The ____________________ of a number is the distance between 0 and the number on a number line.

3. A(n) ____________________ is a line with a scale that is used to show how numbers relate to each other.

4. Two numbers are called ____________________ if their sum is equal to zero.

5. ____________________ is used to describe a set of numbers without actually having to list all the elements.

6. A number line can be used when ____________________. If one number lies to the left of another number on a number line, then that number is less than the other number. Likewise, if one number lies to the right of another number on a number line, then that number is the greater of the two numbers.

7. ____________________ are numbers that can be written with a positive or negative sign.

Objective 1 Classify numbers and graph them on number lines.

List all the sets among the following to which the number belongs: natural numbers, whole numbers, integers, rational numbers, irrational numbers, real numbers.

1. 5 **1.**________________

Name: Date:
Instructor: Section:

2. $\frac{4}{5}$ 2.________________

3. -2.6 3.________________

4. $\sqrt{3}$ 4.________________

5. 11.45 5.________________

Graph each group of rational numbers on a number line.

6. $-2, -1, 0, 2, 4$ 6.________________

7. $3, 5, -1, -3$ 7.________________

8. $\frac{1}{2}, 0, -3, -\frac{5}{2}$ 8.________________

9. $2\frac{1}{3}, 4, 6\frac{2}{3}, 8$ 9.________________

10. $-4.5, -2.3, 1.7, 3.5$ 10.________________

Objective 2 Tell which of two real numbers is less than the other.

Select the smaller number in each pair.

11. $-15, -12$ 11.________________

12. $-.802, -.820$ 12.________________

Name: Date:
Instructor: Section:

13. $\frac{2}{3}, -\frac{1}{2}$ **13.**________________

Decide whether each statement is true *or* false.

14. $-76 < 45$ **14.**________________

15. $-5 > -5$ **15.**________________

16. $-12 > -10$ **16.**________________

Objective 3 Find additive inverses and absolute values of real numbers.

Find the additive inverse of the number.

17. -15 **17.**________________

18. $\frac{5}{8}$ **18.**________________

19. $-2\frac{5}{8}$ **19.**________________

Simplify by removing absolute value symbols.

20. $|-4|$ **20.**________________

21. $-|95|$ **21.**________________

22. $-|-25|$ **22.**________________

23. $-|49-39|$ **23.**________________

Name: Date:
Instructor: Section:

24. $|-7.52|$ **24.**________________

25. $\left|1\frac{1}{2}-2\frac{1}{4}\right|$ **25.**________________

26. $-\left|2\frac{3}{8}-4\frac{3}{4}\right|$ **26.**________________

Objective 4 Interpret the meanings of real numbers from a table of data.

Use the table below of counties in the state of Indiana to answer the questions.

County Name	1997 Estimated Population	1998 Estimated Population	Numeric Population Change 1997 – 1998
Allen	311928	314218	2290
Benton	9661	9725	64
Delaware	117520	116828	−692
Elkhart	170668	172310	1642
Henry	48896	48785	−111
Lake	478536	478323	−213
Monroe	116612	115130	−1482
Union	7320	7263	−57

27. Which county experienced the greatest population growth? **27.**________________

28. Which county experienced the greatest decline in population? **28.**________________

29. By how many more people did the population of Elkhart County change compared with Henry County? **29.**________________

30. By how many more people did the population of Union County change compared with Delaware County? **30.**________________

Name: Date:
Instructor: Section:

Chapter 1 THE REAL NUMBER SYSTEM

1.5 Adding and Subtracting Real Numbers

Learning Objectives

1. Add two numbers with the same sign.
2. Add positive and negative numbers.
3. Use the definition of subtraction.
4. Use the rules for order of operations with real numbers.
5. Interpret words and phrases involving addition and subtraction.
6. Use signed numbers to interpret data.

Key Terms

Use the vocabulary terms listed below to complete each statement in exercises 1-4.

same sign **different signs** **difference** **additive inverses**

1. Two numbers that are the same distance from, but on opposite sides, of 0 on a number line are called ____________________.

2. To add two numbers with ____________________, find the absolute values of the numbers and subtract the smaller absolute value from the larger. Give the answer the sign of the number having the larger absolute value.

3. The answer to a subtraction problem is called the ____________________.

4. To add numbers with the ____________________, add the absolute values of the numbers. The sum has the same sign as the given numbers.

Objective 1 Add two numbers with the same sign.

Find the sum.

1. $8+7$ **1.**________________

2. $9+12$ **2.**________________

3. $-4+(-6)$ **3.**________________

4. $-7+(-11)$ **4.**________________

Name: Date:
Instructor: Section:

Objective 2 Add positive and negative numbers.

Find the sum.

5. $7+(-5)$ **5.**________

6. $9+(-16)$ **6.**________

7. $\frac{7}{12}+\left(-\frac{3}{4}\right)$ **7.**________

8. $-\frac{4}{7}+\frac{3}{5}$ **8.**________

9. $3\frac{5}{8}+\left(-2\frac{1}{4}\right)$ **9.**________

10. $14.1+(-14.1)$ **10.**________

Objective 3 Use the definition of subtraction.

Find each difference.

11. $14-20$ **11.**________

12. $32-36$ **12.**________

13. $-9-4$ **13.**________

14. $13-(-9)$ **14.**________

15. $-4-(-20)$ **15.**________

Name: Date:
Instructor: Section:

16. $4.5-(-2.8)$ **16.**________________

17. $-\frac{3}{10}-\left(-\frac{4}{15}\right)$ **17.**________________

18. $3\frac{3}{4}-\left(-2\frac{1}{8}\right)$ **18.**________________

Objective 4 Use the rules for order of operations with real numbers.

Find the sum.

19. $6+[2+(-7)]$ **19.**________________

20. $-9+[5+(-19)]$ **20.**________________

21. $-2+[-16+(-2)]$ **21.**________________

22. $[(-7)+14]+[(-16)+3]$ **22.**________________

23. $-7.6+[5.2]+(-11.4)$ **23.**________________

24. $-\frac{4}{5}+\left[\frac{1}{4}+\left(-\frac{2}{3}\right)\right]$ **24.**________________

Objective 5 Interpret words and phrases involving addition and subtraction.

Write a numerical expression for the phrase, and then simplify the expression.

25. 10 added to the sum of -2 and -3. **25.**________________

Name: Date:
Instructor: Section:

26. 6 more than -2, increased by 8 **26.**________________

27. -6 subtracted from the sum of 2 and -3 **27.**________________

Objective 6 Use signed numbers to interpret data.

Solve the problem by writing a sum or difference of real numbers and adding or subtracting. No variables are needed.

28. A mountain climber starts to climb at an altitude of 4325 feet. He climbs so that he gains 208 feet in altitude. Then he finds that, because of an obstruction, he must descend 25 feet. Then he climbs 58 feet up. What is his final altitude? **28.**________________

29. At 1:00 A.M., the temperature on the top of Mt. Washington in New Hampshire was $-16°$ F. At 11:00 A.M., the temperature was $25°$ F. What was the rise in temperature? **29.**________________

30. Marlene owes $342.58 on her VISA account. She makes additional purchases which total $173.76. Express her new balance as a negative number. **30.**________________

Name: Date:
Instructor: Section:

Chapter 1 THE REAL NUMBER SYSTEM

1.6 Multiplying and Dividing Real Numbers

Learning Objectives
1 Find the product of a positive number and a negative number.
2 Find the product of two negative numbers.
3 Identify factors of integers.
4 Use the reciprocal of a number to apply the definition of division.
5 Use the rules for order of operations when multiplying and dividing signed numbers.
6 Evaluate expressions involving variables.
7 Interpret words and phrases involving multiplication and division.
8 Translate simple sentences into equations.

Key Terms
Use the vocabulary terms listed below to complete each statement in exercises 1-4.

product **negative** **positive** **multiplicative inverse**

1. The answer to a multiplication problem is called the ___________________.

2. The product or quotient of two numbers with different signs is ___________________.

3. The ___________________ of a nonzero real number a is $\frac{1}{a}$.

4. The product or quotient of two numbers with the same sign is ___________________.

Objective 1 Find the product of a positive number and a negative number.

Find the product.

1. $7(-4)$ **1.**________________

2. $(-80)(4)$ **2.**________________

3. $7(-2.5)$ **3.**________________

Objective 2 Find the product of two negative numbers.

Find the product.

4. $(-3)(-4)$ **4.**________________

Name: Date:
Instructor: Section:

5. $(-13)(-14)$ **5.**________________

6. $\left(-\frac{2}{7}\right)\left(-\frac{14}{5}\right)$ **6.**________________

Objective 3 Identify factors of integers.

Find all the integer factors of the given number.

7. 40 **7.**________________

8. 36 **8.**________________

Objective 4 Use the reciprocal of a number to apply the definition of division.

Use the definition of division to find each quotient.

9. $\frac{-120}{-20}$ **9.**________________

10. $-\frac{3}{16}\div\frac{9}{8}$ **10.**________________

11. $-\frac{27}{35}\div\left(-\frac{9}{5}\right)$ **11.**________________

Objective 5 Use the rules for order of operations when multiplying and dividing signed numbers.

Perform the indicated operations.

12. $(-4)(9)-(-5)(4)$ **12.**________________

13. $-7\left[-4-(-2)(-3)\right]$ **13.**________________

Name: Date:
Instructor: Section:

Simplify the numerators and denominators separately. Then find the quotients.

14. $\frac{9(-4)}{-6-(-2)}$ **14.**________________

15. $\frac{-3-(-4+1)}{-7-(-6)}$ **15.**________________

16. $\frac{-4[8-(-3+7)]}{-6[3-(-2)]-3(-3)}$ **16.**________________

17. $\frac{2^2+4^2}{5^2-3^2}$ **17.**________________

Objective 6 Evaluate expressions involving variables.

Evaluate the following expressions if $x=-3$, $y=2$, *and* $a=4$.

18. $-x+2y-3a$ **18.**________________

19. $(x-2)(4-y)$ **19.**________________

20. $(-2y+4a)-(3x+y)$ **20.**________________

21. $\frac{2x^2-3y}{4a}$ **21.**________________

Name: Date:
Instructor: Section:

Objective 7 Interpret words and phrases involving multiplication and division.

Write a numerical expression for each phrase and simplify.

22. The product of 7 and –2, added to 4 **22.**________________

23. The product of 10 and –2, subtracted from –2 **23.**________________

24. 70% of the sum of 20 and –4 **24.**________________

25. –34 subtracted from two-thirds of the sum of 16 and –10 **25.**________________

26. The sum of –12 and the quotient of 49 and –7 **26.**________________

27. The product of –4 and 7, divided by the sum of –3 and 14 **27.**________________

Objective 8 Translate simple sentences into equations.

Write each statement in symbols, using x as the variable.

28. The quotient of a number and –2 is –9. **28.**________________

29. The product of a number and –1 is 7. **29.**________________

30. When a number is divided by –4, the result is 1. **30.**________________

Name: Date:
Instructor: Section:

Chapter 1 THE REAL NUMBER SYSTEM

1.7 Properties of Real Numbers

Learning Objectives
1 Use the commutative properties.
2 Use the associative properties.
3 Use the identity properties.
4 Use the inverse properties.
5 Use the distributive property.

Key Terms
Use the vocabulary terms listed below to complete each statement in exercises 1-5.

commutative property **associative property** **identity property**

inverse property **distributive property**

1. For any real numbers *a*, *b*, and *c*, the ____________________ states that $a(b+c)=ab+ac$ and $(b+c)a=ba+ca$.

2. The ____________________ states that the sum of 0 and any number equals the number, and the product of 1 and any number equals the number.

3. The ____________________ states that the way in which numbers being added (or multiplied) are grouped does not change the sum (or product).

4. The ____________________ states that a number added to its opposite is 0 and a number multiplied by its reciprocal is 1.

5. The ____________________states that the order of the numbers in an addition problem (or multiplication problem) can be changed without changing the sum (or product).

Objective 1 Use the commutative properties.

Complete each statement. Use a commutative property.

1. $y+4=$ _____ $+\,y$ **1.**________________

2. $5(2)=$ _____ (5) **2.**________________

3. $(ab)(2)=(2)$ _____ **3.**________________

Name: Date:
Instructor: Section:

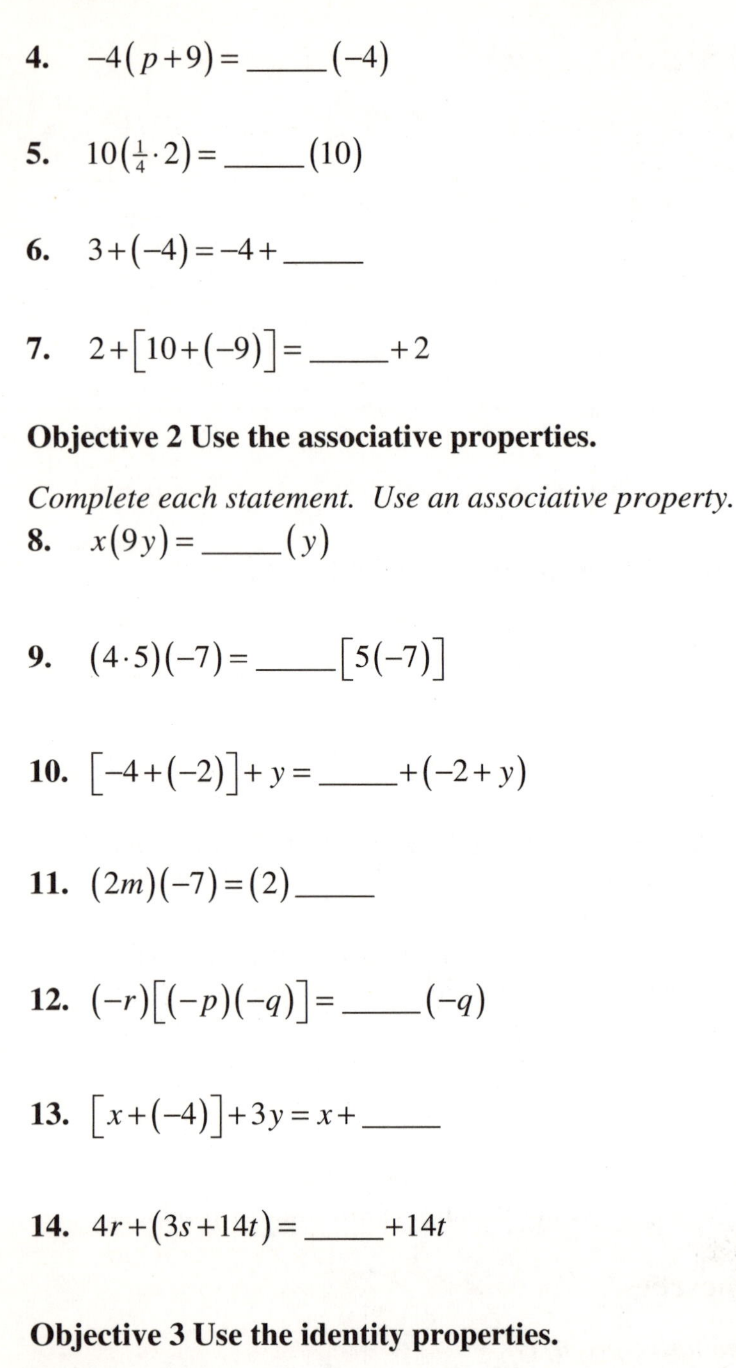

4. $-4(p+9) = \underline{\qquad}(-4)$ 4.____________

5. $10(\frac{1}{4} \cdot 2) = \underline{\qquad}(10)$ 5.____________

6. $3 + (-4) = -4 + \underline{\qquad}$ 6.____________

7. $2 + [10 + (-9)] = \underline{\qquad} + 2$ 7.____________

Objective 2 Use the associative properties.

Complete each statement. Use an associative property.

8. $x(9y) = \underline{\qquad}(y)$ 8.____________

9. $(4 \cdot 5)(-7) = \underline{\qquad}[5(-7)]$ 9.____________

10. $[-4 + (-2)] + y = \underline{\qquad} + (-2 + y)$ 10.____________

11. $(2m)(-7) = (2)\underline{\qquad}$ 11.____________

12. $(-r)[(-p)(-q)] = \underline{\qquad}(-q)$ 12.____________

13. $[x + (-4)] + 3y = x + \underline{\qquad}$ 13.____________

14. $4r + (3s + 14t) = \underline{\qquad} + 14t$ 14.____________

Objective 3 Use the identity properties.

Simplify.

15. $-7 + 0 =$ 15.____________

16. $1(-4) =$ 16.____________

17. $7(1) =$ 17.____________

Name: Date:
Instructor: Section:

Objective 4 Use the inverse properties.

Complete the statements so that they are examples of either an identity property or an inverse property. Identify which property is used.

18. $-4+\_\_\_\_\_=0$ **18.**________________

19. $\_\_\_\_\_+\frac{1}{7}=0$ **19.**________________

20. $1\cdot\_\_\_\_\_=1$ **20.**________________

21. $\frac{2}{7}\cdot\_\_\_\_\_=1$ **21.**________________

22. $-\frac{3}{5}\cdot\_\_\_\_\_=1$ **22.**________________

23. $-14+\_\_\_\_\_=0$ **23.**________________

Objective 5 Use the distributive property.

Use the distributive property to rewrite each expression. In Exercises 24 and 25, simplify the result.

24. $6y+7y$ **24.**________________

25. $10r-4r$ **25.**________________

26. $3(a+b)$ **26.**________________

27. $4c-4d$ **27.**________________

28. $n(2a-4b+6c)$ **28.**________________

29. $-2(5y-9z)$ **29.**________________

30. $-14x+(-14y)$ **30.**________________

Name: Date:
Instructor: Section:

Chapter 1 THE REAL NUMBER SYSTEM

1.8 Simplifying Expressions

Learning Objectives
1 Simplify expressions.
2 Identify terms and numerical coefficients.
3 Identify like terms.
4 Combine like terms.
5 Simplify expressions from word phrases.

Key Terms
Use the vocabulary terms listed below to complete each statement in exercises 1-5.

term **numerical coefficient** **like terms** **unlike terms**

combining like terms

1. The numerical factor in a term is its ____________________.

2. ____________________ is a method of adding or subtracting like terms by using the properties of real numbers.

3. A(n) ____________________ is a number, a variable, or the product or quotient of a number and one or more variables raised to powers.

4. Terms with exactly the same variables raised to exactly the same powers are called ____________________.

5. ____________________ are terms that do not have the same variable or terms with the same variables but whose variables are not raised to the same powers.

Objective 1 Simplify expressions.

Simplify each expression.

1. $14+3y-8$ **1.**________________

2. $4(2x+5)+7$ **2.**________________

3. $-(9-4b)-8$ **3.**________________

Name: Date:
Instructor: Section:

4. $11-(d-2)+(-6)$ **4.**________________

5. $-2(-5x+2)+7$ **5.**________________

6. $4(-6p-2)+2-4$ **6.**________________

Objective 2 Identify terms and numerical coefficients.

Give the numerical coefficient of each term.

7. $4x$ **7.**________________

8. $-7a^2$ **8.**________________

9. $.3a^2b$ **9.**________________

10. $-\frac{5}{9}v^6w^4$ **10.**________________

Objective 3 Identify like terms.

Identify each group of terms as like *or* unlike.

11. $2x, 7x$ **11.**________________

12. $9, -2a$ **12.**________________

13. $-7q^2, 2q^2$ **13.**________________

14. $2w, 4w, -w$ **14.**________________

Name: Date:
Instructor: Section:

15. $4x, -10x^2, -9x^2$ 15.________________

16. $2, -4, 16$ 16.________________

Objective 4 Combine like terms.

Simplify each expression by combining like terms.

17. $12-4x-2-7x$ 17.________________

18. $4a^2-4a^3-2a^2+7a^3$ 18.________________

19. $\frac{7}{10}r+\frac{3}{10}s-\frac{2}{5}r-\frac{4}{5}s$ 19.________________

Use the distributive property and combine like terms to simplify the following expressions.

20. $2(3x+5)$ 20.________________

21. $7r-(2r+4)$ 21.________________

22. $-6(a+2)+4(2a-1)$ 22.________________

23. $2(4x-1)-(5x+2)$ 23.________________

24. $2.5(3y+1)-4.5(2y-3)$ 24.________________

Name: Date:
Instructor: Section:

Objective 5 Simplify expressions from word phrases.

Write each phrase as a mathematical expression and simplify by combining like terms. Use x as the variable.

25. Seven times a number, added to twice the number **25.**________________

26. The sum of seven times a number and 2, subtracted from three times the number **26.**________________

27. The difference between five times a number and 3, added to four times the sum of the number and 2 **27.**________________

28. The sum of ten times a number and 7, subtracted from the difference between 2 and nine times the number **28.**________________

29. Twelve times the difference between 4 and twice a number, subtracted from 10 **29.**________________

30. Four times the difference between twice a number and –10, subtracted from three times the sum of –7 and five times the number **30.**________________

Name: Date:
Instructor: Section:

Chapter 2 LINEAR EQUATIONS AND INEQUALITIES IN ONE VARIABLE

2.1 The Addition Property of Equality

Learning Objectives
1 Identify linear equations.
2 Use the addition property of equality.
3 Simplify and then use the addition property of equality.

Key Terms
Use the vocabulary terms listed below to complete each statement in exercises 1-3.

solution set **equivalent equations** **addition property of equality**

1. The ____________________ states that the same number can be added to (or subtracted from) both sides of an equation to obtain an equivalent equation.

2. The ____________________ is the set of all solutions to a particular equation.

3. ____________________ are equations that have the same solution set.

Objective 1 Identify linear equations.

Tell whether each of the following is a linear equation.

1. $9x+2=0$ **1.**________________

2. $3x^2+4x+3=0$ **2.**________________

3. $7x^2=10$ **3.**________________

4. $3x^3=2x^2+5x$ **4.**________________

5. $\frac{5}{x}-\frac{3}{2}=0$ **5.**________________

6. $4x-2=12x+9$ **6.**________________

Name: Date:
Instructor: Section:

Objective 2 Use the addition property of equality.

Solve each equation by using the addition property of equality. Check each solution.

7. $y-4=16$ **7.**________________

8. $r+9=8$ **8.**________________

9. $4x+2=5x+7$ **9.**________________

10. $6y=7y-1$ **10.**________________

11. $p-\frac{2}{3}=\frac{5}{6}$ **11.**________________

12. $y+4\frac{1}{2}=3\frac{3}{4}$ **12.**________________

13. $\frac{2}{3}t-5=\frac{5}{3}t$ **13.**________________

14. $\frac{9}{8}p-\frac{1}{2}=\frac{1}{8}p$ **14.**________________

Name: Date:
Instructor: Section:

15. $5.7x+12.8=4.7x$ **15.**________________

16. $9.5y-2.4=10.5y$ **16.**________________

Objective 3 Simplify and then use the addition property of equality.

Solve each equation. First simplify each side of the equation as much as possible. Check each solution.

17. $6x-5x+2=-2$ **17.**________________

18. $6x+3x-7x+4=10$ **18.**________________

19. $3(t+3)-(2t+7)=9$ **19.**________________

20. $5x+2(2x+1)-(8x-1-2)=5\frac{1}{4}$ **20.**________________

21. $-4(5g-7)+3(8g-3)=15-4+3g$ **21.**________________

Name: Date:
Instructor: Section:

22. $10x+4x-11x+4-7=2-4x-3+8x$

22.________________

23. $4(3a-2)-7(2+a)=4(a-5)$

23.________________

24. $2(4t+6)-3(2t-3)=-3(3t-4)+5+10t$

24.________________

25. $-7(1+2b)-6(3-5b)=5(4+3b)-45$

25.________________

26. $8(2-4b)+3(5-b)=4(1-9b)+22$

26.________________

27. $\frac{8}{5}t+\frac{1}{3}=\frac{5}{6}+\frac{3}{5}t-\frac{1}{6}$

27.________________

28. $\frac{5}{12}+\frac{7}{6}s-\frac{1}{6}=\frac{5}{6}s+\frac{1}{4}-\frac{2}{3}s$

28.________________

29. $3.6p+4.8+4.0p=8.6p-3.1+.7$

29.________________

30. $.03x+.6+.09x-.9=2.1$

30.________________

Name: Date:
Instructor: Section:

Chapter 2 LINEAR EQUATIONS AND INEQUALITIES IN ONE VARIABLE

2.2 The Multiplication Property of Equality

Learning Objectives
1 Use the multiplication property of equality.
2 Combine terms in equations, and then use the multiplication property of equality.

Key Terms
Use the vocabulary terms listed below to complete each statement in exercises 1-3.

multiplication property of equality **reciprocal** **coefficient**

1. The ____________________ states that the same nonzero number can be multiplied by (or divided into) both sides of an equation to obtain an equivalent equation.

2. A ____________________ is the numerical factor of a term.

3. To eliminate a fractional coefficient, it is usually easier to multiply by the ____________________ of the fraction.

Objective 1 Use the multiplication property of equality.

Solve each equation and check your solution.

1. $8x = 24$ 1.________________

2. $-3w = 42$ 2.________________

3. $-16a = -48$ 3.________________

4. $\frac{b}{5} = 4$ 4.________________

5. $\frac{3p}{7} = -6$ 5.________________

6. $\frac{b}{-2} = 21$ 6.________________

Name: Date:
Instructor: Section:

7. $\frac{3}{4}r=-27$ 7.________________

8. $-\frac{7}{2}t=-4$ 8.________________

9. $\frac{6}{7}y=\frac{2}{3}$ 9.________________

10. $.9x=5.4$ 10.________________

11. $2.1a=9.03$ 11.________________

12. $7.5p=-61.5$ 12.________________

13. $-2.7v=-17.28$ 13.________________

14. $-5.9y=-21.24$ 14.________________

Objective 2 Combine terms in equations, and then use the multiplication property of equality.

Solve each equation and check your solution.

15. $4r+3r=63$ 15.________________

16. $3x+6x=72$ 16.________________

17. $7y-2y=45$ 17.________________

18. $10a-7a=-24$ 18.________________

Name: Date:
Instructor: Section:

19. $4v+3v+7v=98$ **19.**________________

20. $8f+4f-3f=72$ **20.**________________

21. $-y=3.9$ **21.**________________

22. $-t=-26$ **22.**________________

23. $-h=\frac{7}{4}$ **23.**________________

24. $3b-4b=8$ **24.**________________

25. $9p-10p=-18$ **25.**________________

26. $3w-7w=20$ **26.**________________

27. $7q-10q=-24$ **27.**________________

28. $4x-8x+2x=16$ **28.**________________

29. $2f+3f-7f=48$ **29.**________________

30. $-11h-6h+14h=-21$ **30.**________________

Name: Date:
Instructor: Section:

Chapter 2 LINEAR EQUATIONS AND INEQUALITIES IN ONE VARIABLE

2.3 More on Solving Linear Equations

Learning Objectives
1 Learn and use the four steps for solving a linear equation.
2 Solve equations with fractions or decimals as coefficients.
3 Solve equations with no solution or infinitely many solutions.
4 Write expressions for two related unknown quantities.

Key Terms
Use the vocabulary terms listed below to complete each statement in exercises 1-6.

simplify each side separately **LCD** **conditional equation** **identity**

contradiction **empty set**

1. A(n) ____________________ is true for some replacements of the variable and false for others.

2. You can clear fractions from an equation by multiplying both sides of the equation by the ____________________ of all the fractions.

3. A(n) ____________________ is an equation that is never true. It has no solution.

4. When solving a linear equation, the first step is to ____________________.

5. The ____________________, denoted by { } or Ø, is the set containing no elements.

6. A(n) ____________________ is an equation that is true for all replacements of the variable. It has an infinite number of solutions.

Objective 1 Learn and use the four steps for solving a linear equation.

Solve each equation and check your solution.

1. $7t+6=11t-4$ **1.**________________

2. $7x+11=9x+25$ **2.**________________

Name: Date:
Instructor: Section:

3. $7j+1=10j-29$ **3.**________________

4. $4+x=-(x+6)$ **4.**________________

5. $4(z-2)-(3z-1)=2z-6$ **5.**________________

6. $3(x+4)=6-2(x-8)$ **6.**________________

7. $4w-5w+3(w-7)=-4(w+4)+7$ **7.**________________

8. $3a-6a+4(a-4)=-2(a+2)$ **8.**________________

9. $4r-3(3r-2)=8-3(r-4)$ **9.**________________

Name: Date:
Instructor: Section:

Objective 2 Solve equations with fractions or decimals as coefficients.

Solve each equation.

10. $\frac{1}{5}(z-5)=\frac{1}{3}(z+2)$ **10.**________________

11. $\frac{3}{8}x-\frac{1}{3}x=\frac{1}{12}$ **11.**________________

12. $\frac{2}{3}y-\frac{1}{4}y=-\frac{5}{12}y+\frac{1}{2}$ **12.**________________

13. $\frac{5}{6}(r-2)-\frac{2}{9}(r+4)=\frac{7}{18}$ **13.**________________

14. $\frac{1}{2}r+\frac{5}{14}r=r-\frac{4}{7}$ **14.**________________

15. $.90x=.40(30)+.15(100)$ **15.**________________

16. $.35(20)+.45y=.125(200)$ **16.**________________

17. $.12x+.24(x-5)=.56x$ **17.**________________

Name: Date:
Instructor: Section:

18. $.45a - .35(20 - a) = .02(50)$

18.________________

Objective 3 Solve equations with no solution or infinitely many solutions.

Solve each equation.

19. $3(6x - 7) = 2(9x - 6)$

19.________________

20. $4x + 15 = 4(x - 6)$

20.________________

21. $6y - 3(y + 2) = 3(y - 2)$

21.________________

22. $-1 - (2 + y) = -(-4 + y)$

22.________________

23. $8k + 14 = 2(k + 2) + 3(2k + 1)$

23.________________

24. $4(2p - 3) - 3(3p + 1) = -18 - p + 3$

24.________________

Name: Date:
Instructor: Section:

25. $7y-11=6(2y+3)-5y$ **25.**________________

26. $8(2d-4)-3(7d+8)=-5(d+2)$ **26.**________________

27. $4(4-4h)+2(3+3h)=12+5(2-2h)$ **27.**________________

Objective 4 Write expressions for two related unknown quantities.

Write an expression for the two related unknown quantities.

28. Two numbers have a sum of 36. One is m. Find the other number. **28.**________________

29. The product of two numbers is 17. One number is p. What is the other number? **29.**________________

30. A cashier has q dimes. Find the value of the dimes in cents. **30.**________________

Name: Date:
Instructor: Section:

Chapter 2 LINEAR EQUATIONS AND INEQUALITIES IN ONE VARIABLE

2.4 An Introduction to Applications of Linear Equations

Learning Objectives

1 Learn the six steps for solving applied problems.
2 Solve problems involving unknown numbers.
3 Solve problems involving sums of quantities.
4 Solve problems involving supplementary and complementary angles.
5 Solve problems involving consecutive integers.

Key Terms

Use the vocabulary terms listed below to complete each statement in exercises 1-6.

solving an applied problem **degree** **complementary** **supplementary**

straight angle **consecutive integers**

1. ____________________ angles are angles whose measures have a sum of 90°.

2. One ____________________ is a basic unit of measure for angles equal to $\frac{1}{360}$ of a complete revolution.

3. ____________________ angles are angles whose measures have a sum of 180°.

4. You should use the six-step method when ____________________.

5. A ____________________ measures 180°.

6. Two integers that differ by 1 are called ____________________.

Objective 2 Solve problems involving unknown numbers.

Write an equation for each of the following and then solve the problem. Use x as the variable.

1. If 4 is added to 3 times a number, the result is 7. Find the number. **1.**________________

Name: Date:
Instructor: Section:

2. If 2 is subtracted from four times a number, the result is 3 more than six times the number. What is the number?

2.________________

3. If –2 is multiplied by the difference between 4 and a number, the result is 24. Find the number.

3.________________

4. Six times the difference between a number and 4 equals the product of the number and –2. Find the number.

4.________________

5. When the difference between a number and 4 is multiplied by –3, the result is two more than –5 times the number. Find the number.

5.________________

6. If four times a number is added to 7, the result is five less than six times the number. Find the number.

6.________________

7. If a number is subtracted from 83, the result is 19 more than 37. Find the number.

7.________________

8. If seven times a number is added to 3, the result is two less than eight times the number. Find the number.

8.________________

Name: Date:
Instructor: Section:

Objective 3 Solve problems involving sums of quantities.

Solve each problem.

9. A rope 116 inches long is cut into three pieces. The middle-sized piece is 10 inches shorter than twice the shortest piece. The longest piece is $\frac{5}{3}$ as long as the shortest piece. What is the length of the shortest piece?

9.________________

10. George and Al were opposing candidates in the school board election. George received 21 more votes than Al, with 439 votes cast. How many votes did Al receive?

10.________________

11. On a psychology test, the highest grade was 38 points more than the lowest grade. The sum of the two grades was 142. Find the lowest grade.

11.________________

12. Mount McKinley in Alaska is 5910 feet higher than Mount Rainier in Washington. Together, their heights total 34,730 feet. How high is each mountain?

12.________________

13. Charles bought five general admission tickets and four student tickets for a movie. He paid $35.25. If each student ticket cost $3.50, how much did each general admission ticket cost?

13.________________

Name: Date:
Instructor: Section:

14. Penny is making punch for a party. The recipe requires twice as much orange juice as cranberry juice and 8 times as much ginger ale as cranberry juice. If she plans to make 176 ounces of punch, how much of each ingredient should she use?

14.________________

15. Pablo, Faustino, and Mark swim at a public pool each day for exercise. One day Pablo swam five more than three times as many laps as Mark, and Faustino swam four times as many laps as Mark. If the men swam 29 laps altogether, how many laps did each one swim?

15.________________

16. Linda wishes to build a rectangular dog pen using 52 feet of fence and the back of her house, which is 36 feet long to enclose the pen. How wide will the dog pen be if the pen is 36 feet long?

16.________________

Objective 4 Solve problems involving supplementary and complementary angles.

Solve each problem.

17. Find the measure of an angle if the measure of the angle is 8° less than three times the measure of its supplement.

17.________________

18. Find the measure of an angle whose supplement measures 20° more than twice its complement.

18.________________

Name: Date:
Instructor: Section:

19. Find the measure of an angle such that the difference between the measure of its supplement and twice the measure of its complement is 49°.

19.________________

20. Find the measure of an angle whose complement is 9° more than twice its measure.

20.________________

21. Find the measure of an angle whose supplement measures 6° more than 7 times its complement.

21.________________

22. Find the measure of an angle such that the difference between the measures of an angle and its complement is 20°.

22.________________

23. Find the measure of an angle if its supplement measures 4° less than three times its complement.

23.________________

24. Find the measure of an angle such that the difference between the measure of the angle and the measure of its complement is 28°.

24.________________

Name: Date:
Instructor: Section:

Objective 5 Solve problems involving consecutive integers.

Solve each problem.

25. Find two consecutive even integers whose sum is 154.

25.________________

26. Find two consecutive even integers such that the smaller, added to twice the larger, is 292.

26.________________

27. Find two consecutive integers such that the larger, added to three times the smaller, is 109.

27.________________

28. Find two consecutive odd integers such that if three times the smaller is added to twice the larger, the sum is 69.

28.________________

29. Find two consecutive odd integers such that the larger, added to eight times the smaller, equals 119.

29.________________

30. Find three consecutive odd integers whose sum is 363.

30.________________

Name: Date:
Instructor: Section:

Chapter 2 LINEAR EQUATIONS AND INEQUALITIES IN ONE VARIABLE

2.5 Formulas and Applications from Geometry

Learning Objectives
1 Solve a formula for one variable, given the values of the other variables.
2 Use a formula to solve an applied problem.
3 Solve problems involving vertical angles and straight angles.
4 Solve a formula for a specified variable.

Key Terms
Use the vocabulary terms listed below to complete each statement in exercises 1-5.

formula **area** **perimeter** **vertical angles**

solving a literal equation

1. A(n) ____________________ is an equation in which letters are used to describe relationships.

2. ____________________ is a measure of the surface covered by a two-dimensional (flat) figure.

3. When two intersecting lines are drawn, the angles that lie opposite each other have the same measure and are called ____________________.

4. The ____________________ of a two-dimensional figure is the measure of the distance around the outside edges of the figure – that is, the sum of the lengths of its sides.

5. The process of solving a formula for a specified variable is called ____________________.

Objective 1 Solve a formula for one variable, given the values of the other variables.

In the following exercises, a formula is given, along with the values of all but one of the variables in the formula. Find the value of the variable that is not given.

1. $V = LWH$; $L = 2$, $W = 4$, $H = 3$ 1.________________

2. $A = \frac{1}{2}bh$; $b = 8$, $h = 2.5$ 2.________________

Name: Date:
Instructor: Section:

3. $V = \frac{1}{3}Bh$; $B = 27$, $V = 63$ **3.**________________

4. $A = \pi r^2$; $r = 3$, $\pi = 3.14$ **4.**________________

5. $I = prt$; $I = 288$, $r = .04$, $t = 3$ **5.**________________

6. $C = \frac{5}{9}(F - 32)$; $F = 104$ **6.**________________

7. $A = \frac{1}{2}(b + B)h$; $b = 6$, $B = 16$, $A = 132$ **7.**________________

8. $V = \frac{4}{3}\pi r^3$; $r = 3$, $\pi = 3.14$ **8.**________________

Objective 2 Use a formula to solve an applied problem.

Use a formula to write an equation for each of the following applications; then solve the application.

9. Find the length of a rectangular garden if its perimeter is 96 feet and its width is 12 feet. **9.**________________

10. Find the height of a triangular banner whose area is 48 square inches and base is 12 inches. **10.**________________

Name: Date:
Instructor: Section:

11. A water tank is a right circular cylinder. The tank has a radius of 6 meters and a volume of 1356.48 cubic meters. Find the height of the tank. (Use 3.14 as an approximation for π.)

11.________________

12. A tent has the shape of a right pyramid. The volume is 200 cubic feet and the height is 12 feet. Find the area of the floor of the tent.

12.________________

13. A spherical balloon has a radius of 9 centimeters. Find the amount of air required to fill the balloon. (Use 3.14 as an approximation for π.)

13.________________

14. Linda invests \$5000 at 6% simple interest and earns \$450. How long did Linda invest her money?

14.________________

15. Find the height of an ice cream cone if the diameter is 6 centimeters and the volume is 37.68 cubic centimeters. (Use 3.14 as an approximation for π.)

15.________________

Name: Date:
Instructor: Section:

Objective 3 Solve problems involving vertical angles and straight angles.

Find the measure of each marked angle.

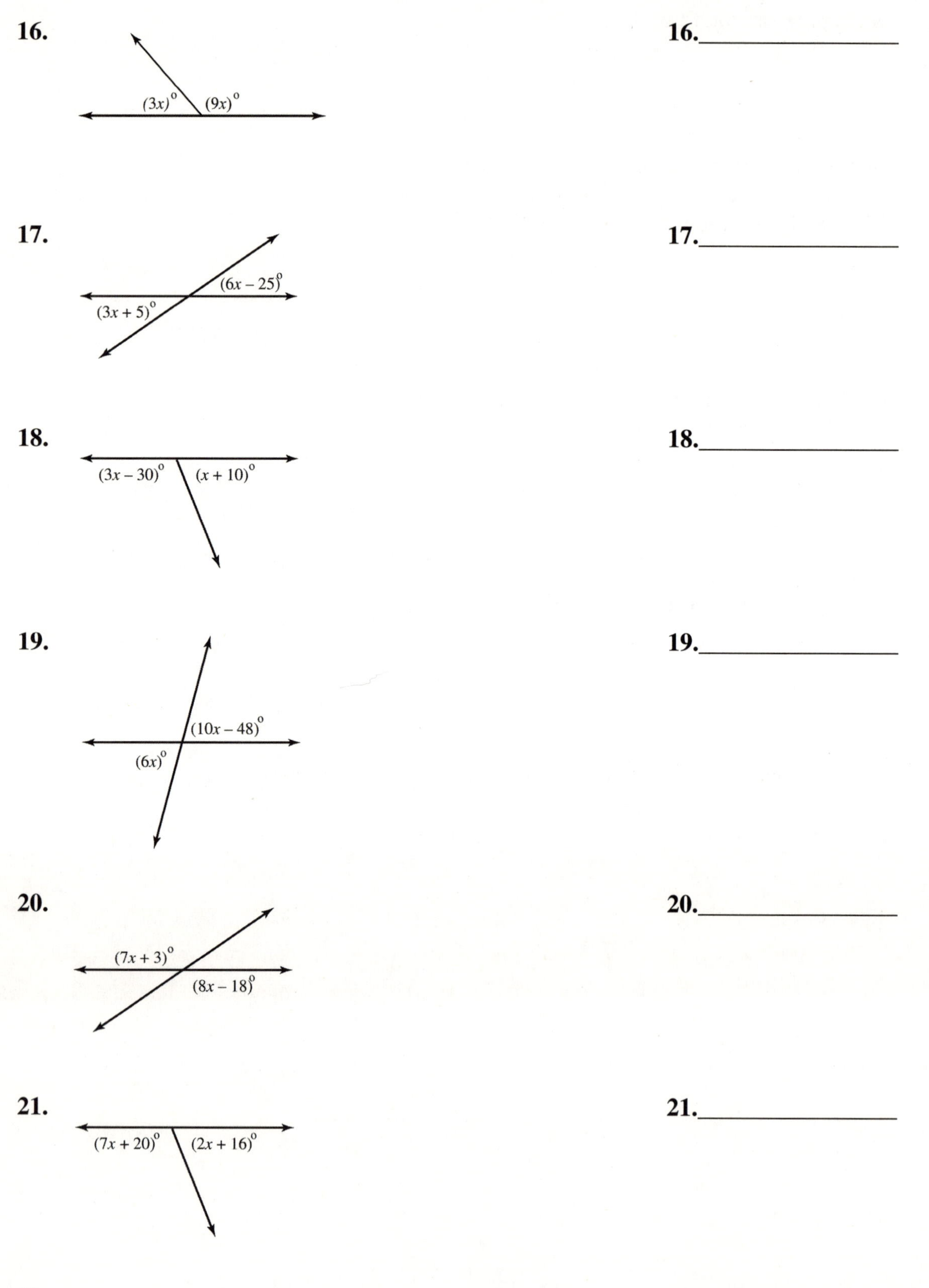

Name: Date:
Instructor: Section:

22.

$(4x+7)^\circ$ $(11x-22)^\circ$

22.________________

Objective 4 Solve a formula for a specified variable.

Solve each formula for the specified variable.

23. $V = LWH$ for H

23.________________

24. $A = p + prt$ for p

24.________________

25. $S = \dfrac{a}{1-r}$ for r

25.________________

26. $S = 2\pi rh + 2\pi r^2$ for h

26.________________

27. $a_n = a_1 + (n-1)d$ for n

27.________________

28. $V = \pi r^2 h$ for h

28.________________

Name: Date:
Instructor: Section:

29. $C = \frac{5}{9}(F - 32)$ for F **29.**________________

30. $S = (n - 2)180$ for n **30.**________________

Name: Date:
Instructor: Section:

Chapter 2 LINEAR EQUATIONS AND INEQUALITIES IN ONE VARIABLE

2.6 Ratios and Proportions

Learning Objectives
1 Write ratios.
2 Solve proportions.
3 Solve applied problems by using proportions.

Key Terms
Use the vocabulary terms listed below to complete each statement in exercises 1-6.

ratio **proportion** **terms** **means** **extremes** **cross products**

1. The ____________________ of the proportion $\frac{a}{b}=\frac{c}{d}$ are the a and d terms.

2. The ____________________ of the proportion $\frac{a}{b}=\frac{c}{d}$ are the b and c terms.

3. A ____________________ is a comparison of two quantities with the same units.

4. A ____________________ is a statement that two ratios are equal.

5. The ____________________ in the proportion $\frac{a}{b}=\frac{c}{d}$ are ad and bc.

6. The ____________________ of the proportion $\frac{a}{b}=\frac{c}{d}$ are a, b, c, and d.

Objective 1 Write ratios.

Write a ratio for each word phrase. Write fractions in lowest terms.

1. 14 marbles to 18 marbles **1.**________________

2. 10 days to 2 weeks **2.**________________

3. 9 dollars to 48 quarters **3.**________________

4. 23 yards to 10 feet **4.**________________

5. 5 months to 2 years **5.**________________

Name: Date:
Instructor: Section:

6. 145 cents to $9 **6.**____________

Zagara's supermarket was surveyed and the following prices were charged for items in various sizes. Find the best buy (based on price per unit) for each of the following items.

7. Rice: **7.**____________
1-pound box: $1.29
2-pound box: $2.31
3-pound box: $3.32
5-pound box: $4.44

8. Tomato catsup **8.**____________
14-ounce size: $ .93
32-ounce size: $1.92
44-ounce size: $2.59
64-ounce size: $3.45

9. Trash bags **9.**____________
10-count box: $1.25
15-count box: $1.60
20-count box: $1.99
25-count box: $2.49

10. Corn oil **10.**____________
18-ounce bottle: $1.27
32-ounce bottle: $2.40
48-ounce bottle: $3.19
64-ounce bottle: $4.43

11. Applesauce **11.**____________
8-ounce jar: $ .59
16-ounce jar: $ .96
24-ounce jar: $1.31
48-ounce jar: $1.99

Name: Date:
Instructor: Section:

12. Freeze-dried coffee
2-ounce size: $2.21
4-ounce size: $3.78
7-ounce size: $4.20
10-ounce size: $6.14

12.________________

Objective 2 Solve proportions.

Solve each equation.

13. $\frac{25}{3}=\frac{125}{x}$

13.________________

14. $\frac{4}{r}=\frac{30}{12}$

14.________________

15. $\frac{z}{20}=\frac{25}{125}$

15.________________

16. $\frac{5}{m}=\frac{12}{5}$

16.________________

17. $\frac{m}{5}=\frac{m-2}{2}$

17.________________

18. $\frac{w+4}{6}=\frac{w+10}{8}$

18.________________

Name: Date:
Instructor: Section:

19. $\frac{x+8}{x-9}=\frac{1}{4}$

19.________________

20. $\frac{6y-4}{y}=\frac{11}{5}$

20.________________

21. $\frac{4}{z+1}=\frac{2}{z+7}$

21.________________

22. $\frac{3x+4}{x-2}=\frac{1}{3}$

22.________________

Objective 3 Solve applied problems by using proportions.

Solve the following problems involving proportions.

23. Ginny can type 8 pages of her term paper in 30 minutes. How long will it take her to type the paper if it has 20 pages?

23.________________

24. If 6 typewriter ribbons cost $40.50, how much will 4 ribbons cost?

24.________________

25. On a road map, 6 inches represents 50 miles. How many inches would represent 125 miles?

25.________________

26. A certain lawn mower uses 5 tanks of gas to cut 18 acres of lawn. How many acres could be cut using 12 tanks of gas?

26.________________

Name: Date:
Instructor: Section:

27. If 3 ounces of medicine must be mixed with 10 ounces of water, how many ounces of medicine must be mixed with 15 ounces of water?

27.________________

28. A certain lawn mower uses 7 tanks of gas to cut 15 acres of lawn. How many tanks of gas are needed to cut 30 acres of lawn?

28.________________

29. If 12 rolls of tape cost $4.60, how much will 15 rolls cost?

29.________________

30. If four pounds of fertilizer will cover 50 square feet of garden, how many pounds would be needed for 125 square feet?

30.________________

Name: Date:
Instructor: Section:

Chapter 2 LINEAR EQUATIONS AND INEQUALITIES IN ONE VARIABLE

2.7 Further Applications of Linear Equations

Learning Objectives
1 Use percent in solving problems involving rates.
2 Solve problems involving mixtures.
3 Solve problems involving simple interest.
4 Solve problems involving denominations of money.
5 Solve problems involving distance, rate, and time.

Key Terms
Use the vocabulary terms listed below to complete each statement in exercises 1-4.

mixture problem **simple interest** **denomination** **distance problems**

1. The formula $I = prt$ is used to calculate ____________________, where p is the principal, r is the rate, and t is time (usually in years).

2. A(n) ____________________ typically involves mixing different concentrations of a substance.

3. Most ____________________ are solved using the formula $d = rt$.

4. The ____________________ of a coin or bill gives the monetary value of the coin or bill.

Objective 1 Use percent in solving problems involving rates.

Solve the problem.

1. How much pure alcohol is in 50 liters of a 45% alcohol solution?

1.________________

2. How much pure antifreeze is in 72 liters of a 22% antifreeze solution?

2.________________

3. If $10,000 is invested for one year at 8% simple interest, how much interest is earned?

3.________________

Name: Date:
Instructor: Section:

4. How much interest is earned if \$7800 is invested at 12% simple interest for one year?

4.______________

5. What is the monetary value of 63 nickels?

5.______________

6. What is the monetary value of 47 quarters?

6.______________

Objective 2 Solve problems involving mixtures.

Solve the problem.

7. A pharmacist has 2 liters of a solution containing 30% alcohol. If he wants to have a solution containing 44% alcohol, how much pure alcohol must he add?

7.______________

8. A car radiator contains 4 gallons of a coolant which is a mixture of antifreeze and water. If the coolant in the radiator is 30% antifreeze, how much coolant must be added with 80% antifreeze to have a 50% solution?

8.______________

9. A cereal manufacturer has 2500 pounds of a cereal mixture containing 25% corn, 40% sugar, 25% oat flour, and a 10% combination of wheat-starch and salt. If he wants to increase the percentage of corn content in the cereal mixture to 40%, how many pounds of corn should he add?

9.______________

10. A chemist has two acid solutions. One is a 60% solution and the other a 30% solution. How many liters of each should she mix to obtain 10 liters of 51% acid solution?

10.______________

Name: Date:
Instructor: Section:

11. How many liters of water must be added to 2 liters of pure alcohol to obtain a 10% alcohol solution?

11.________________

12. A merchant has 25 pounds of cashews worth \$4.30 per pound. He wishes to mix the cashews with peanuts worth \$1.80 per pound to have a nut mixture he can sell for \$2.80 per pound. How many pounds of peanuts should he use?

12.________________

Objective 3 Solve problems involving simple interest.

Solve the problem. Assume that simple interest is being paid.

13. August Zarcone has an annual interest income of \$3390 from two investments. He has \$10,000 more invested at 8% than he has invested at 6%. Find the amount invested at each rate.

13.________________

14. Felicia Whitcomb has some money invested at 5%, and \$5000 more than this amount invested at 9%. Her total annual interest income is \$1430. Find the amount invested at each rate.

14.________________

15. Georgia Levy has 3 times as much money invested in 8% bonds as she has in stocks paying $9\frac{1}{2}$%. How much does she have invested in each if her yearly income from the investments is \$5695?

15.________________

Name: Date:
Instructor: Section:

16. A total of \$2000 is invested for one year, part at $7\frac{1}{2}\%$ and the remainder at $8\frac{1}{2}\%$. If \$156 interest is earned, how much is invested at $7\frac{1}{2}\%$?

16.________________

17. Noah Ouellette has \$1000 more invested at 9% than he has invested at 11%. If the annual income for the two investments is \$1290, find how much he has invested at each rate.

17.________________

18. Adam Costello received an inheritance of \$13,500. He wishes to divide the amount between investments at 4% and 7% to receive an average return of 6% on the two investments. How much should he invest at each rate?

18.________________

Objective 4 Solve problems involving denominations of money.

Solve the problem.

19. A collection of coins consisting of nickels and dimes has a value of \$5.80. Find the number of nickels and dimes in the collection if there are 22 more dimes than nickels.

19.________________

20. A cashier has \$645 in ten-dollar bills and five-dollar bills. There are 90 bills in all. How many of each bill does the cashier have?

20.________________

Name: Date:
Instructor: Section:

21. Total receipts from the sale of 300 tickets to a school musical were $1130. If student tickets cost $3 each and adult tickets $5 each, how many student tickets were sold?

21.______________

22. A stamp collector buys some 20¢ stamps and some 35¢ stamps, paying $9.35 for them. The number of 35¢ stamps is one more than the number of 20¢ stamps. Find the number of 35¢ stamps she buys.

22.______________

23. Merideth has $10.35 in nickels, dimes, and quarters. If she has six more dimes than nickels and twice as many quarters as nickels, how many of each kind of coin does she have?

23.______________

24. Tickets for a performance of *The Nutcracker* ballet cost $14 for adults and $12 for students and senior citizens. If total receipts from the sale of 650 tickets were $3646, how many adult tickets were sold?

24.______________

Objective 5 Solve problems involving distance, rate, and time.

Solve the problem, using $d = rt$, $r = \frac{d}{t}$, or $t = \frac{d}{r}$, *as necessary.*

25. A driver averaged 48 miles per hour and took 6 hours to travel from Chicago to St. Louis. What is the distance between Chicago and St. Louis?

25.______________

26. An Amtrak train traveled from Seattle to San Francisco, averaging 54 miles per hour. The distance between the two cities is 810 miles. How long did the trip take?

26.______________

Name: Date:
Instructor: Section:

27. In the 1988 Olympics, the 400-meter women's relay was won by the United States team of Brown, Echols, Griffith-Joyner, and Ashford in a time of 42 seconds. Find the relay team's average speed in meters per second. (Round to the nearest hundredth.)

27._______________

Solve the problem.

28. Ron and Doug leave the same point at the same time traveling in cars going in opposite directions. Ron travels at 40 miles per hour and Doug travels at 60 miles per hour. In how many hours will they be 350 miles apart?

28._______________

29. A boat goes 3 miles upstream in the same time it takes the boat to go 5 miles downstream. If the rate of the current is 2 miles per hour, what is the speed of the boat in still water?

29._______________

30. Elly and Sam are jogging. Elly runs from point A to point B in one hour at 2 miles per hour faster than Sam does. If Sam takes $\frac{1}{2}$ hour more time than Elly to go the same distance, find the distance between points A and B.

30._______________

Name: Date:
Instructor: Section:

Chapter 2 LINEAR EQUATIONS AND INEQUALITIES IN ONE VARIABLE

2.8 Solving Linear Inequalities

Learning Objectives

1. Graph intervals on a number line.
2. Use the addition property of inequality.
3. Use the multiplication property of inequality.
4. Solve linear inequalities by using both properties in inequality.
5. Solve applied problems by using inequalities.
6. Solve linear inequalities with three parts.

Key Terms

Use the vocabulary terms listed below to complete each statement in exercises 1-7.

interval **interval notation** **negative infinity**

linear inequality in one variable **addition property of inequality**

multiplication property of inequality **three part inequality**

1. A(n) ____________________can be written in the form $Ax+B<C$, $Ax+B\leq C$, $Ax+B>C$, or $Ax+B\geq C$, where *A*, *B*, and *C* are real numbers, with $A\neq 0$.

2. ____________________ is used to indicate all real numbers to the left of a specific location on a number line.

3. The ____________________ states that the same number can be added to (or subtracted from) both sides of an inequality without changing the solution set.

4. ____________________ is a simplified notation that uses parentheses ( ) and/or brackets [ ] to describe an interval on a number line.

5. An inequality that says that one number is between two other numbers is called a ____________________.

6. The ____________________ states that both sides of an inequality may be multiplied (or divided) by a positive number without changing the direction of the inequality symbol. Multiplying (or dividing) by a negative number reverses the inequality symbol.

7. A(n) ____________________ is a portion of a number line.

Name: Date:
Instructor: Section:

Objective 1 Graph intervals on a number line.

Graph each inequality on a number line.

1. $x \geq 3$ **1.**________________

2. $7 < a$ **2.**________________

3. $-4 \leq x < 4$ **3.**________________

4. $-3 < a \leq 2$ **4.**________________

Objective 2 Use the addition property of inequality.

Solve each inequality and graph the solutions.

5. $j + 6 \leq 11$ **5.**________________

6. $y - 7 > -12$ **6.**________________

7. $5a + 3 \leq 6a$ **7.**________________

8. $6 + 3x < 4x + 4$ **8.**________________

9. $9 + 8b > 9b + 11$ **9.**________________

Name: Date:
Instructor: Section:

Objective 3 Use the multiplication property of inequality.

Solve each inequality and graph the solutions.

10. $2x \leq 10$ **10.**_______________

11. $-2s > 4$ **11.**_______________

12. $\frac{1}{2}r > 5$ **12.**_______________

13. $4k \geq -16$ **13.**_______________

14. $-5t \leq -35$ **14.**_______________

Objective 4 Solve linear inequalities by using both properties in inequality.

Solve each inequality.

15. $4(y-3)+2>3(y-2)$ **15.**_______________

16. $7(2-x)-3 \leq -2(x-4)-x$ **16.**_______________

Name: Date:
Instructor: Section:

Solve each inequality and graph the solutions.

17. $7m-8\geq 5m$ **17.**________________

18. $5p-5-p>7p-2$ **18.**________________

19. $5(x+4)+2x<2(3x-1)+8$ **19.**________________

20. $3-\frac{1}{4}z\leq 2+\frac{3}{8}z$ **20.**________________

Objective 5 Solve applied problems by using inequalities.

Use an inequality to solve each problem.

21. Faustino sold two antique desks for $280 and $305. How much should he charge for the third in order to average at least $300 per desk? **21.**________________

22. Find every number such that one third the sum of that number and 24 is less than or equal to 10. **22.**________________

Name: Date:
Instructor: Section:

23. Lauren has grades of 98 and 86 on her first two chemistry quizzes. What must she score on her third quiz to have an average of at least 91 on the three quizzes?

23.________________

24. Nina has a budget of \$230 for gifts for this year. So far she has bought gifts costing \$47.52, \$38.98, and \$26.98. If she has three more gifts to buy, find the average amount she can spend on each gift and still stay within her budget.

24.________________

25. Ruth tutors mathematics in the evenings in an office for which she pays \$600 per month rent. If rent is her only expense and she charges each student \$40 per month, how many students must she teach to make a profit of at least \$1600 per month?

25.________________

Objective 6 Solve linear inequalities with three parts.

Solve each inequality and graph the solutions.

26. $4 \le x - 3 \le 7$

26.________________

27. $1 < 2r - 3 \le 5$

27.________________

Name: Date:
Instructor: Section:

28. $-2 < 5t + 3 \leq 10$ **28.**_______________

29. $-10 \leq 4t - 2 < 6$ **29.**_______________

30. $-5 \leq 3x - 8 < 6$ **30.**_______________

Name: Date:
Instructor: Section:

Chapter 3 LINEAR EQUATIONS IN TWO VARIABLES

3.1 Reading Graphs; Linear Equations in Two Variables

Learning Objectives

1 Interpret graphs.
2 Write a solution as an ordered pair.
3 Decide whether a given ordered pair is a solution of a given equation.
4 Complete ordered pairs for a given equation.
5 Complete a table of values.
6 Plot ordered pairs.

Key Terms

Use the vocabulary terms listed below to complete each statement in exercises 1-10.

bar graph **line graph** **linear equation in two variables** **ordered pair**

***x*-axis** ***y*-axis** **rectangular coordinate system** **quadrant** **origin**

coordinates

1. A(n) ___________________ is a pair of numbers written with parentheses in which the order of the numbers is important.

2. A(n) ___________________ is one of the four regions in the plane determined by a rectangular coordinate system.

3. The vertical line in a rectangular coordinate system is called the ___________________.

4. A(n) ___________________ is a series of line segments that connect points representing data.

5. The numbers in an ordered pair are called the ___________________ of the corresponding point.

6. The horizontal line in a rectangular coordinate system is called the ___________________.

7. A point at which the x-axis and y-axis of a rectangular coordinate system intersect is called the ___________________.

8. A(n) ___________________ is a series of bars arranged either vertically or horizontally to show comparisons of data.

Name: Date:
Instructor: Section:

9. The x-axis and y-axis placed at a right angle at their zero points form a
____________________.

10. A(n) ____________________ is an equation that can be written in the form $Ax + By = C$, where A, B, and C are real numbers and A and B are not both zero.

Objective 1 Interpret graphs.

The graphs below show the total number of degrees awarded by Jefferson University for the years 1990 – 1995 and the distribution of degrees awarded over this period. Use these graphs to answer the questions in Exercises 1 – 3.

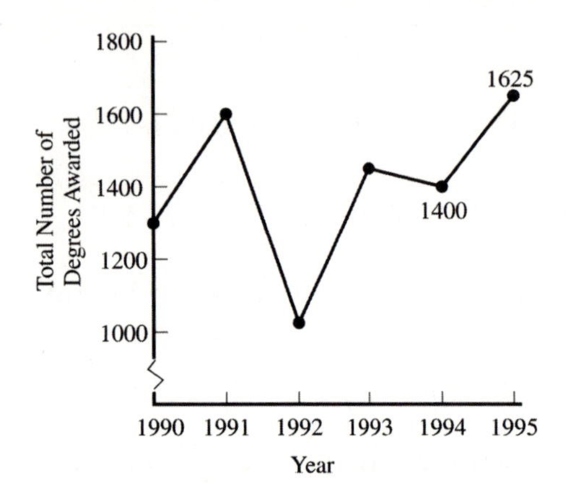

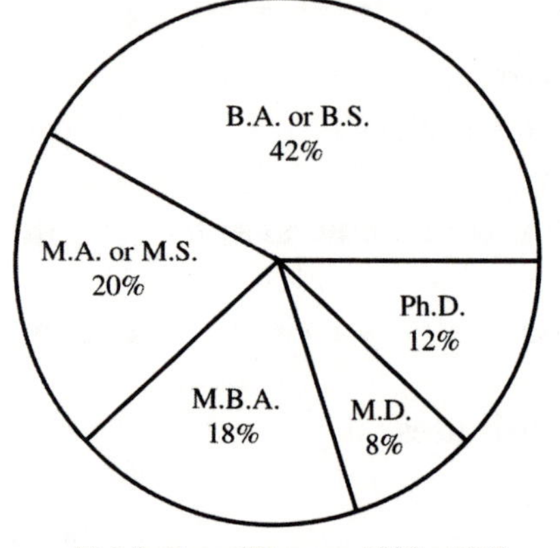

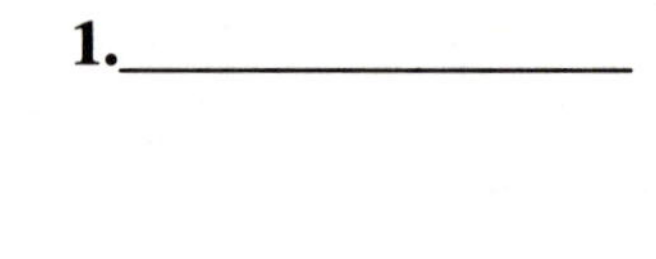
Distribution of Degrees, 1990 – 1995

1. Between which two years did the total number of degrees awarded show the greatest decline? **1.**________________

2. About how many more students received M.B.A. degrees in 1995 than 1994? **2.**________________

3. Between which two years did the total number of degrees awarded show the smallest change? **3.**________________

Name: Date:
Instructor: Section:

The graphs below show the usage of a mathematics help center by subject and by day of the week. Use these graphs to answer the questions in Exercises 4 – 6.

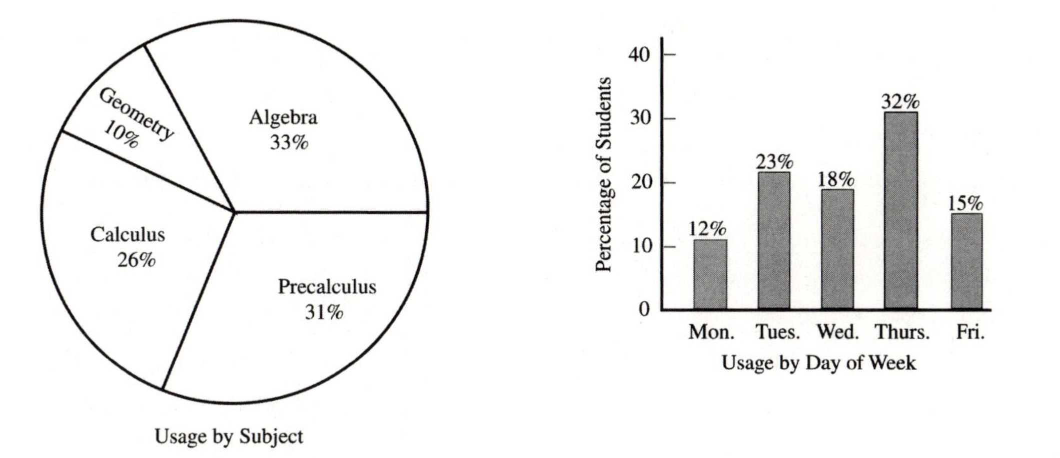

4. On which day was the center used one-and-a-half times as much as Monday? **4.**_______________

5. If 4500 students used the math center during the school year, how many of these students were not enrolled in geometry or pre-calculus courses? Assume each student is enrolled in only one math class. **5.**_______________

6. Which day had the greatest decrease in usage of the center as compared to the previous day? **6.**_______________

Objective 2 Write a solution as an ordered pair.

Write each solution as an ordered pair.

7. $x = 4$ and $y = 7$ **7.**_______________

Name: Date:
Instructor: Section:

8. $x=-2$ and $y=-3$ **8.**________________

9. $y=\frac{1}{3}$ and $x=0$ **9.**________________

Objective 3 Decide whether a given ordered pair is a solution of a given equation.

Decide whether the given ordered pair is a solution of the given equation.

10. $5x-2y=6;\ (2,-2)$ **10.**________________

11. $2x-3y=1;\ (0,\frac{1}{3})$ **11.**________________

12. $2x=3y;\ (3,2)$ **12.**________________

13. $x=1-2y;\ (0,-\frac{1}{2})$ **13.**________________

Objective 4 Complete ordered pairs for a given equation.

For each of the given equations, complete the ordered pairs beneath it.

14. $y=2x-5$ **14.**________________

(a) $(2,\ \ )$

(b) $(0,\ \ )$

(c) $(\ \ ,3)$

(d) $(\ \ ,-7)$

(e) $(\ \ ,9)$

15. $y=3+2x$ **15.**________________

(a) $(-4,\ \ )$

(b) $(2,\ \ )$

(c) $(\ \ ,0)$

(d) $(-2,\ \ )$

Name: Date:
Instructor: Section:

(e) $(\quad , -7)$

16. $5x + 4y = 10$ **16.**________________

(a) $(2, \quad)$

(b) $(4, \quad)$

(c) $(\quad , 3)$

(d) $(0, \quad)$

(e) $(\quad , 0)$

17. $x = -2$ **17.**________________

(a) $(\quad , -2)$

(b) $(\quad , 0)$

(c) $(\quad , 19)$

(d) $(\quad , 3)$

(e) $(\quad , -\frac{2}{3})$

Objective 5 Complete a table of values.

Complete the table of ordered pairs for each equation.

18. $4x + y = 6$ **18.**________________

x	2		1
y		4	

19. $3x + 2y = 4$ **19.**________________

x	0		4
y		0	

20. $y - 4 = 0$ **20.**________________

x	−6	0	6
y			

21. $3x - 4y = -6$ **21.**________________

x	y
0	
	0
2	

Name: Date:
Instructor: Section:

22. $4x+3y=12$ **22.**____________

x	y
0	
	0
	−1

23. $y-4=0$ **23.**____________

x	y
−4	
0	
6	

Objective 6 Plot ordered pairs.

Plot the following ordered pairs on a coordinate system.

24. $(7,1)$ **24.**____________

25. $(-2,4)$ **25.**____________

26. $(-2,-7)$ **26.**____________

27. $(4,-2)$ **27.**____________

28. $(0,-4)$ **28.**____________

29. $(-5,0)$ **29.**____________

30. $(0,0)$ **30.**____________

Name: Date:
Instructor: Section:

Chapter 3 LINEAR EQUATIONS IN TWO VARIABLES

3.2 Graphing Linear Equations in Two Variables

Learning Objectives
1 Graph linear equations by plotting ordered pairs.
2 Find intercepts.
3 Graph linear equations of the form $Ax + By = 0$.
4 Graph linear equations of the form $y = k$ or $x = k$.
5 Use a linear equation to model data.

Key Terms
Use the vocabulary terms listed below to complete each statement in exercises 1-6.

graph of a linear equation	***x*-intercept**	***y*-intercept**
line through the origin	**horizontal line**	**vertical line**

1. The graph of a ____________________ has an equation of the form $y = k$.

2. The ____________________ in two variables is a straight line.

3. A point where a graph intersects the x-axis is called a(n) ____________________.

4. The graph of a ____________________ has an equation of the form $x = k$.

5. A point where a graph intersects the y-axis is called a(n) ____________________.

6. A ____________________ passes through the point (0, 0).

Name: Date:
Instructor: Section:

Objective 1 Graph linear equations by plotting ordered pairs.

Complete the ordered pairs for each equation. Then graph the equation by plotting the points and drawing a line through them.

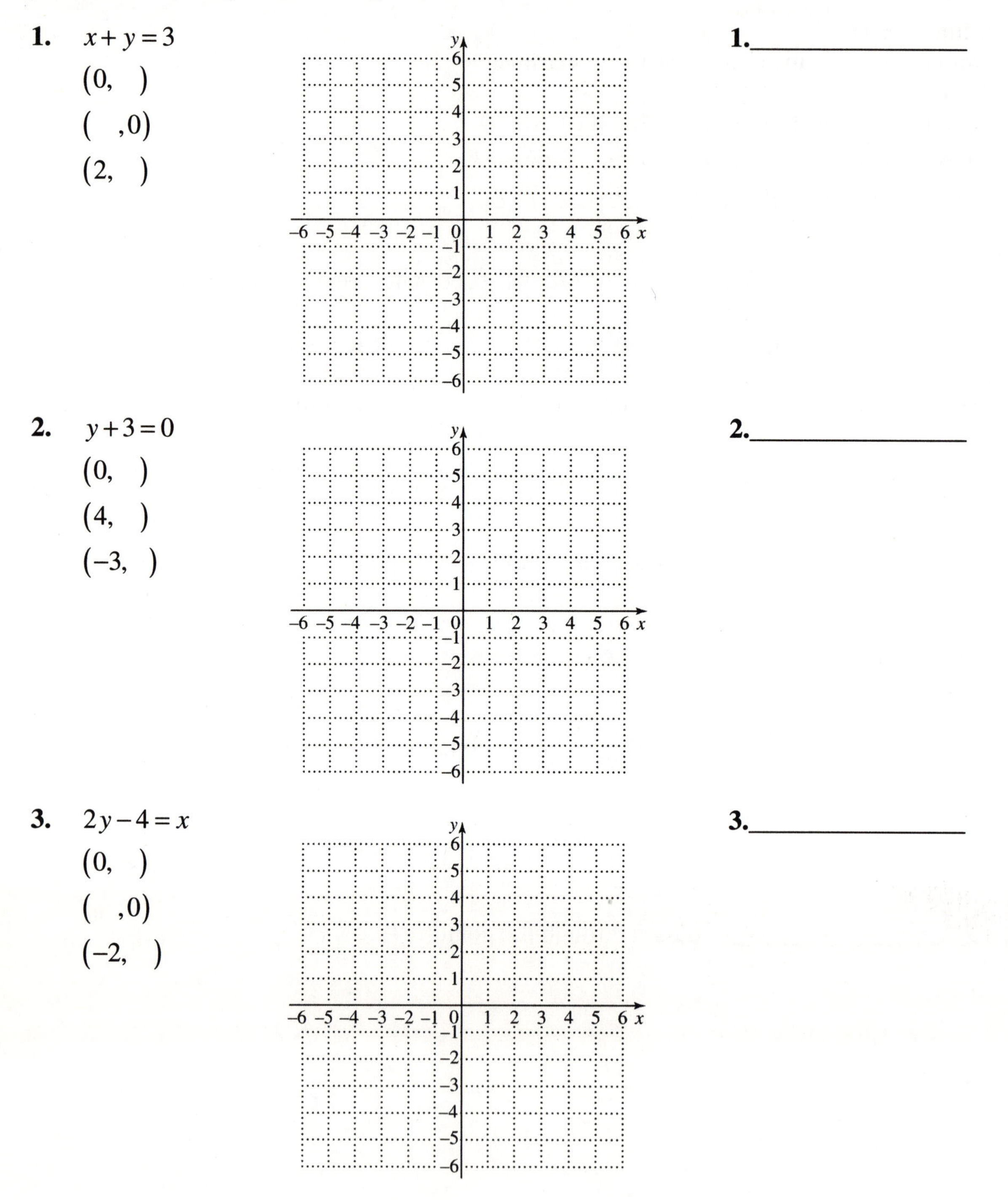

1. $x + y = 3$

(0,)

(,0)

(2,)

1.________________

2. $y + 3 = 0$

(0,)

(4,)

(−3,)

2.________________

3. $2y - 4 = x$

(0,)

(,0)

(−2,)

3.________________

Name: Date:
Instructor: Section:

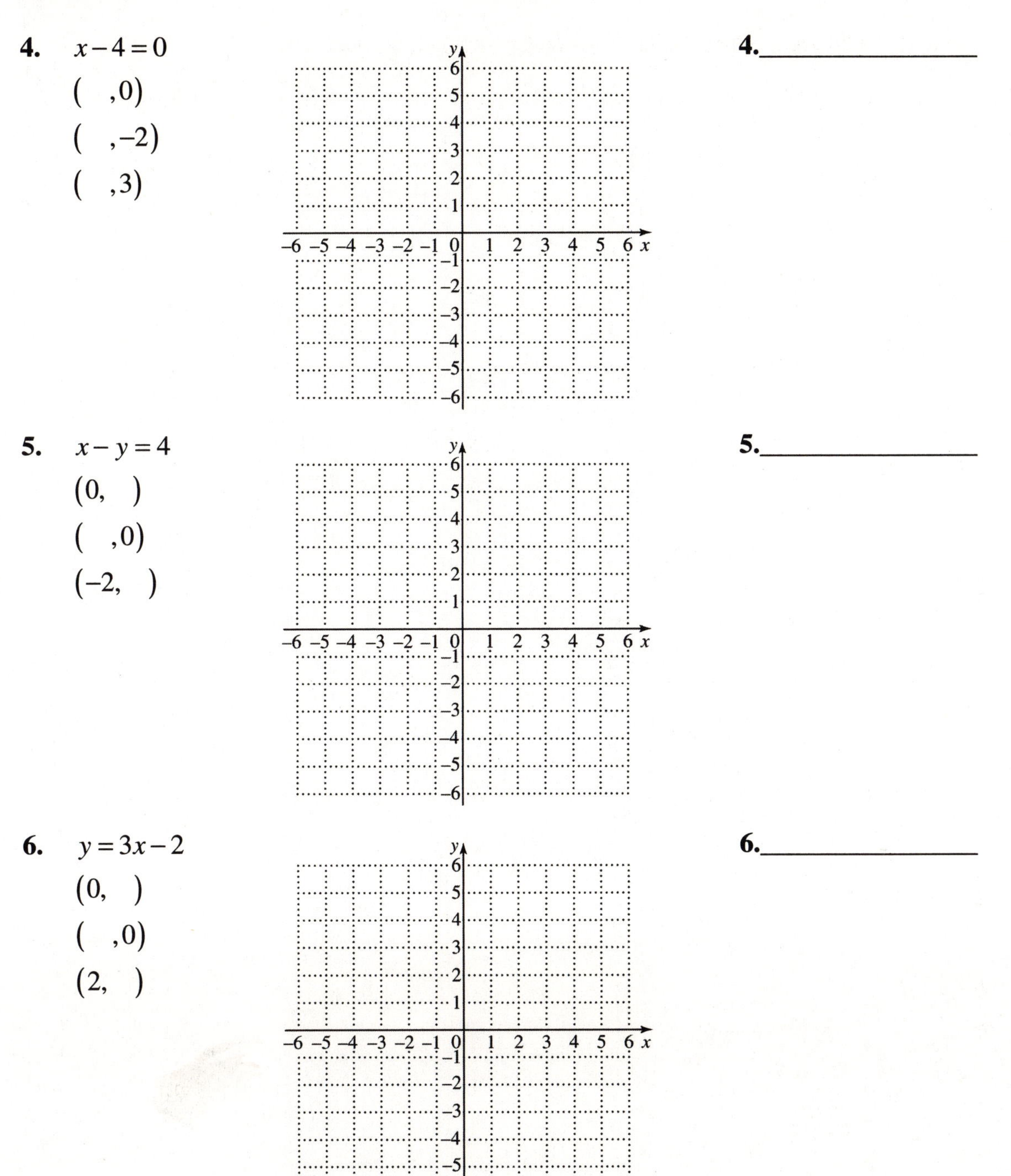

4. $x-4=0$

$(\quad ,0)$

$(\quad ,-2)$

$(\quad ,3)$

4.________________

5. $x-y=4$

$(0, \quad)$

$(\quad ,0)$

$(-2, \quad)$

5.________________

6. $y=3x-2$

$(0, \quad)$

$(\quad ,0)$

$(2, \quad)$

6.________________

Name: Date:
Instructor: Section:

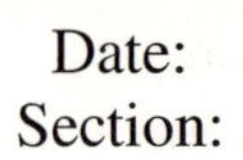

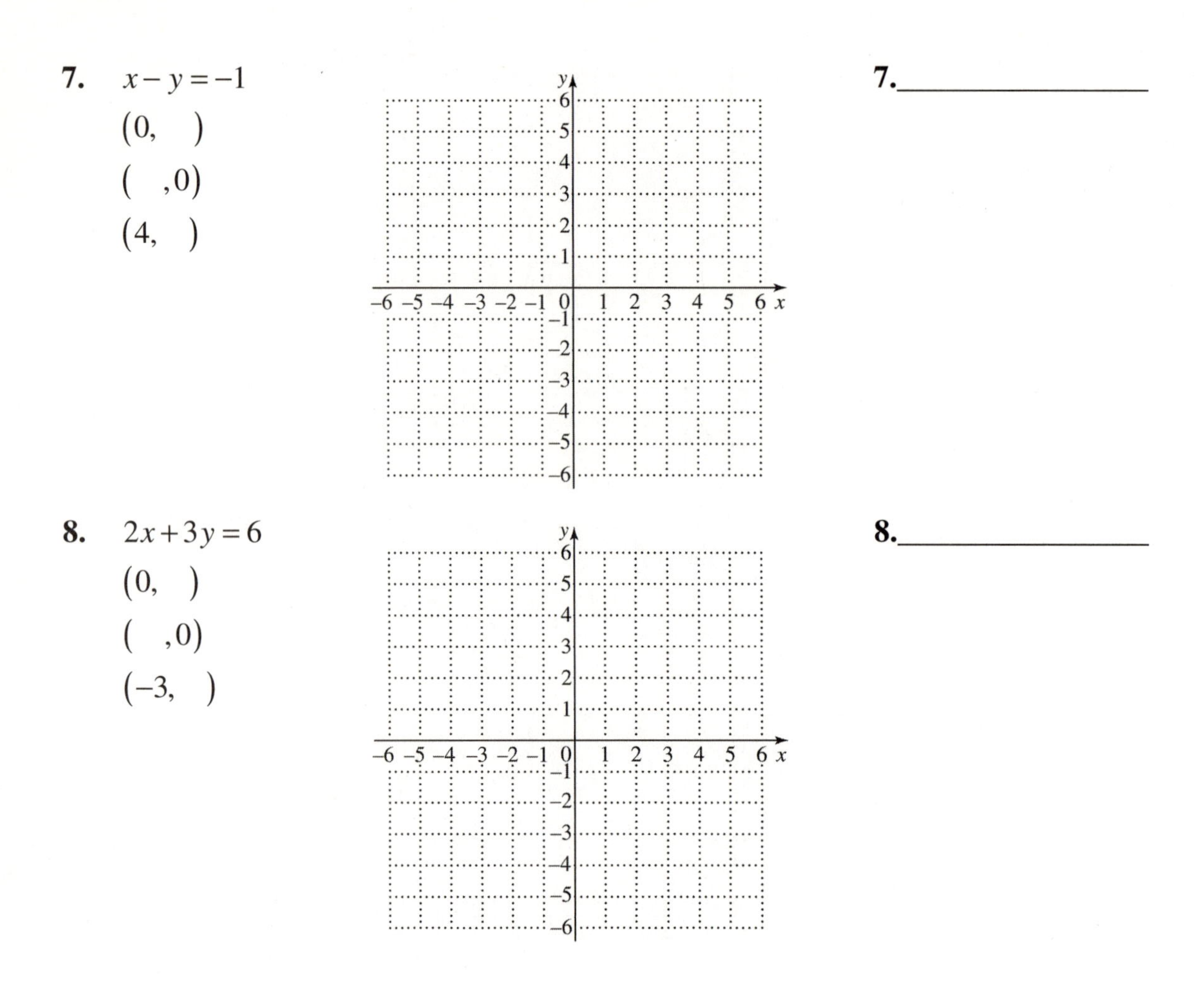

7. $x - y = -1$

$(0, \quad)$

$(\quad ,0)$

$(4, \quad)$

7.________________

8. $2x + 3y = 6$

$(0, \quad)$

$(\quad ,0)$

$(-3, \quad)$

8.________________

Objective 2 Find intercepts.

Find the intercepts for the graph of each equation.

9. $-5x + 2y = 10$

9.________________

10. $3x + 2y = 12$

10.________________

11. $2x + 4y = 0$

11.________________

12. $4x + 5y = 8$

12.________________

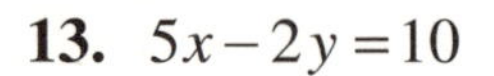

13. $5x - 2y = 10$

13.________________

Name: Date:
Instructor: Section:

14. $4x+3y=9$ **14.**________________

15. $3x+2y=-2$ **15.**________________

16. $5x-3y=12$ **16.**________________

17. $2x+9y=-9$ **17.**________________

18. $3x+4y=9$ **18.**________________

Objective 3 Graph linear equations of the form $Ax+By=0$.

Graph each equation.

19. $3x+y=6$ **19.**________________

20. $6x+5y=15$ **20.**________________

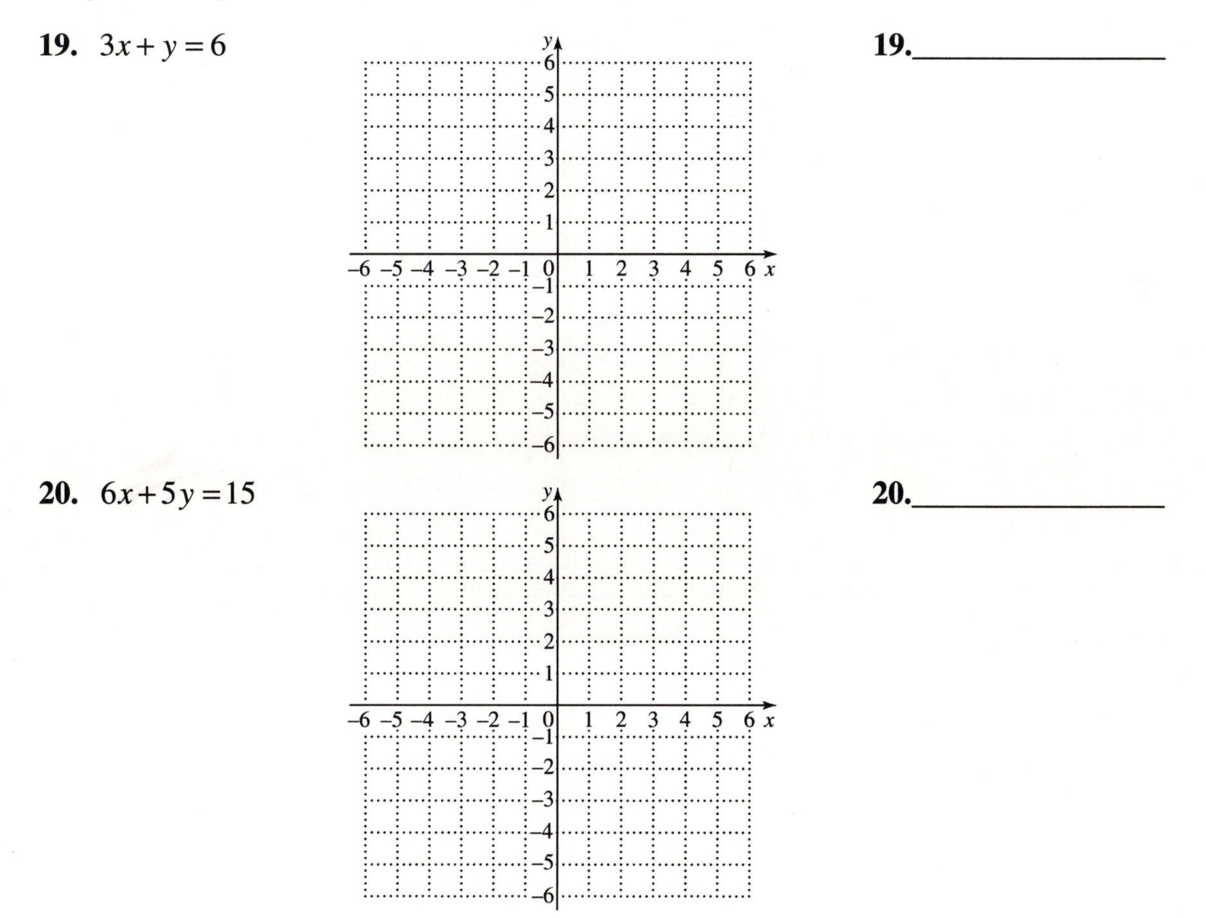

Name: Date:
Instructor: Section:

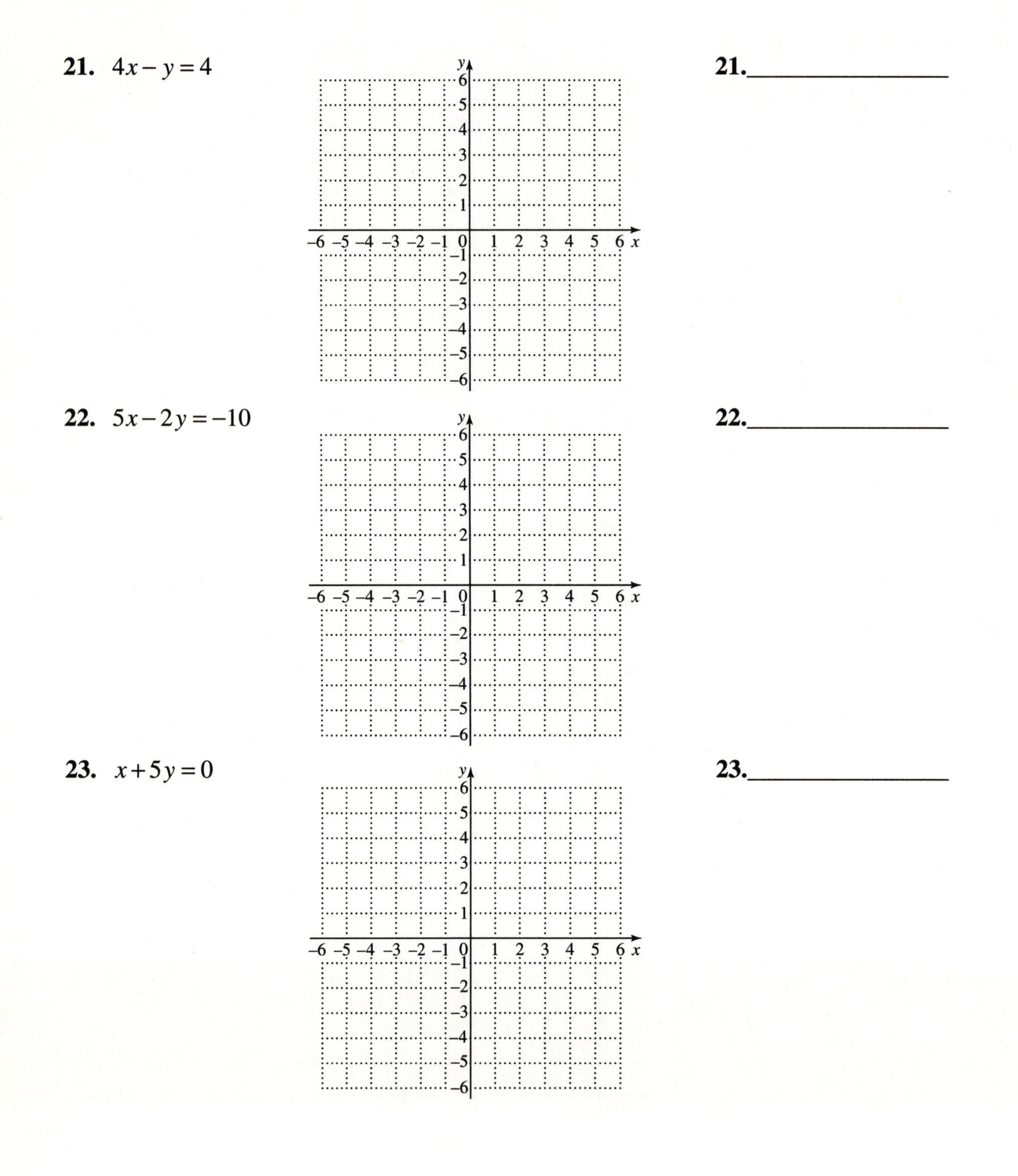

21. $4x - y = 4$

21.________________

22. $5x - 2y = -10$

22.________________

23. $x + 5y = 0$

23.________________

Name: Date:
Instructor: Section:

Objective 4 Graph linear equations of the form $y = k$ or $x = k$.

Graph each equation.

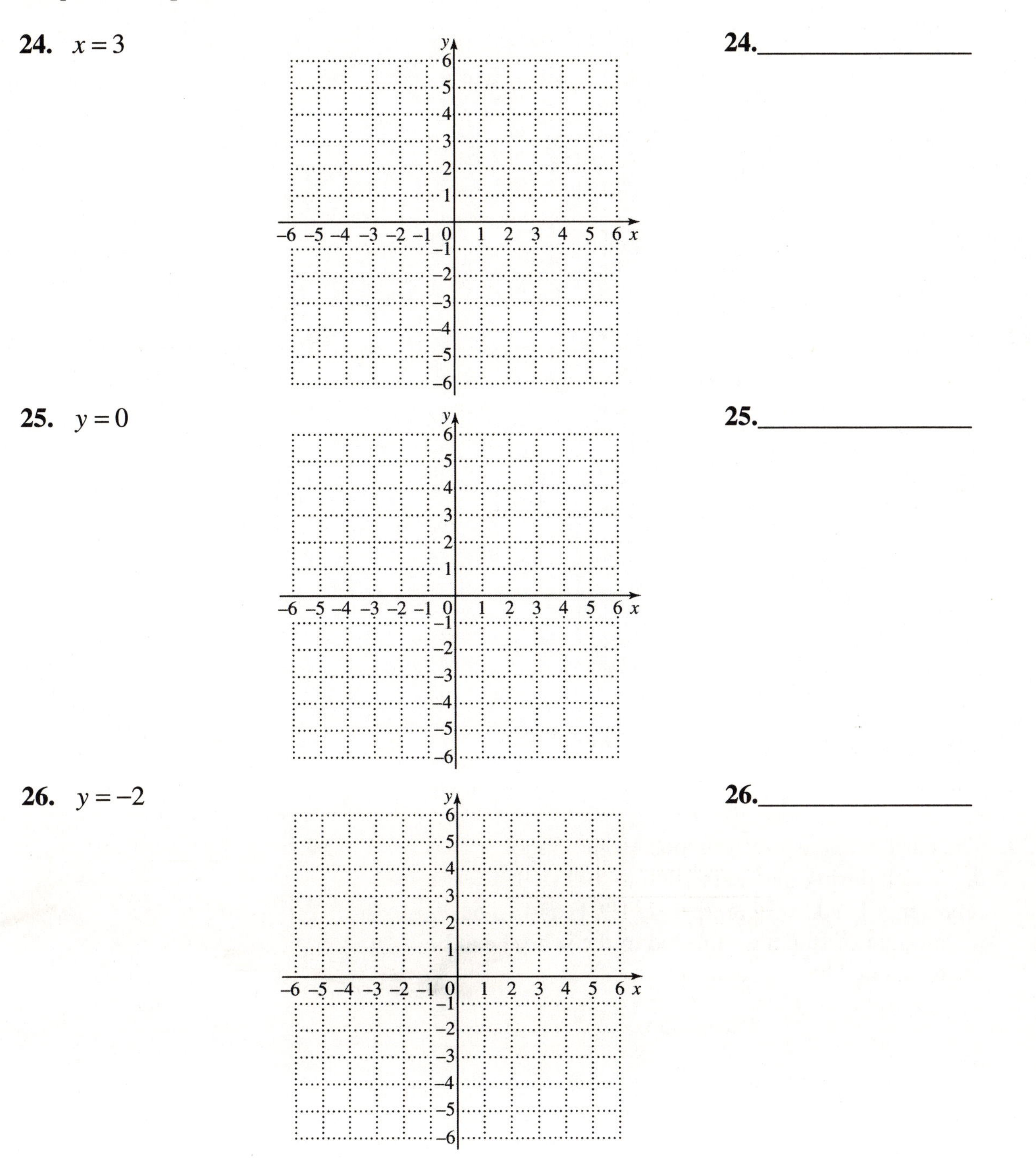

24. $x = 3$

24.________________

25. $y = 0$

25.________________

26. $y = -2$

26.________________

Name: Date:
Instructor: Section:

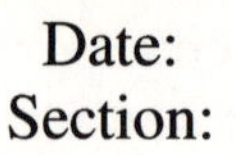

27. $x-1=0$ **27.**________________

28. $y+3=0$ **28.**________________

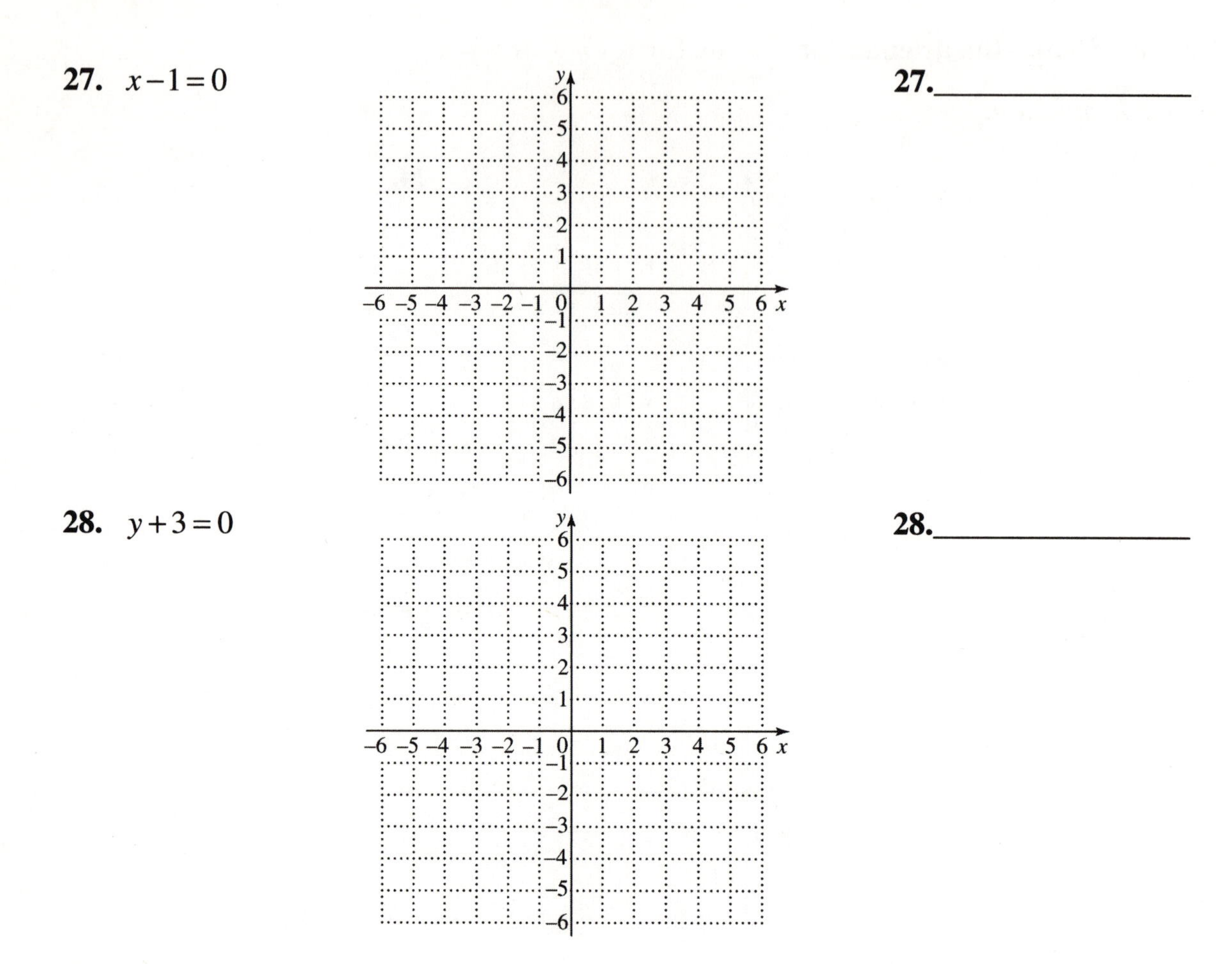

Objective 5 Use a linear equation to model data.

Simplify each expression.

29. The enrollment at Lincolnwood High School decreased during the years 1990 to 1995. If $x=0$ represents 1990, $x=1$ represents 1991, and so on, the number of students enrolled in the school can be approximated by the equation **29.**________________

$$y=-85x+2435.$$

Use this equation to approximate the number of students in each year from 1990 through 1995.

Name: Date:
Instructor: Section:

30. Suppose that the demand and price for a certain model of calculator are related by the equation

$$y = 45 - \tfrac{3}{5}x,$$

where y is the price (in dollars) and x is the demand (in thousands of calculators). Assuming that this model is valid for a demand up to 50,000 calculators, find the price at each of the following levels of demand.

30.________________

(a) 0 calculators

(b) 5000 calculators

(c) 20,000 calculators

(d) 45,000 calculators

Name: Date:
Instructor: Section:

Chapter 3 LINEAR EQUATIONS IN TWO VARIABLES

3.3 The Slope of a Line

Learning Objectives
1. Find the slope of a line, given two points.
2. Find the slope from the equation of a line.
3. Use slopes to determine whether two lines are parallel, perpendicular, or neither.

Key Terms
Use the vocabulary terms listed below to complete each statement in exercises 1-10.

rise **run** **slope** **subscript notation** **positive slope**

negative slope **horizontal line** **vertical line** **parallel** **perpendicular**

1. A line with ____________________ falls from left to right.

2. The ratio of the change in y to the change in x along a line is called the ____________________ of the line.

3. The ____________________ is the horizontal change between two points on a line – that is, the change in x-values.

4. A ____________________ has undefined slope.

5. ____________________ is a way of indicating nonspecific values, such as x_1 and x_2.

6. A line with ____________________ rises from left to right.

7. Two lines that are ____________________ have the same slope.

8. The ____________________ is the vertical change between two points on a line – that is, the change in y-values.

9. A ____________________ has slope equal to zero.

10. Two lines that are ____________________ have slopes whose product is -1.

Name: Date:
Instructor: Section:

Objective 1 Find the slope of a line, given two points.

Find the slope of each line.

1. Through $(4,3)$ and $(3,5)$ **1.**________________

2. Through $(2,3)$ and $(6,7)$ **2.**________________

3. Through $(-3,2)$ and $(7,4)$ **3.**________________

4. Through $(5,-2)$ and $(2,7)$ **4.**________________

5. Through $(2,-4)$ and $(-3,-1)$ **5.**________________

6. Through $(7,2)$ and $(-7,3)$ **6.**________________

7. Through $(-7,-7)$ and $(2,-7)$ **7.**________________

8. Through $(-4,-4)$ and $(-2,-2)$ **8.**________________

9. Through $(-4,6)$ and $(-4,-1)$ **9.**________________

10. Through $(2,-7)$ and $(-2,1)$ **10.**________________

Objective 2 Find the slope from the equation of a line.

Find the slope of each line.

11. $y=-5x$ **11.**________________

Name: Date:
Instructor: Section:

12. $y = \frac{1}{2}x + 5$ **12.**________________

13. $y = -\frac{2}{5}x - 4$ **13.**________________

14. $y = -\frac{4}{7}x + 9$ **14.**________________

15. $4y = 3x + 7$ **15.**________________

16. $2x + 7y = 7$ **16.**________________

17. $4x - 3y = 0$ **17.**________________

18. $y = -4$ **18.**________________

19. $x = 0$ **19.**________________

20. $3x = 4y$ **20.**________________

Objective 3 Use slopes to determine whether two lines are parallel, perpendicular, or neither.

In each pair of equations, give the slope of each line, and then determine whether the two lines are parallel, perpendicular, *or* neither.

21. $y = -5x - 2$ **21.**________________

$y = 5x + 11$

22. $y = 4x + 4$ **22.**________________

$y = 3 - \frac{1}{4}x$

Name: Date:
Instructor: Section:

23. $-x+y=-7$
$x-y=-3$

23.________________

24. $2x+2y=7$
$2x-2y=5$

24.________________

25. $4x+2y=8$
$x+4y=-3$

25.________________

26. $9x+3y=2$
$x-3y=5$

26.________________

27. $4x+2y=7$
$5x+3y=11$

27.________________

28. $8x+2y=7$
$x=3-y$

28.________________

29. $y+4=0$
$y-7=0$

29.________________

30. $y=9$
$x=0$

30.________________

Name: Date:
Instructor: Section:

Chapter 3 LINEAR EQUATIONS IN TWO VARIABLES

3.4 Equations of a Line

Learning Objectives
1 Write an equation of a line by using its slope and *y*-intercepts.
2 Graph a line by using its slope and a point on the line.
3 Write an equation of a line by using its slope and any point on the line.
4 Write an equation of a line by using two points on the line.
5 Find an equation of a line that fits a data set.

Key Terms
Use the vocabulary terms listed below to complete each statement in exercises 1-3.

slope-intercept form **point-slope form** **standard form**

1. A linear equation is written in ____________________ if it is in the form $y = mx + b$, where m is the slope and $(0, b)$ is the *y*-intercept.

2. A linear equation is written in ____________________ if it is in the form $y - y_1 = m(x - x_1)$, where m is the slope and (x_1, y_1) is a point on the line.

3. A linear equation in two variables written in the form $Ax + By = C$, with A and B both not 0, is in ____________________.

Objective 1 Write an equation of a line by using its slope and *y*-intercepts.

Write an equation in slope-intercept form for each of the following lines.

1. $m = \frac{2}{3}$; $b = -4$ **1.**________________

2. $m = -2$; $b = 0$ **2.**________________

3. $m = -7$; $b = -2$ **3.**________________

4. Slope $\frac{-1}{2}$; *y*-intercept $(0, -3)$ **4.**________________

Name: Date:
Instructor: Section:

5. Slope –4; y-intercept $(0,0)$ **5.**________________

6. Slope 0; y-intercept $(0,-4)$ **6.**________________

Objective 2 Graph a line by using its slope and a point on the line.

Graph the line passing through the given point and having the given slope.

7. $(4,-2)$; $m=-1$ **7.**________________

8. $(-3,-2)$; $m=\frac{2}{3}$ **8.**________________

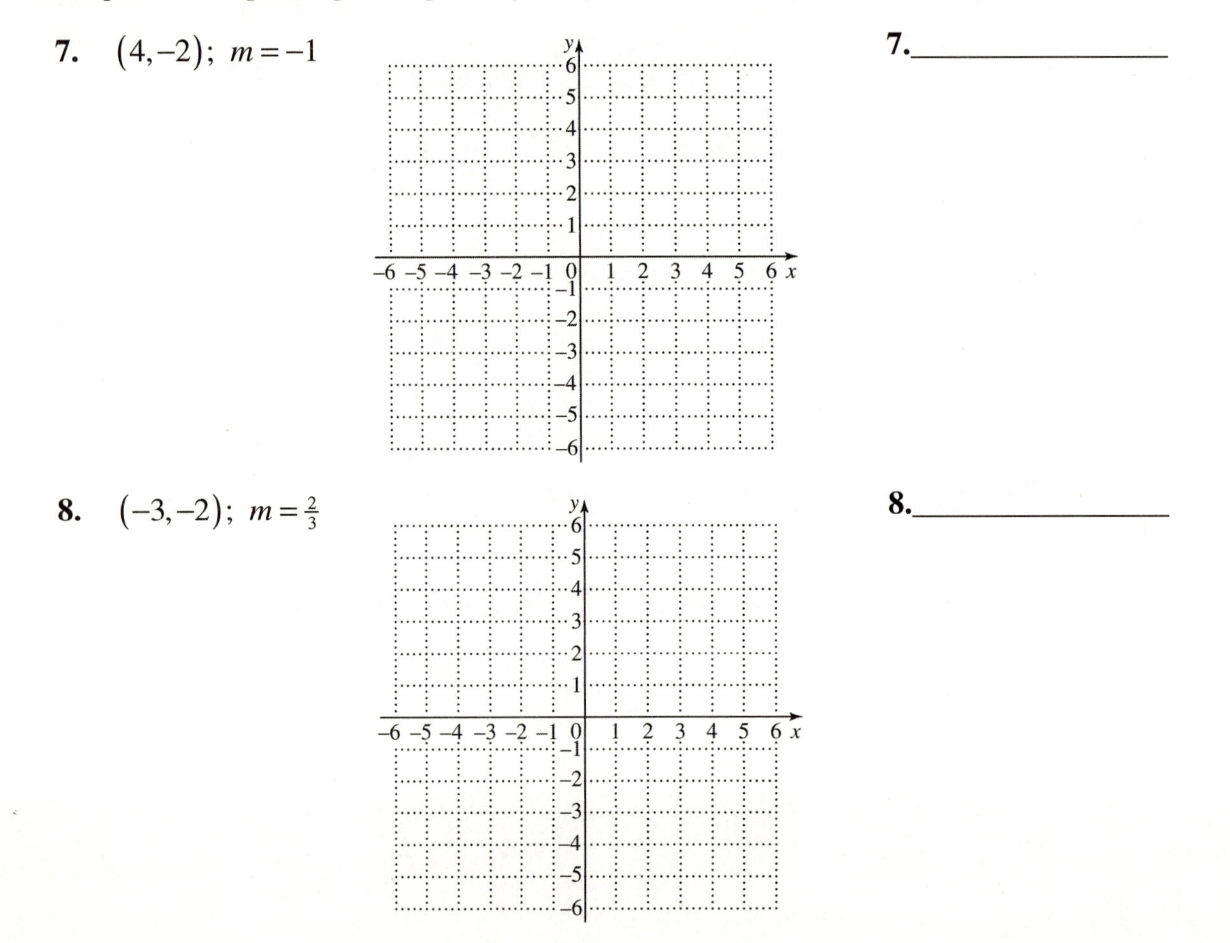

Name: Date:
Instructor: Section:

9. $(2,4)$;
undefined slope

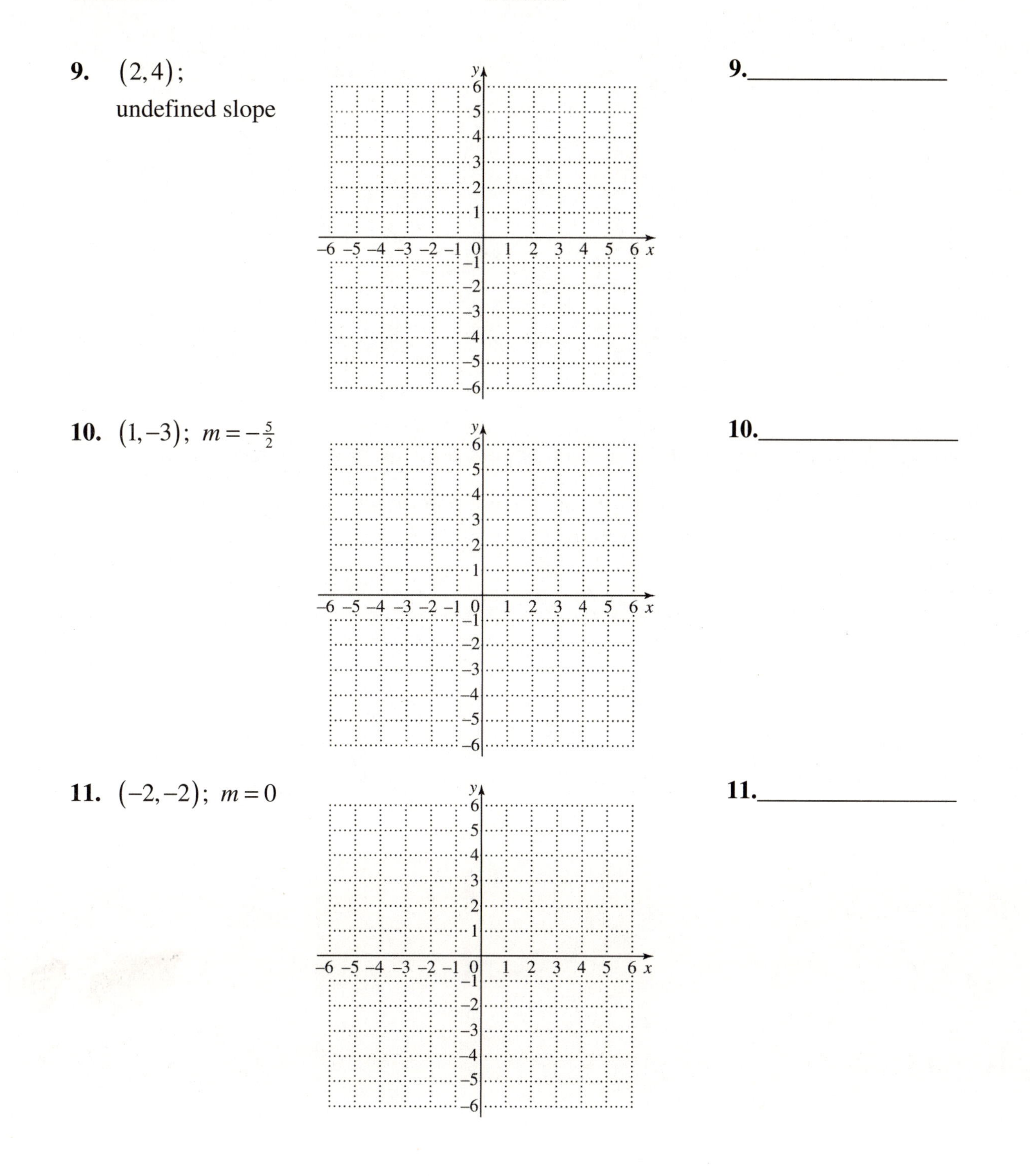

9.________________

10. $(1,-3)$; $m=-\frac{5}{2}$

10.________________

11. $(-2,-2)$; $m=0$

11.________________

Name: Date:
Instructor: Section:

12. $(3,-1)$; $m=2$

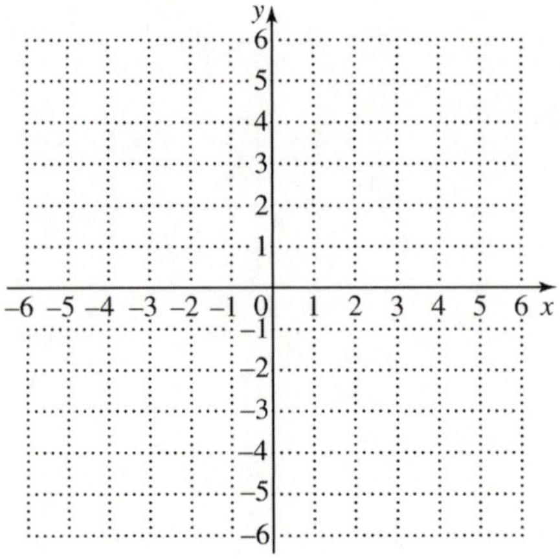

12.________________

Objective 3 Write an equation of a line by using its slope and any point on the line.

Write an equation for the line passing through the given point and having the given slope. Write the equations in the standard form $Ax+By=C$.

13. $(5,4)$; $m=\frac{1}{3}$

13.________________

14. $(-2,4)$; $m=2$

14.________________

15. $(-3,-1)$; $m=-\frac{2}{3}$

15.________________

16. $(-4,-7)$; $m=\frac{3}{4}$

16.________________

17. $(0,0)$; $m=0$

17.________________

Name: Date:
Instructor: Section:

18. $(-4,-2)$; undefined slope

18.________________

19. $(2,-2)$; $m=-1$

19.________________

Objective 4 Write an equation of a line by using two points on the line.

Write an equation for the line passing through each pair of points. Write the equations in standard form $Ax+By=C$.

20. $(2,3)$ and $(7,5)$

20.________________

21. $(3,-4)$ and $(2,7)$

21.________________

22. $(1,-2)$ and $(-2,8)$

22.________________

23. $(-3,4)$ and $(-3,-7)$

23.________________

24. $(7,2)$ and $(-2,-4)$

24.________________

Name: Date:
Instructor: Section:

25. $(2,6)$ and $(-4,6)$ **25.**________________

26. $\left(\frac{1}{2},\frac{2}{3}\right)$ and $\left(-\frac{3}{2},2\right)$ **26.**________________

Objective 5 Find an equation of a line that fits a data set.

The total expenditures (in millions of dollars) for the purchase of memorabilia collectibles is given below. Use the information in the chart to answer questions 27 – 30.

Year	x	Millions of dollars (y)
1993	0	84
1994	1	101
1995	2	123
1996	3	136
1997	4	160
1998	5	181
1999	6	196

27. Use the data from 1994 and 1999 to find the slope of the line that approximates this information. Then use the slope to find the equation of the line in slope-intercept form. **27.**________________

28. To see how well the equation in exercise 27 approximates the ordered pairs (x, y) in the table of data, let $x = 4$ (for 1997) and find y. **28.**________________

Name: Date:
Instructor: Section:

29. Use the data from 1995 and 1998 to find the slope of the line that approximates this information. Then use the slope to find the equation of the line in slope-intercept form.

29.________________

30. To see how well the equation in exercise 29 approximates the ordered pairs (x, y) in the table of data, let $x = 3$ (for 1996) and find y.

30.________________

Name: Date:
Instructor: Section:

Chapter 4 EXPONENTS AND POLYNOMIALS

4.1 The Product Rule and Power Rules for Exponents

Learning Objectives
1 Use exponents.
2 Use the product rule for exponents.
3 Use the rule $(a^m)^n = a^{mn}$.
4 Use the rule $(ab)^m = a^m b^m$.
5 Use the rule $\left(\frac{a}{b}\right)^m = \frac{a^m}{b^m}$.
6 Use combinations of rules.
7 Use the rules for exponents in a geometry application.

Key Terms
Use the vocabulary terms listed below to complete each statement in exercises 1-4.

product rule for exponents **power raised to a power**
product raised to a power **quotient raised to a power**

1. The ____________________ states that the product of two terms with the same base is equal the base raised to the sum of the exponents on each term.

2. The rule $\left(\frac{a}{b}\right)^m = \frac{a^m}{b^m}$ describes how to evaluate a ____________________.

3. The rule $(a^m)^n = a^{mn}$ describes how to evaluate a ____________________.

4. The rule $(ab)^m = a^m b^m$ describes how to evaluate a ____________________.

Objective 1 Use exponents.

Write each expression in exponential form and evaluate.

1. $2\cdot 2\cdot 2\cdot 2\cdot 2\cdot 2$ 1.________________

2. $(-\frac{2}{5})(-\frac{2}{5})(-\frac{2}{5})(-\frac{2}{5})$ 2.________________

Name: Date:
Instructor: Section:

Write each expression in exponential form.

3. $(-2y)(-2y)(-2y)(-2y)(-2y)$ **3.**________________

4. $(.5st)(.5st)(.5st)(.5st)$ **4.**________________

Evaluate each exponential expression. Name the base and the exponent.

5. $(-4)^4$ **5.**________________

6. -6^2 **6.**________________

Objective 2 Use the product rule for exponents.

Use the product rule to simplify each expression, if possible. Write each answer in exponential form.

7. $4^7 \cdot 4^4$ **7.**________________

8. $(-5)^{10} \cdot (-5)^4$ **8.**________________

Multiply.

9. $10n^4 \cdot 10n \cdot n^7$ **9.**________________

10. $(-4x^3)(8x^{12})$ **10.**________________

Name: Date:
Instructor: Section:

In each of the following exercises, first add the given terms. Then start over and multiply them.

11. $-3m^6, \ -4m^6$ **11.**______________

12. $-5a^3, \ 4a^3, \ -a^3$ **12.**______________

Objective 3 Use the rule $\left(a^m\right)^n = a^{mn}$.

Simplify each expression. Write all answers in exponential form.

13. $\left(3^4\right)^3$ **13.**______________

14. $\left(9^5\right)^9$ **14.**______________

15. $\left[(-2)^7\right]^5$ **15.**______________

16. $-\left(21^5\right)^4$ **16.**______________

Objective 4 Use the rule $(ab)^m = a^m b^m$.

Simplify each expression.

17. $\left(yz^4\right)^3$ **17.**______________

18. $5\left(ab^3\right)^3$ **18.**______________

19. $\left(-2r^3s\right)^6$ **19.**______________

Name: Date:
Instructor: Section:

20. $\left(4c^3d^4\right)^3$

20.________________

Objective 5 Use the rule $\left(\frac{a}{b}\right)^m = \frac{a^m}{b^m}$**.**

Simplify each expression. Assume all variables represent nonzero real numbers.

21. $\left(\frac{-2a}{b^2}\right)^7$

21.________________

22. $\left(-\frac{2x}{5}\right)^3$

22.________________

23. $\left(\frac{x}{2y}\right)^4$

23.________________

24. $\left(\frac{4}{g}\right)^3$

24.________________

Objective 6 Use combinations of rules.

Simplify. Write all answers in exponential form.

25. $\left(\frac{3b^2}{11}\right)^4$

25.________________

Name: Date:
Instructor: Section:

26. $(5x^2y^3)^7(5xy^4)^4$

26.________________

27. $\left(\frac{7a^2b^3}{2}\right)^7$

27.________________

28. $\left(\frac{km^4p^2}{3n^4}\right)^7 \quad (n \neq 0)$

28.________________

Objective 7 Use the rules for exponents in a geometry application.

Find an expression that represents the area of the figure described.

29. A rectangle with length $5x^4$ and width $2x^3$

29.________________

30. A circle with radius $4a^5$

30.________________

Name: Date:
Instructor: Section:

Chapter 4 EXPONENTS AND POLYNOMIALS

4.2 Integer Exponents and the Quotient Rule

Learning Objectives
1 Use 0 as an exponent.
2 Use negative numbers as exponents.
3 Use the quotient rule for exponents.
4 Use combinations of rules.

Key Terms
Use the vocabulary terms listed below to complete each statement in exercises 1-3.

zero exponent **negative exponent** **quotient rule for exponents**

1. A number raised to a ____________________ can be rewritten as the reciprocal of the number raised to the opposite of the exponent.

2. A nonzero number raised to a ____________________ has a value of 1.

3. The ____________________ states that the quotient of two terms with the same base is equal the base raised to the difference of the exponents on each term.

Objective 1 Use 0 as an exponent.

Evaluate each expression.

1. 7^0 **1.**________________

2. $-(-8)^0$ **2.**________________

3. -12^0 **3.**________________

4. $(-5)^0-(-5)^0$ **4.**________________

5. $-12^0+(-12)^0$ **5.**________________

6. $\frac{0^8}{8^0}$ **6.**________________

7. $-r^0 \quad (r \neq 0)$ **7.**________________

Name: Date:
Instructor: Section:

Objective 2 Use negative numbers as exponents.

Evaluate each expression.

8. 4^{-2} **8.**________________

9. $\left(\frac{2}{7}\right)^{-1}$ **9.**________________

10. $(-2)^{-4}$ **10.**________________

11. $10^{-2}+5^{-2}$ **11.**________________

Simplify by using the definition of negative exponents. Write each expression with only positive exponents. Assume all variables represent nonzero real numbers

12. a^{-4} **12.**________________

13. $\dfrac{2}{r^{-7}}$ **13.**________________

14. $\dfrac{2x^{-4}}{3y^{-7}}$ **14.**________________

Objective 3 Use the quotient rule for exponents.

Use the quotient rule to simplify each expression. Write answers with only positive exponents. Assume that all variables represent nonzero real numbers.

15. $\dfrac{2^9}{2^5}$ **15.**________________

16. $\dfrac{(-2)^8}{(-2)^3}$ **16.**________________

Name: Date:
Instructor: Section:

17. $\dfrac{2^4 \cdot x^2}{2^5 \cdot x^8}$ **17.**________________

18. $\dfrac{12x^9 y^5}{12^4 x^3 y^7}$ **18.**________________

19. $\dfrac{a^4 b^3}{a^{-2} b^{-3}}$ **19.**________________

20. $\dfrac{3^{-1} m^{-4} p^6}{3^4 m^{-1} p^{-2}}$ **20.**________________

21. $\dfrac{8b^{-3} c^4}{8^{-4} b^{-7} c^{-3}}$ **21.**________________

Objective 4 Use combinations of rules.

Simplify each expression. Write answers with only positive exponents. Assume that all variables represent nonzero real numbers.

22. $\dfrac{(7^2)^6}{7^6}$ **22.**________________

23. $x^{-3} \cdot x^3 \cdot x^9$ **23.**________________

Name: Date:
Instructor: Section:

24. $\left(2^{-4}\right)^{4}$

24.________________

25. $\left(5w^{2}y^{2}\right)^{-2}\left(4wy^{-3}\right)^{2}$

25.________________

26. $\dfrac{(2y)^{-4}}{(3y)^{-2}}$

26.________________

27. $\dfrac{c^{10}\left(c^{2}\right)^{3}}{\left(c^{3}\right)^{3}\left(c^{2}\right)^{-9}}$

27.________________

28. $\dfrac{\left(a^{-1}b^{-2}\right)^{-4}\left(ab^{2}\right)^{6}}{\left(a^{3}b\right)^{-2}}$

28.________________

29. $\left(\dfrac{k^{3}t^{4}}{k^{2}t^{-1}}\right)^{-4}$

29.________________

30. $\dfrac{\left(3^{-2}x^{-5}y\right)^{-4}\left(2x^{2}y^{-4}\right)^{2}}{\left(2x^{-2}y^{2}\right)^{-2}}$

30.________________

Name: Date:
Instructor: Section:

Chapter 4 EXPONENTS AND POLYNOMIALS

4.3 An Applications of Exponents: Scientific Notation

Learning Objectives
1 Express numbers in scientific notation.
2 Convert numbers in scientific notation to numbers without exponents.
3 Use scientific notation in calculations.

Key Terms
Use the vocabulary terms listed below to complete each statement in exercises 1-3.

scientific notation **left** **right**

1. A number written in ___________________ has the form $a \times 10^n$, where $1 \le |a| < 10$ and n is an integer.

2. To write a number in the form $a \times 10^n$, where $1 \le |a| < 10$ and $n \ge 1$, without exponents, move the decimal to the ___________________ n places.

3. To write a number in the form $a \times 10^n$, where $1 \le |a| < 10$ and $n \le -1$, without exponents, move the decimal to the ___________________ n places.

Objective 1 Express numbers in scientific notation.

Write each number in scientific notation.

1. 325 **1.**_______________

2. 4579 **2.**_______________

3. 23,651 **3.**_______________

4. 209,907 **4.**_______________

5. 429,600,000,000 **5.**_______________

6. .0257 **6.**_______________

Name: Date:
Instructor: Section:

7. .246 **7.**________________

8. .00000413 **8.**________________

9. .00426 **9.**________________

10. –.00047 **10.**________________

Objective 2 Convert numbers in scientific notation to numbers without exponents.

Write each number without exponents.

11. 2.5×10^{4} **11.**________________

12. 7.2×10^{7} **12.**________________

13. -2.45×10^{6} **13.**________________

14. 4.045×10^{0} **14.**________________

15. 4.5×10^{7} **15.**________________

16. 6.4×10^{-3} **16.**________________

17. 7.24×10^{-4} **17.**________________

18. 4.007×10^{-2} **18.**________________

19. 4.752×10^{-1} **19.**________________

Name: Date:
Instructor: Section:

20. -9.11×10^{-4} **20.**________________

Objective 3 Use scientific notation in calculations.

Perform the indicated operations with the numbers in scientific notation, and then write your answers without exponents.

21. $(7\times10^{7})\times(3\times10^{0})$ **21.**________________

22. $(3\times10^{6})\times(4\times10^{-2})\times(2\times10^{-1})$ **22.**________________

23. $(2.3\times10^{4})\times(1.1\times10^{-2})$ **23.**________________

24. $(2.3\times10^{-4})\times(3.1\times10^{-2})$ **24.**________________

25. $\dfrac{4.6\times10^{-3}}{2.3\times10^{-1}}$ **25.**________________

26. $\dfrac{8.5\times10^{-3}}{1.7\times10^{-7}}$ **26.**________________

27. $\dfrac{9.39\times10^{1}}{3\times10^{3}}$ **27.**________________

Name: Date:
Instructor: Section:

28. $\left(3\times10^{4}\right)\times\left(4\times10^{2}\right)\div\left(2\times10^{3}\right)$

28.________________

29. $\dfrac{\left(7.5\times10^{6}\right)\times\left(4.2\times10^{-5}\right)}{\left(6\times10^{4}\right)\times\left(2.5\times10^{-3}\right)}$

29.________________

30. $\dfrac{\left(2.1\times10^{-3}\right)\times\left(4.8\times10^{4}\right)}{\left(1.6\times10^{6}\right)\times\left(7\times10^{-6}\right)}$

30.________________

Name: Date:
Instructor: Section:

Chapter 4 EXPONENTS AND POLYNOMIALS

4.4 Adding and Subtracting Polynomials; Graphing Simple Polynomials

Learning Objectives
1 Identify terms and coefficients.
2 Add like terms.
3 Know the vocabulary for polynomials.
4 Evaluate polynomials.
5 Add and subtract polynomials.
6 Graph equations defined by polynomials of degree 2.

Key Terms
Use the vocabulary terms listed below to complete each statement in exercises 1-8.

term **coefficient** **descending powers** **degree of a term**

degree of a polynomial **monomial** **binomial** **trinomial**

1. A ___________________ is a polynomial with only one term.

2. The ___________________ is the sum of the exponents on the variables in the term.

3. The ___________________ is the greatest degree of any of the terms in the polynomial.

4. A ___________________ is the numerical factor of a term.

5. A ___________________ is a polynomial with exactly two terms.

6. A ___________________ is a number, variable, or the product or quotient of a number and one or more variables raised to powers.

7. A ___________________ is a polynomial with exactly three terms.

8. A polynomial in one variable is written in ___________________ of the variable if the exponents on the terms of the polynomial decrease from left to right.

Objective 1 Identify terms and coefficients.

For each of the following, determine the number of terms in the polynomial and name the coefficients of the terms.

1. $-7y^4$ **1.**________________

Name: Date:
Instructor: Section:

2. $-4b^3 + 2b^2 - 3b$ **2.**____________

3. $9x^3 + 3x^3 - 4x + 2$ **3.**____________

4. $-\frac{2}{5}y^2 + \frac{4}{3}y$ **4.**____________

Objective 2 Add like terms.

In each polynomial, add like terms whenever possible. Write the result in descending powers of the variable.

5. $-6s^3 + 12s^3$ **5.**____________

6. $7z^3 - 4z^3 + 5z^3 - 11z^3$ **6.**____________

7. $4y^4 - 7y^2 + 4 - 7y^3 + 9y^4 - 2y^2 - 3y$ **7.**____________

8. $-12x^3 - 2x^2 + 4x - 7x^3$ **8.**____________

9. $-\frac{1}{2}r^3 + \frac{1}{3}r + \frac{1}{4}r^3 - \frac{1}{3}r$ **9.**____________

Objective 3 Know the vocabulary for polynomials.

Choose one or more of the following descriptions for each expression: (a) polynomial, (b) polynomial written in descending order, (c) not a polynomial.

10. $-2w^3 + 9w^2 + 4w - 10$ **10.**____________

Name: Date:
Instructor: Section:

11. $3a^5 - 4a^3 - 3a^{-1} + 5$ **11.**________________

For each polynomial, first simplify, if possible, and write the resulting polynomial in descending powers of the variable. Then give the degree of this polynomial, and tell whether it is a monomial, *a* binomial, *a* trinomial, *or* none of these.

12. $7m^2 + 3m + m^4$ **12.**________________

13. $3n^8 - n^2 - 2n^8$ **13.**________________

Objective 4 Evaluate polynomials.

Find the value of the polynomial when (a) $x = 3$ and (b) $x = -2$.

14. $7x + 5$ **14.**________________

15. $3x^2 - 4x + 1$ **15.**________________

16. $-x^2 + 5x - 9$ **16.**________________

17. $2x^3 + 4x^2 - 7x + 3$ **17.**________________

18. $5x^4 - 2x^2 + 8x - 1$ **18.**________________

Name: Date:
Instructor: Section:

Objective 5 Add and subtract polynomials.

Add.

19. $\begin{array}{r} 5m^4+2m^3-4 \\ \underline{-3m^4+5m^3-3} \end{array}$

19.________________

20. $\begin{array}{l} 5w^4+2m^2-6m+6 \\ \underline{2w^4-2m^2+7m+8} \end{array}$

20.________________

21. $\begin{array}{l} 7p^4 \qquad\quad +5p^2-2p+\ 8 \\ \underline{2p^4-3p^3-8p^2-3p+14} \end{array}$

21.________________

22. $(3x^2+2x^4-3)+(8x^3-5x^4-6x^2)$

22.________________

23. $(-4a^5-6a^3+5a+2)+(4a^3-7a-6)$

23.________________

Subtract.

24. $\begin{array}{l} 6m^3-5m \\ \underline{7m^3-3m} \end{array}$

24.________________

25. $\begin{array}{r} 2m^2-5m+1 \\ \underline{-2m^2-5m+3} \end{array}$

25.________________

26. $\begin{array}{l} 2z^5+z^4-2z^3 \qquad\qquad\quad +5 \\ \underline{\qquad\ \ 2z^4 \qquad -7z^2+8z-2} \end{array}$

26.________________

Name: Date:
Instructor: Section:

27. $\left(2x^3-4x^2+3x+10\right)-\left(6x^3-4x+2\right)$

27.________________

28. $\left(8p^2+7p-2\right)-\left(3p^2-3p+7\right)-\left(2p^2-3\right)$

28.________________

Objective 6 Graph equations defined by polynomials of degree 2.

Select several values for x; then find the corresponding y-values, and graph.

29. $y=x^2-1$

29.________________

30. $y=2-x^2$

30.________________

Name: Date:
Instructor: Section:

Chapter 4 EXPONENTS AND POLYNOMIALS

4.5 Multiplying Polynomials

Learning Objectives
1 Multiply a monomial and a polynomial.
2 Multiply two polynomials.
3 Multiply binomials by the FOIL method.

Key Terms
Use the vocabulary terms listed below to complete each statement in exercises 1-2.

multiply two polynomials **FOIL method**

1. To ___________________ , multiply each term of the first polynomial by each term of the second polynomial and add the products.

2. The ___________________ is a method used for multiplying two binomials.

Objective 1 Multiply a monomial and a polynomial.

Find each product.

1. $(3y^3)(4y^2)$ **1.**________________

2. $7z(5z^3+2)$ **2.**________________

3. $-2x^4(3+6x+2x^2)$ **3.**________________

4. $-6z(z^5+3z^3+4z+2)$ **4.**________________

5. $-3y^2(2y^3+3y^2-4y+11)$ **5.**________________

6. $2m(3+7m^2+3m^3)$ **6.**________________

Name: Date:
Instructor: Section:

7. $7b^2\left(-5b^2+1-4b\right)$ **7.**________________

8. $-4r^4\left(2r^2-3r+2\right)$ **8.**________________

9. $-3r^2s^3\left(8r^2s^2-4rs+2rs^2\right)$ **9.**________________

Objective 2 Multiply two polynomials.

Find the product.

10. $(x+3)(x+9)$ **10.**________________

11. $(3p+4)(p+2)$ **11.**________________

12. $(3m-5)(2m+4)$ **12.**________________

13. $(y+4)\left(y^2-4y+16\right)$ **13.**________________

14. $(r+3)\left(2r^2-3r+5\right)$ **14.**________________

15. $\left(2m^2+1\right)\left(3m^3+2m^2-4m\right)$ **15.**________________

Name: Date:
Instructor: Section:

Find each product, using the vertical method of multiplication.

16. $(2x+3)(2x^2-3x+2)$ **16.**________________

17. $(2y-3)(3y^3+2y^2-y+2)$ **17.**________________

18. $(2x^2+3x+2)(4x^3+2x+3)$ **18.**________________

19. $(2r+s)(2r+s)$ **19.**________________

20. $(3x^2+x)(2x^2+3x-4)$ **20.**________________

21. $(2x^2-x+2)(2x^2-x+2)$ **21.**________________

Name: Date:
Instructor: Section:

Objective 3 Multiply binomials by the FOIL method.

Use the FOIL method to find each product.

22. $(4m+3)(m-7)$ **22.**________________

23. $(3x+2y)(2x-3y)$ **23.**________________

24. $(5a-b)(4a+3b)$ **24.**________________

25. $(11k-4)(11k+4)$ **25.**________________

26. $(3+4a)(1+2a)$ **26.**________________

27. $(2v^2+w^2)(v^2-3w^2)$ **27.**________________

28. $(2y+.1)(2y-.5)$ **28.**________________

29. $(x-\frac{1}{3})(x-\frac{4}{3})$ **29.**________________

30. $(z+\frac{4}{5})(z-\frac{2}{5})$ **30.**________________

Name: Date:
Instructor: Section:

Chapter 4 EXPONENTS AND POLYNOMIALS

4.6 Special Products

Learning Objectives
1 Square binomials.
2 Find the product of the sum and difference of two terms.
3 Find greater powers of binomials.

Key Terms
Use the vocabulary terms listed below to complete each statement in exercises 1-2.

square of a binomial **product of the sum and difference of two terms**

1. The __________________ is a binomial consisting of the square of the first term minus the square of the second term.

2. The __________________ is a trinomial consisting of the square of the first term of the binomial, plus twice the product of the two terms, plus the square of the last term of the binomial.

Objective 1 Square binomials.

Find each square by using the pattern for the square of a binomial.

1. $(z+3)^2$ **1.**______________

2. $(a+2b)^2$ **2.**______________

3. $(5y-3)^2$ **3.**______________

4. $(2m+5)^2$ **4.**______________

5. $(5m+3n)^2$ **5.**______________

Name: Date:
Instructor: Section:

6. $(5+2y)^2$ 6.________________

7. $(2p+3q)^2$ 7.________________

8. $(2m-3p)^2$ 8.________________

9. $(4y-.7)^2$ 9.________________

10. $(4x-\frac{1}{4}y)^2$ 10.________________

11. $(3a+\frac{1}{2}b)^2$ 11.________________

Objective 2 Find the product of the sum and difference of two terms.

Find each product by using the pattern for the sum and difference of two terms.

12. $(z-6)(z+6)$ 12.________________

13. $(12+x)(12-x)$ 13.________________

14. $(7x-3y)(7x+3y)$ 14.________________

15. $(4p+7q)(4p-7q)$ 15.________________

Name: Date:
Instructor: Section:

16. $(2+3x)(2-3x)$ 16.________________

17. $(9-4y)(9+4y)$ 17.________________

18. $(x+.2)(x-.2)$ 18.________________

19. $(7m-\frac{3}{4})(7m+\frac{3}{4})$ 19.________________

20. $(2a+\frac{4}{3}b)(2a-\frac{4}{3}b)$ 20.________________

21. $(y^2+2)(y^2-2)$ 21.________________

22. $(5m^4-7n^3)(5m^4+7n^3)$ 22.________________

Objective 3 Find greater powers of binomials.

Find each product.

23. $(x+2)^3$ 23.________________

24. $(a-3)^3$ 24.________________

Name: Date:
Instructor: Section:

25. $(2x-3)^3$

25.________________

26. $(2x+1)^3$

26.________________

27. $(k+2)^4$

27.________________

28. $(x+2y)^4$

28.________________

29. $(3b-2)^4$

29.________________

30. $(4s+3t)^4$

30.________________

Name: Date:
Instructor: Section:

Chapter 4 EXPONENTS AND POLYNOMIALS

4.7 Dividing Polynomials

Learning Objectives
1 Divide a polynomial by a monomial.
2 Divide a polynomial by a polynomial.

Key Terms
Use the vocabulary terms listed below to complete each statement in exercises 1-3.

dividing a polynomial by a monomial **dividing a polynomial by a polynomial**

long division

1. When ____________________ divide each term of the polynomial by the monomial.

2. We use the method of ____________________ to divide a polynomial by a polynomial.

3. When ____________________ each polynomial must be written in descending powers.

Objective 1 Divide a polynomial by a monomial.

Divide each polynomial by $3m^2$.

1. $6m^3 + 9m^2$ **1.**________________

2. $12m^4 - 9m^3 + 6m^2$ **2.**________________

3. $15m^5 - 9m^3$ **3.**________________

4. $27m^3 + 18m^2 - 6m$ **4.**________________

5. $-54m^3 + 30m^2 + 6m$ **5.**________________

Name: Date:
Instructor: Section:

6. $3m^2 - 3$ **6.**________________

Perform each division.

7. $\dfrac{6p^4 + 18p^7}{3p^3}$ **7.**________________

8. $\dfrac{12x^6 + 18x^5 + 30x^3}{6x^2}$ **8.**________________

9. $(8y^7 - 9y^2) \div (4y)$ **9.**________________

10. $(6m^5 - 4m^3 + 12m) \div (4m)$ **10.**________________

11. $(m^2 + 3m - 12) \div (2m)$ **11.**________________

12. $\dfrac{8y^7 + 9y^6 - 11y - 12}{y^3}$ **12.**________________

13. $\dfrac{12z^5 + 8z^4 - 6z^3 + 5z}{3z^3}$ **13.**________________

14. $\dfrac{6y^5 - 3y^4 + 9y^2 + 27}{-3y}$ **14.**________________

Name: Date:
Instructor: Section:

15. $\dfrac{14y^2-14y+70}{-7y^2}$

15.________________

Objective 2 Divide a polynomial by a polynomial.

Perform each division.

16. $\dfrac{x^2-x-6}{x-3}$

16.________________

17. $\dfrac{18a^2-9a-5}{3a+1}$

17.________________

18. $\left(x^2+16x+64\right)\div\left(x+8\right)$

18.________________

19. $\left(2a^2-11a+16\right)\div\left(2a+3\right)$

19.________________

20. $\left(9w^2+12w+4\right)\div\left(3w+2\right)$

20.________________

Name: Date:
Instructor: Section:

21. $\dfrac{5w^2 - 22w + 4}{w - 4}$

21.________________

22. $\dfrac{9m^2 - 18m + 16}{3m - 4}$

22.________________

23. $\dfrac{4x^2 - 25}{2x - 5}$

23.________________

24. $\dfrac{2z^3 - 7z^2 + 3z + 2}{2z + 3}$

24.________________

25. $\left(27p^4 - 36p^3 - 6p^2 + 26p - 24\right) \div (3p - 4)$

25.________________

26. $\left(3x^3 - 11x^2 + 25x - 25\right) \div \left(x^2 - 3x - 5\right)$

26.________________

Name: Date:
Instructor: Section:

27. $\dfrac{6x^4-12x^3+13x^2-5x-1}{2x^2+3}$

27.________________

28. $\dfrac{2a^4+5a^2+3}{2a^2+3}$

28.________________

29. $\dfrac{y^3+1}{y+1}$

29.________________

30. $\dfrac{6x^5+7x^4-7x^3+7x+4}{3x+2}$

30.________________

Name: Date:
Instructor: Section:

Chapter 5 FACTORING AND APPLICATIONS

5.1 The Greatest Common Factor; Factoring by Grouping

Learning Objectives
1. Find the greatest common factor of a list of terms.
2. Factor out the greatest common factor.
3. Factor by grouping.

Key Terms
Use the vocabulary terms listed below to complete each statement in exercises 1-5.

factoring **factored form** **common factor** **greatest common factor**

factoring by grouping

1. The ____________________ of a list of integers or expressions is the largest common factor of those integers or expressions.

2. A polynomial is written in ____________________ if it is written as a product.

3. The process of writing a polynomial as a product is called ____________________.

4. An integer or expression that is a factor of two or more integers or expressions is called a ____________________.

5. ____________________ is a method for grouping terms of a polynomial in such a way that the polynomial can be factored even though its greatest common factor is 1.

Objective 1 Find the greatest common factor of a list of terms.
Find the greatest common factor for each group of numbers.

1. 36, 18, 24 **1.**________________

2. 108, 48, 84 **2.**________________

3. 17, 23, 40 **3.**________________

Name: Date:
Instructor: Section:

4. 70, 126, 42, 56 **4.**________________

5. 84, 280, 112 **5.**________________

Find the greatest common factor for each list of terms.

6. $18b^3,\ 36b^6,\ 45b^4$ **6.**________________

7. $7m^4,\ 12m^5,\ 21m^9$ **7.**________________

8. $y^7z^2,\ y^4z^8,\ z^3$ **8.**________________

9. $6k^2m^4n^5,\ 8k^3m^7n^4,\ k^4m^8n^7$ **9.**________________

10. $45a^7y^4,\ 75a^3y^2,\ 90a^2y,\ 30a^4y^3$ **10.**________________

11. $9xy^4, 72x^4y^7,\ 27xy^2,\ 108x^2y^5$ **11.**________________

Objective 2 Factor out the greatest common factor.

Complete the factoring.

12. $84 = 4(\quad)$ **12.**________________

13. $-18y^8 = -3y^5(\quad)$ **13.**________________

Name: Date:
Instructor: Section:

14. $-75a^4y^2 = 25a^3y^2(\quad)$ **14.**________________

Factor out the greatest common factor.

15. $26r + 39t$ **15.**________________

16. $45xy + 18x + 27x^3y$ **16.**________________

17. $24ab - 8a^2 + 40ac$ **17.**________________

18. $15a^7 - 25a^3 - 40a^4$ **18.**________________

19. $9y^2 - 7$ **19.**________________

20. $56x^2y^4 - 24xy^3 + 32xy^2$ **20.**________________

21. $3(a+b) - x(a+b)$ **21.**________________

22. $x^2(r-4s) + z^2(r-4s)$ **22.**________________

Name: Date:
Instructor: Section:

Objective 3 Factor by Grouping.

Factor each polynomial by grouping.

23. $x^2 + 2x + 5x + 10$ **23.**________________

24. $x^4 + 2x^2 + 5x^2 + 10$ **24.**________________

25. $3x^2 - 9x + 12x - 36$ **25.**________________

26. $xy - 2x - 2y + 4$ **26.**________________

27. $2a^3 - 3a^2b + 2ab^2 - 3b^3$ **27.**________________

28. $12x^3 - 4xy - 3x^2y^2 + y^3$ **28.**________________

29. $2x^4 + 4x^2y^2 + 3x^2y + 6y^3$ **29.**________________

30. $12x^2 + 4xy - 6xy - 2y^2$ **30.**________________

Name: Date:
Instructor: Section:

Chapter 5 FACTORING AND APPLICATIONS

5.2 Factoring Trinomials

Learning Objectives
1 Factor trinomials with a coefficient of 1 for the squared term.
2 Factor such trinomials after factoring out the greatest common factor.

Key Terms
Use the vocabulary terms listed below to complete each statement in exercises 1-2.

coefficient **prime polynomial**

1. A __________________ is a polynomial that cannot be factored into factors having only integer coefficients.

2. A __________________ is the numerical factor of a term.

Objective 1 Factor trinomials with a coefficient of 1 for the squared term.

List all pairs of integers with the given product. Then find the pair whose sum is given.

1. Product: 42; sum: 17 **1.**________________

2. Product: 28; sum: -11 **2.**________________

3. Product: –64; sum: 12 **3.**________________

4. Product: –54; sum –3 **4.**________________

Complete the factoring.

5. $x^2+7x+12=(x+3)(\quad)$ **5.**________________

6. $x^2+3x-28=(x-4)(\quad)$ **6.**________________

Name: Date:
Instructor: Section:

7. $x^2 + 4x + 4 = (x+2)(\quad)$ **7.**__________

8. $x^2 - x - 30 = (x+5)(\quad)$ **8.**__________

Factor completely. If a polynomial cannot be factored, write prime.

9. $x^2 + 11x + 18$ **9.**__________

10. $x^2 - 11x + 28$ **10.**__________

11. $x^2 - x - 2$ **11.**__________

12. $x^2 + 14x + 49$ **12.**__________

13. $x^2 - 2x - 35$ **13.**__________

14. $x^2 - 8x - 33$ **14.**__________

15. $x^2 + 6x + 5$ **15.**__________

16. $x^2 - 15xy + 56y^2$ **16.**__________

Name: Date:
Instructor: Section:

17. $x^2 - 4xy - 21y^2$ 17.________________

18. $m^2 - 2mn - 3n^2$ 18.________________

Objective 2 Factor such trinomials after factoring out the greatest common factor.

Factor completely.

19. $2x^2 + 10x - 28$ 19.________________

20. $3h^3k - 21h^2k - 54hk$ 20.________________

21. $4a^2 - 24b + 5$ 21.________________

22. $3p^6 + 18p^5 + 24p^4$ 22.________________

23. $2a^3b - 10a^2b^2 + 12ab^3$ 23.________________

24. $3y^3 + 9y^2 - 12y$ 24.________________

25. $5r^2 + 35r + 60$ 25.________________

Name: Date:
Instructor: Section:

26. $3xy^2 - 24xy + 36x$ **26.**________________

27. $10k^6 + 70k^5 + 100k^4$ **27.**________________

28. $x^5 - 3x^4 + 2x^3$ **28.**________________

29. $2x^2y^2 - 2xy^3 - 12y^4$ **29.**________________

30. $a^2b - 12ab^2 + 35b^3$ **30.**________________

Name: Date:
Instructor: Section:

Chapter 5 FACTORING AND APPLICATIONS

5.3 More on Factoring Trinomials

Learning Objectives
1 Factor trinomials by grouping when the coefficient of the squared term is not 1.
2 Factor trinomials by using the FOIL method.

Key Terms
Use the vocabulary terms listed below to complete each statement in exercises 1-2.

squared term of a trinomial **binomial factor**

1. A factor containing only two terms is called a ____________________.

2. The ____________________ is the term in which the variable is raised to the second power.

Objective 1 Factor trinomials by grouping when the coefficient of the squared term is not 1.

Factor by grouping.

1. $8b^2+18b+9$ **1.**________________

2. $3x^2+13x+14$ **2.**________________

3. $15a^2+16a+4$ **3.**________________

4. $6n^2+11n+4$ **4.**________________

5. $3b^2+8b+4$ **5.**________________

6. $3m^2-5m-12$ **6.**________________

7. $3p^3+8p^2+4p$ **7.**________________

Name: Date:
Instructor: Section:

8. $8m^2 + 26mn + 6n^2$ **8.**________________

9. $7a^2b + 18ab + 8b$ **9.**________________

10. $2s^2 + 5st - 3t^2$ **10.**________________

11. $9c^2 + 24cd + 12d^2$ **11.**________________

12. $25a^2 + 30ab + 9b^2$ **12.**________________

13. $9r^2 + 12r - 5$ **13.**________________

14. $12a^3 + 26a^2b + 12ab^2$ **14.**________________

Objective 2 Factor trinomials by using the FOIL method.

Complete the factoring.

15. $2x^2 + 5x - 3 = (2x - 1)(\quad)$ **15.**________________

16. $6x^2 + 19x + 10 = (3x + 2)(\quad)$ **16.**________________

17. $16x^2 + 4x - 6 = (4x + 3)(\quad)$ **17.**________________

18. $24y^2 - 17y + 3 = (3y - 1)(\quad)$ **18.**________________

Complete each trinomial by trial and error (using FOIL backwards).

19. $10x^2 + 19x + 6$ **19.**________________

Name: Date:
Instructor: Section:

20. $4y^2 + 3y - 10$ **20.**________________

21. $2a^2 + 13a + 6$ **21.**________________

22. $8q^2 + 10q + 3$ **22.**________________

23. $8m^2 - 10m - 3$ **23.**________________

24. $14b^2 + 3b - 2$ **24.**________________

25. $15q^2 - 2q - 24$ **25.**________________

26. $9w^2 + 12wz + 4z^2$ **26.**________________

27. $10c^2 - cd - 2d^2$ **27.**________________

28. $6x^2 + xy - 12y^2$ **28.**________________

29. $18x^2 - 27xy + 4y^2$ **29.**________________

30. $12y^2 + 11y - 15$ **30.**________________

Name: Date:
Instructor: Section:

Chapter 5 FACTORING AND APPLICATIONS

5.4 Special Factoring Techniques

Learning Objectives

1 Factor a difference of squares.
2 Factor a perfect square trinomial.
3 Factor a difference of cubes.
4 Factor a sum of cubes.

Key Terms

Use the vocabulary terms listed below to complete each statement in exercises 1-4.

difference of squares **perfect square trinomial** **difference of cubes**

sum of cubes

1. The ____________________ can be factored as a product of the sum and difference of two terms.

2. A ____________________ can be factored as $(x+y)(x^2-xy+y^2)$.

3. A ____________________ is a trinomial that can be factored as the square of a binomial.

4. A ____________________ can be factored as $(x-y)(x^2+xy+y^2)$.

Objective 1 Factor a difference of squares.

Factor each binomial completely. If a binomial cannot be factored, write prime.

1. x^2-49 **1.**________________

2. $100r^2-9s^2$ **2.**________________

3. $9j^2-\frac{16}{49}$ **3.**________________

4. $36-121d^2$ **4.**________________

5. $9m^4-1$ **5.**________________

Name: Date:
Instructor: Section:

6. $m^4n^2 - m^2$ 6.________________

Objective 2 Factor a perfect square trinomial.

Factor each trinomial completely. It may be necessary to factor out the greatest common factor first.

7. $y^2 + 6y + 9$ 7.________________

8. $m^2 - 8m + 16$ 8.________________

9. $z^2 - \frac{4}{3}z + \frac{4}{9}$ 9.________________

10. $64p^4 + 48p^2q^2 + 9q^4$ 10.________________

11. $-16x^2 - 48x - 36$ 11.________________

12. $-12a^2 + 60ab - 75b^2$ 12.________________

Objective 3 Factor a difference of cubes.

Find each difference. Write each answer in lowest terms.

13. $a^3 - 1$ 13.________________

14. $b^3 - 27$ 14.________________

15. $c^3 - 216$ 15.________________

16. $125z^3 - 8$ 16.________________

Name: Date:
Instructor: Section:

17. $c^9 - d^6$

17.________________

18. $125m^3 - 8p^3$

18.________________

19. $64x^3 - 27y^3$

19.________________

20. $8m^3 - \frac{1}{27}$

20.________________

21. $1000a^3 - 27b^3$

21.________________

Objective 4 Factor a sum of cubes.

Find each difference. Write each answer in lowest terms.

22. $y^3 + 27$

22.________________

23. $m^3 + 64$

23.________________

24. $n^3 + 216$

24.________________

25. $8b^3 + 1$

25.________________

Name: Date:
Instructor: Section:

26. $343d^3 + 27$ **26.**________________

27. $t^6 + 1$ **27.**________________

28. $64x^3 + 27y^3$ **28.**________________

29. $216m^3 + 125p^3$ **29.**________________

30. $27t^3 + \frac{1}{64}$ **30.**________________

Name: Date:
Instructor: Section:

Chapter 5 FACTORING AND APPLICATIONS

5.5 Solving Quadratic Equations by Factoring

Learning Objectives
1 Solve quadratic equations by factoring.
2 Solve other equations by factoring.

Key Terms

Use the vocabulary terms listed below to complete each statement in exercises 1-3.

quadratic equation **standard form** **zero-factor property**

1. The ____________________ states that if two numbers have a product of 0, then at least one of the numbers is 0.

2. A quadratic equation written in the form $ax^2 + bx + c = 0$, where $a \neq 0$, is in ____________________.

3. A ____________________ is an equation that can be written in the form $ax^2 + bx + c = 0$, where a, b, and c are real numbers, with $a \neq 0$.

Objective 1 Solve quadratic equations by factoring.

Solve each equation. Check your answers.

1. $(y+9)(2y-3)=0$ **1.**________________

2. $(3k+4)(5k-7)=0$ **2.**________________

3. $b^2 - 49 = 0$ **3.**________________

4. $2x^2 - 3x - 20 = 0$ **4.**________________

5. $x^2 - 2x - 63 = 0$ **5.**

Name: Date:
Instructor: Section:

6. $8r^2 = 24r$ 6.__________

7. $3x^2 - 7x - 6 = 0$ 7.__________

8. $3 - 5x = 8x^2$ 8.__________

9. $9x^2 + 12x + 4 = 0$ 9.__________

10. $25x^2 = 20x$ 10.__________

11. $9y^2 = 16$ 11.__________

12. $12x^2 + 7x - 12 = 0$ 12.__________

13. $14x^2 - 17x - 6 = 0$ 13.__________

14. $c(5c + 17) = 12$ 14.__________

15. $3x(x + 3) = (x + 2)^2 - 1$ 15.__________

Name: Date:
Instructor: Section:

Objective 2 Solve other equations by factoring.

Solve each equation.

16. $3x(x+7)(x-2)=0$ **16.**______________

17. $x(2x^2-7x-15)=0$ **17.**______________

18. $z(4z^2-9)=0$ **18.**______________

19. $z^3-49z=0$ **19.**______________

20. $25a=a^3$ **20.**______________

21. $x^3+2x^2-8x=0$ **21.**______________

22. $2m^3+m^2-6m=0$ **22.**______________

23. $(4x^2-9)(x-2)=0$ **23.**______________

24. $z^4+8z^3-9z^2=0$ **24.**______________

Name: Date:
Instructor: Section:

25. $3z^3+z^2-4z=0$ **25.**________________

26. $(y^2-5y+6)(y^2-36)=0$ **26.**________________

27. $15x^2=x^3+56x$ **27.**________________

28. $(y-7)(2y^2+7y-15)=0$ **28.**________________

29. $(x-\frac{3}{2})(2x^2+11x+15)=0$ **29.**________________

30. $(y-1)(y^2-25)=0$ **30.**________________

Name: Date:
Instructor: Section:

Chapter 5 FACTORING AND APPLICATIONS

5.6 Applications of Quadratic Functions

Learning Objectives
1 Solve problems involving geometric figures.
2 Solve problems involving consecutive integers.
3 Solve problems by using the Pythagorean formula.
4 Solve problems by using given quadratic models.

Key Terms
Use the vocabulary terms listed below to complete each statement in exercises 1-4.

consecutive integers **consecutive odd integers** **consecutive even integers**
hypotenuse

1. ____________________ are odd integers that are next to each other.

2. Two integers that differ by 1 are ____________________.

3. The ____________________ is the longest side in a right triangle. It is the side opposite the right angle.

4. ____________________ are even integers that are next to each other.

Objective 1 Solve problems involving geometric figures.

Solve the problem.

1. The length of a rectangle is 8 centimeters more than the width. The area is 153 square centimeters. Find the length and width of the rectangle.

1.________________

2. The length of a rectangle is three times its width. If the width were increased by 4 and the length remained the same, the resulting rectangle would have an area of 231 square inches. Find the dimensions of the original rectangle.

2.________________

Name: Date:
Instructor: Section:

3. The area of a rectangular room is 252 square feet. Its width is 4 feet less than its length. Find the length and width of the room.

3.________________

4. Two rectangles with different dimensions have the same area. The length of the first rectangle is three times its width. The length of the second rectangle is 4 meters more than the width of the first rectangle, and its width is 2 meters more than the width of the first rectangle. Find the lengths and widths of the two rectangles.

4.________________

5. Each side of one square is 1 meter less than twice the length of each side of a second square. If the difference between the areas of the two squares is 16 meters, find the lengths of the sides of the two squares.

5.________________

6. The area of a triangle is 42 square centimeters. The base is 2 centimeters less than twice the height. Find the base and height of the triangle.

6.________________

7. A rectangular bookmark is 6 centimeters longer than it is wide. Its area is numerically 3 more than its perimeter. Find the length and width of the bookmark.

7.________________

8. A book is three times as long as it is wide. Find the length and width of the book in inches if its area is numerically 128 more than its perimeter

8.________________

Name: Date:
Instructor: Section:

9. The volume of a box is 192 cubic feet. If the length of the box is 8 feet and the width is 2 feet more than the height, find the width of the box.

9.________________

10. Mr. Fixxall is building a box which will have a volume of 60 cubic meters. The height of the box will be 4 meters, and the length will be 2 meters more than the width. Find the width of the box.

10.________________

Objective 2 Solve problems involving consecutive integers.

Solve the problem.

11. The product of two consecutive integers is four less than four times their sum. Find the integers.

11.________________

12. Find two consecutive integers such that the square of their sum is 169.

12.________________

13. Find two consecutive integers such that the sum of the squares of the two integers is 3 more than the opposite (additive inverse) of the smaller integer.

13.________________

14. The product of two consecutive even integers is 24 more than three times the larger integer. Find the integers.

14.________________

15. Find all possible pairs of consecutive odd integers whose sum is equal to their product decreased by 47.

15.________________

Name: Date:
Instructor: Section:

16. Find two consecutive positive even integers whose product is 6 more than three times its sum.

16.________________

Objective 3 Solve problems by using the Pythagorean formula.

Solve the problem.

17. The hypotenuse of a right triangle is 4 inches longer than the shorter leg. The longer leg is 4 inches shorter than twice the shorter leg. Find the length of the shorter leg.

17.________________

18. A flag is shaped like a right triangle. The hypotenuse is 6 meters longer than twice the length of the shortest side of the flag. If the length of the other side is 2 meters less than the hypotenuse, find the lengths of the sides of the flag.

18.________________

19. A field has a shape of a right triangle with one leg 10 meters longer than twice the length of the other leg. The hypotenuse is 4 meters longer than three times the length of the shorter leg. Find the dimensions of the field.

19.________________

20. A train and a car leave a station at the same time, the train traveling due north and the car traveling west. When they are 100 miles apart, the train has traveled 20 miles farther than the car. Find the distance each has traveled.

20.________________

21. The hypotenuse of a right triangle is 1 foot larger than twice the shorter leg. The longer leg is 7 feet larger than the shorter leg. Find the length of the longer leg.

21.________________

Name: Date:
Instructor: Section:

22. Mark is standing directly beneath a kite attached to a string which Nina is holding, with her hand touching the ground. The height of the kite at that instant is 12 feet less than twice the distance between Mark and Nina. The length of the kite string is 12 feet more than that distance. Find the length of the kite string.

22.________________

23. A 30-foot ladder is leaning against a building. The distance from the bottom of the ladder to the building is 6 feet less than the distance from the top of the ladder to the ground. How far is the bottom of the ladder from the building?

23.________________

24. A field is in the shape of a right triangle. The shorter leg measures 45 meters. The hypotenuse measures 45 meters less than twice the longer leg. Find the dimensions of the lot.

24.________________

25. Two ships left a dock at the same time. When they were 25 miles apart, the ship that sailed due south had gone 10 miles less than twice the distance traveled by the ship that sailed due west. Find the distance traveled by the ship that sailed due south.

25.________________

26. A ladder is leaning against a building. The distance from the bottom of the ladder to the building is 8 feet less than the length of the ladder. How high up the side of the building is the top of the ladder if that distance is 4 feet less than the length of the ladder?

26.________________

Name: Date:
Instructor: Section:

Objective 4 Solve problems by using given quadratic formulas.

Use the quadratic model to answer the questions.

The equation $y = -.04x^2 + .93x + 21$ *was developed to model fuel economy trends within the automobile industry starting in 1978. Suppose that an automotive engineer is revising the model to project fuel economy trends into the 21st century. She develops the following formula:*

$$y = -.02x^2 + 1.19x + 27$$

and determines that x is coded so that x = 0 represents 1999.

27. Calculate the expected miles per gallon in 2005. Round your answer to the nearest tenth.

27.________________

28. Calculate the expected miles per gallon in 2049. Round your answer to the nearest tenth.

28.________________

Use the quadratic model to answer the questions.

29. If a ball is thrown upward from ground level with an initial velocity of 80 feet per second, its height h (in feet) t seconds later is given by the equation

$$h = -16t^2 + 80t$$

After how many seconds is the height 100 feet?

29.________________

30. An object is propelled upward from a height of 16 feet with an initial velocity of 48 feet per second. Its height h (in feet) t seconds later is given by the equation

$$h = -16t^2 + 48t + 16$$

After how many seconds is the height 48 feet?

30.________________

Name: Date:
Instructor: Section:

Chapter 6 RATIONAL EXPRESSIONS AND APPLICATIONS

6.1 The Fundamental Property of Rational Expressions

Learning Objectives
1. Find the numerical value of a rational expression.
2. Find the values of the variable for which a rational expression is undefined.
3. Write rational expressions in lowest terms.
4. Recognize equivalent forms of rational expressions.

Key Terms
Use the vocabulary terms listed below to complete each statement in exercises 1-3.

rational expression **lowest terms** **fundamental property of rational expressions**

1. A __________________ is the quotient of two polynomials with denominator not 0.

2. The __________________ states that $\frac{PK}{QK} = \frac{P}{Q}$ for $Q \neq 0$ and $K \neq 0$.

3. A rational expression is in __________________ if the greatest common factor of its numerator and denominator is 1.

Objective 1 Find the numerical value of a rational expression.

Find the numerical value of each expression when (a) $x = 4$ *and (b)* $x = -3$.

1. $\frac{3x^2 - 2x}{2x}$ 1.__________________

2. $\frac{(-4 - x)^2}{x + 3}$ 2.__________________

3. $\frac{2x + 5}{4x^2 - 25}$ 3.__________________

4. $\frac{3x^2}{3x + 2}$ 4.__________________

5. $\frac{-3x}{x^2 - x - 12}$ 5.__________________

Name: Date:
Instructor: Section:

6. $\dfrac{2x^2-4}{x^2-2}$ 6.________________

7. $\dfrac{2x-5}{2+x-x^2}$ 7.________________

8. $\dfrac{-2x^2}{2x^2-x+2}$ 8.________________

Objective 2 Find the values of the variable for which a rational expression is undefined.

Find all values for which the following expressions are undefined.

9. $\dfrac{9}{4x}$ 9.________________

10. $\dfrac{4x^2}{x+7}$ 10.________________

11. $\dfrac{x-4}{4x^2-16x}$ 11.________________

12. $\dfrac{4x+3}{x^2+x-12}$ 12.________________

13. $\dfrac{5x}{x^2-25}$ 13.________________

14. $\dfrac{z-3}{z^2+9}$ 14.________________

15. $\dfrac{x+5}{x^3+9x^2+18x}$ 15.________________

Name: Date:
Instructor: Section:

16. $\dfrac{y+2}{y^4-16}$ 16.________________

Objective 3 Write rational expressions in lowest terms.

Write each rational expression in lowest terms.

17. $\dfrac{24a^6b^9}{8a^2b^3}$ 17.________________

18. $\dfrac{15ab^3c^9}{-24ab^2c^{10}}$ 18.________________

19. $\dfrac{16r^2-4s^2}{8r+4s}$ 19.________________

20. $\dfrac{b^2+2b-15}{b^2+9b+20}$ 20.________________

21. $\dfrac{2x^2+9x-5}{2x^2+3x-2}$ 21.________________

22. $\dfrac{16-x^2}{2x-8}$ 22.________________

23. $\dfrac{3y^2-13y-10}{2y^2-9y-5}$ 23.________________

24. $\dfrac{6r^2-7rs-10s^2}{r^2+3rs-10s^2}$ 24.________________

Name: Date:
Instructor: Section:

25. $\dfrac{5x^2-17xy-12y^2}{x^2-7xy+12y^2}$

25.______________

26. $\dfrac{vw-5v+3w-15}{vw-5v-2w+10}$

26.______________

Objective 4 Recognize equivalent forms of rational expressions.

Write four equivalent forms of the following rational expressions.

27. $-\dfrac{2x-3}{x+2}$

27.______________

28. $\dfrac{4x+1}{5x-3}$

28.______________

29. $-\dfrac{2x-1}{3x+5}$

29.______________

30. $-\dfrac{2x+6}{x-5}$

30.______________

Name: Date:
Instructor: Section:

Chapter 6 RATIONAL EXPRESSIONS AND APPLICATIONS

6.2 Multiplying and Dividing Rational Expressions

Learning Objectives
1 Multiply rational expressions.
2 Divide rational expressions.

Key Terms
Use the vocabulary terms listed below to complete each statement in exercises 1-2.

multiply **divide**

1. To ____________________ rational expressions, multiply the numerators and multiply the denominators.

2. To ____________________ rational expressions, multiply the first rational expression by the reciprocal of the second rational expression.

Objective 1 Multiply rational expressions.

Multiply. Write each answer in lowest terms.

1. $\frac{8m^4n^3}{3} \cdot \frac{5}{4mn^2}$ **1.**________________

2. $\frac{6}{9y+36} \cdot \frac{4y+16}{9}$ **2.**________________

3. $\frac{12-4z}{6} \cdot \frac{9}{4z-12}$ **3.**________________

4. $\frac{x^2+x-12}{x^2+7x+10} \cdot \frac{x^2+3x-10}{x^2+2x-8}$ **4.**________________

Name: Date:
Instructor: Section:

5. $\dfrac{a^2-3a+2}{a^2-1}\cdot\dfrac{a^2+2a-3}{a^2+a-6}$

5.________________

6. $\dfrac{3m^2-m-10}{2m^2-7m-4}\cdot\dfrac{4m^2-1}{6m^2+7m-5}$

6.________________

7. $\dfrac{2x^2+5x-12}{x^2-2x-24}\cdot\dfrac{x^2-9x+18}{9-4x^2}$

7.________________

8. $\dfrac{3x^2-12}{x^2-x-6}\cdot\dfrac{x^2-6x+9}{2x-4}$

8.________________

9. $\dfrac{2x+1}{16-x^2}\cdot\dfrac{x-4}{4x+2}$

9.________________

10. $\dfrac{4r+4p}{8z^2}\cdot\dfrac{36z^6}{r^2+rp}$

10.________________

11. $\dfrac{3x+12}{6x-30}\cdot\dfrac{x^2-x-20}{x^2-16}$

11.________________

12. $\dfrac{m^2-16}{m-3}\cdot\dfrac{m^2-9}{m+4}$

12.________________

Name: Date:
Instructor: Section:

13. $\frac{9a-18}{6a+12} \cdot \frac{6a+12}{30-15a}$

13.________________

14. $\frac{x^2-4}{2x^2-2} \cdot \frac{x-x^2}{2x^2+4x}$

14.________________

Objective 2 Divide rational expressions.

Divide. Write each answer in lowest terms.

15. $\frac{9}{7m^3} \div \frac{15}{28m^6}$

15.________________

16. $\frac{6z^3}{9zw} \div \frac{z^7}{21zw^2}$

16.________________

17. $\frac{b-7}{16} \div \frac{7-b}{8}$

17.________________

18. $\frac{2x+2y}{8z} \div \frac{x^2\left(x^2-y^2\right)}{24}$

18.________________

19. $\frac{4m-12}{2m+10} \div \frac{m^2-9}{m^2-25}$

19.________________

Name: Date:
Instructor: Section:

20. $\frac{6a(a+3)}{3a+1} \div \frac{a^2(a+3)}{9a^2-1}$

20.________________

21. $\frac{m^2+2mn+n^2}{m^2+m} \div \frac{m^2-n^2}{m^2-1}$

21.________________

22. $\frac{9a^2-1}{9a^2-6a+1} \div \frac{3a^2-11a-4}{12a^2+5a-3}$

22.________________

23. $\frac{12k^2-5k-3}{9k^2-1} \div \frac{16k^2-9}{12k^2+13k+3}$

23.________________

24. $\frac{2z^2-11z-21}{z^2-5z-14} \div \frac{4z^2-9}{z^2-6z-16}$

24.________________

25. $\frac{y^2+yz-12z^2}{y^2+yz-20z^2} \div \frac{y^2+9yz+20z^2}{y^2-yz-30z^2}$

25.________________

26. $\frac{4(b-3)(b+2)}{b^2+3b+2} \div \frac{b^2-6b+9}{b^2+4b+4}$

26.________________

27. $\frac{27-3k^2}{3k^2+8k-3} \div \frac{k^2-6k+9}{6k^2-19k+3}$

27.________________

Name: Date:
Instructor: Section:

28. $\frac{2y^2-21y-11}{2-8y^2} \div \frac{y^2-12y+11}{4y^2+14y-8}$

28.________________

29. $\frac{y^2+7y+10}{3y+6} \div \frac{y^2+2y-15}{4y-4}$

29.________________

30. $\frac{2k^2+5k-12}{2k^2+k-3} \div \frac{k^2+8x+16}{2k^2+11k+12}$

30.________________

Name: Date:
Instructor: Section:

Chapter 6 RATIONAL EXPRESSIONS AND APPLICATIONS

6.3 Least Common Denominators

Learning Objectives
1 Find the least common denominator for a group of fractions.
2 Rewrite rational expressions with given denominators.

Key Terms
Use the vocabulary terms listed below to complete each statement in exercises 1-2.

least common denominator **greatest**

1. To find the least common denominator, start by factoring each denominator into prime factors and listing each different denominator factor the __________________ number of times it appears in any denominators.

2. The __________________ is the simplest expression that is divisible by all of the denominators in all of the expressions.

Objective 1 Find the least common denominator for a group of fractions.

Find the least common denominator for each list of rational expressions.

1. $\frac{5}{12}, \frac{9}{16}$ 1.________________

2. $\frac{5}{9}, \frac{7}{15}$ 2.________________

3. $\frac{8}{12}, \frac{7}{20}, \frac{11}{18}$ 3.________________

4. $\frac{9}{14}, \frac{17}{20}, \frac{7}{15}$ 4.________________

5. $\frac{5}{8ab^2}, \frac{7}{6a^2b}$ 5.________________

Name: Date:
Instructor: Section:

6. $\frac{13}{36b^4}, \frac{17}{27b^2}$

6.________________

7. $\frac{4}{5r-25}, \frac{7}{15r^3}$

7.________________

8. $\frac{15}{7t-28}, \frac{21}{6t-24}$

8.________________

9. $\frac{7}{x-y}, \frac{3}{y-x}$

9.________________

10. $\frac{4}{a^2-b^2}, \frac{8}{b^2-a^2}$

10.________________

11. $\frac{3m}{2m^2+9m-5}, \frac{4}{m^2+5m}$

11.________________

12. $\frac{v-4}{3v^4-6v^3}, \frac{v+2}{v^2+2v-8}$

12.________________

13. $\frac{3z+1}{z^4+2z^3-8z^2}, \frac{5z+2}{z^3+8z^2+16z}$

13.________________

14. $\frac{3p+2}{p^2-9}, \frac{2p+7}{p^2-p-12}$

14.________________

15. $\frac{11q-3}{2q^2-q-10}, \frac{21-q}{2q^2-9q+10}$

15.________________

Name: Date:
Instructor: Section:

16. $\dfrac{17r}{9r^2-6r-8}, \dfrac{-13r}{9r^2-9r-4}$

16.________________

17. $\dfrac{m+2}{m^3-2m^2}, \dfrac{3-m}{m^2+5m-14}$

17.________________

Objective 2 Rewrite rational expressions with given denominators.

Rewrite each rational expression with the indicated denominator. Give the numerator of the new fraction.

18. $\dfrac{5}{6}=\dfrac{?}{18}$

18.________________

19. $\dfrac{4}{r}=\dfrac{?}{7r}$

19.________________

20. $\dfrac{7m}{8n}=\dfrac{?}{24n^6}$

20.________________

21. $\dfrac{11a+1}{2a-6}=\dfrac{?}{8a-24}$

21.________________

22. $\dfrac{-3y}{4y+12}=\dfrac{?}{4(y+3)^2}$

22.________________

23. $\dfrac{8z}{3z+3}=\dfrac{?}{12z^2+15z+3}$

23.________________

24. $\dfrac{5}{2r+8}=\dfrac{?}{2(r+4)(r^2+2r-8)}$

24.________________

Name: Date:
Instructor: Section:

25. $\dfrac{9}{y^2-4}=\dfrac{?}{(y+2)^2(y-2)}$ 25.________

26. $\dfrac{2}{7p-35}=\dfrac{?}{14p^3-70p^2}$ 26.________

27. $\dfrac{3}{5r-10}=\dfrac{?}{50r^2-100r}$ 27.________

28. $\dfrac{3}{k^2+3k}=\dfrac{?}{k^3+10k^2+21k}$ 28.________

29. $\dfrac{3x+1}{x^2-4}=\dfrac{?}{2x^2-8}$ 29.________

30. $\dfrac{2p^2}{p-9}=\dfrac{?}{p^2-81}$ 30.________

Name: Date:
Instructor: Section:

Chapter 6 RATIONAL EXPRESSIONS AND APPLICATIONS

6.4 Adding and Subtracting Rational Expressions

Learning Objectives
1 Add rational expressions having the same denominator.
2 Add rational expressions having different denominators.
3 Subtract rational expressions.

Key Terms
Use the vocabulary terms listed below to complete each statement in exercises 1-3.

same denominator **different denominators** **parentheses**

1. To add rational expressions with ____________________, rewrite each expression with the LCD of the rational expressions and add.

2. When subtracting rational expressions, be sure to use ____________________ after the subtraction sign.

3. To add rational expressions with the ____________________, add the numerators and keep the same denominator.

Objective 1 Add rational expressions having the same denominator.

Find each sum. Write each answer in lowest terms.

1. $\dfrac{4}{3w^2}+\dfrac{7}{3w^2}$ 1.________________

2. $\dfrac{x}{x^2-4}+\dfrac{7x}{x^2-4}$ 2.________________

3. $\dfrac{b}{b^2-4}+\dfrac{2}{b^2-4}$ 3.________________

4. $\dfrac{3m+4}{2m^2-7m-15}+\dfrac{m+2}{2m^2-7m-15}$ 4.________________

5. $\dfrac{4m}{m^2+3m+2}+\dfrac{8}{m^2+3m+2}$ 5.________________

6. $\dfrac{6x}{(2x+1)^2}+\dfrac{3}{(2x+1)^2}$ 6.________________

7. $\dfrac{2y-5}{2y^2-5y-3}+\dfrac{2-y}{2y^2-5y-3}$ 7.________________

Name: Date:
Instructor: Section:

8. $\frac{2x}{9x^2-25y^2}+\frac{x-5y}{9x^2-25y^2}$ **8.**__________

Objective 2 Add rational expressions having different denominators.

Find each sum. Write each answer in lowest terms.

9. $\frac{x}{3}-\frac{2}{5}$ **9.**__________

10. $\frac{3z-2}{5z+20}+\frac{2z+1}{3z+12}$ **10.**__________

11. $\frac{2}{a^2-4}+\frac{3}{a+2}$ **11.**__________

12. $\frac{-4}{h+1}+\frac{2h}{1-h^2}$ **12.**__________

13. $\frac{2y+9}{y^2+6y+8}+\frac{y+3}{y^2+2y-8}$ **13.**__________

14. $\frac{7}{x-5}+\frac{4}{x+5}$ **14.**__________

15. $\frac{2s+3}{3s^2-14s+8}+\frac{4s+5}{2s^2-5s-12}$ **15.**__________

Name: Date:
Instructor: Section:

16. $\dfrac{5p-2}{2p^2+9p+9}+\dfrac{p+7}{6p^2+13p+6}$

16.________________

17. $\dfrac{1-3x}{4x^2-1}+\dfrac{3x-5}{2x^2+5x+2}$

17.________________

18. $\dfrac{3}{6x^2+x-2}+\dfrac{-1}{3x^2+8x+4}$

18.________________

Objective 3 Subtract rational expressions.

Find each difference. Write each answer in lowest terms.

19. $\dfrac{6p}{p-4}-\dfrac{p+20}{p-4}$

19.________________

20. $\dfrac{2x}{x^2+3x-10}-\dfrac{x+2}{x^2+3x-10}$

20.________________

21. $\dfrac{4x}{9x^2-16y^2}-\dfrac{x+4y}{9x^2-16y^2}$

21.________________

22. $\dfrac{7}{x+4}-\dfrac{5}{3x+12}$

22.________________

23. $\dfrac{5-10s}{6}-\dfrac{5-5s}{9}$

23.________________

Name: Date:
Instructor: Section:

24. $\frac{6}{x-y}-\frac{4+y}{y-x}$ **24.**________________

25. $\frac{6}{2q^2+5q+2}-\frac{5}{2q^2-3q-2}$ **25.**________________

26. $\frac{m}{m^2-4}-\frac{1-m}{m^2+4m+4}$ **26.**________________

27. $\frac{4y}{y^2+4y+3}-\frac{3y+1}{y^2-y-2}$ **27.**________________

28. $\frac{3}{n^2-16}-\frac{6n}{n^2+8n+16}$ **28.**________________

29. $\frac{4x-1}{2x^2+5x-3}-\frac{x+3}{6x^2+x-2}$ **29.**________________

30. $\frac{6z}{(z+1)^2}-\frac{2z+3}{z^2-1}$ **30.**________________

Name: Date:
Instructor: Section:

Chapter 6 RATIONAL EXPRESSIONS AND APPLICATIONS

6.5 Complex Fractions

Learning Objectives
1 Simplify a complex fraction by writing it as a division problem (Method 1).
2 Simplify a complex fraction by multiplying by the least common denominator (Method 2).

Key Terms
Use the vocabulary terms listed below to complete each statement in exercises 1-3.

complex fraction **method 1** **method 2**

1. ____________________ is the procedure used to simplify complex fractions by (i) writing both the numerator and denominator as single fractions, (ii) changing the complex fraction to a division problem, and (iii) performing the indicated division.

2. ____________________ is the procedure used to simplify complex fractions by (i) finding the LCD of all fractions within the complex fraction and (ii) multiplying both the numerator and the denominator of the complex fraction by this LCD.

3. A ____________________ is a rational expression with one or more fractions in the numerator, or denominator, or both.

Objective 1 Simplify a complex fraction by writing it as a division problem (Method 1).

Simplify each complex fraction by writing it as a division problem.

1. $\dfrac{-\frac{3}{5}}{\frac{9}{10}}$ **1.**________________

2. $\dfrac{\frac{3}{4}-\frac{1}{2}}{\frac{1}{4}+\frac{5}{8}}$ **2.**________________

3. $\dfrac{\dfrac{49m^3}{18n^5}}{\dfrac{21m}{27n^2}}$ **3.**________________

Name: Date:
Instructor: Section:

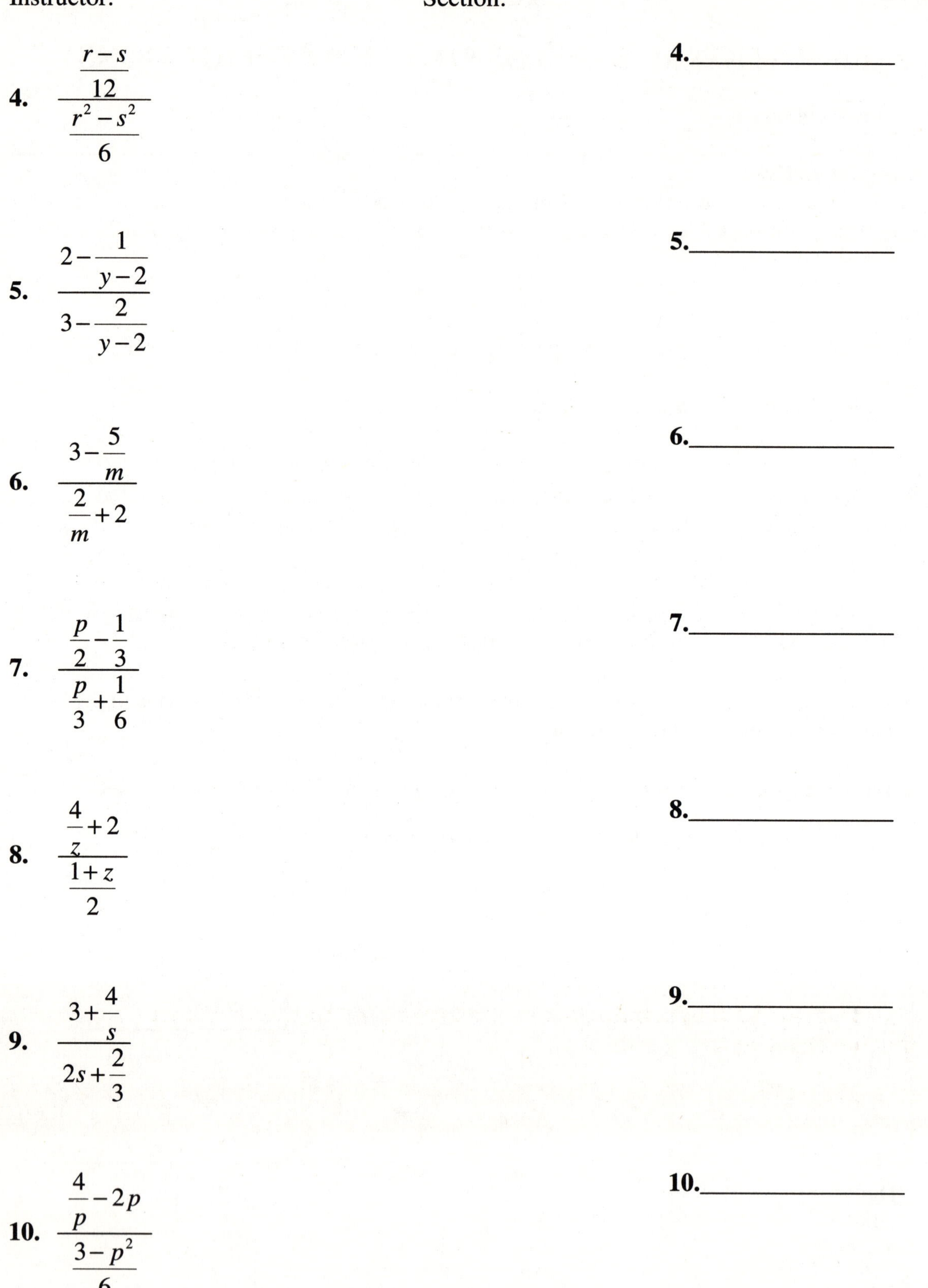

4. $\dfrac{\dfrac{r-s}{12}}{\dfrac{r^2-s^2}{6}}$ 4.________________

5. $\dfrac{2-\dfrac{1}{y-2}}{3-\dfrac{2}{y-2}}$ 5.________________

6. $\dfrac{3-\dfrac{5}{m}}{\dfrac{2}{m}+2}$ 6.________________

7. $\dfrac{\dfrac{p}{2}-\dfrac{1}{3}}{\dfrac{p}{3}+\dfrac{1}{6}}$ 7.________________

8. $\dfrac{\dfrac{4}{z}+2}{\dfrac{1+z}{2}}$ 8.________________

9. $\dfrac{3+\dfrac{4}{s}}{2s+\dfrac{2}{3}}$ 9.________________

10. $\dfrac{\dfrac{4}{p}-2p}{\dfrac{3-p^2}{6}}$ 10.________________

Name: Date:
Instructor: Section:

11. $\dfrac{\dfrac{a+2}{a-2}}{\dfrac{1}{a^2-4}}$

11.________________

12. $\dfrac{\dfrac{2}{a+2}-4}{\dfrac{1}{a+2}-3}$

12.________________

13. $\dfrac{\dfrac{3}{w-4}-\dfrac{3}{w+4}}{\dfrac{1}{w+4}+\dfrac{1}{w^2-16}}$

13.________________

14. $\dfrac{\dfrac{5}{rs^2}-\dfrac{2}{rs}}{\dfrac{3}{rs}-\dfrac{4}{r^2s}}$

14.________________

15. $\dfrac{\dfrac{3a+4}{a}}{\dfrac{1}{a}+\dfrac{2}{5}}$

15.________________

Objective 2 Simplify a complex fraction by multiplying by the least common denominator (Method 2).

Simplify each complex fraction by multiplying by the least common denominator.

16. $\dfrac{\frac{9}{20}}{-\frac{11}{25}}$

16.________________

17. $\dfrac{\frac{1}{2}+\frac{3}{8}}{\frac{3}{4}-\frac{9}{8}}$

17.________________

Name: Date:
Instructor: Section:

18. $\dfrac{\frac{a^2}{b}}{\frac{a^2}{b^2}}$

18.________________

19. $\dfrac{\frac{16r^2}{11s^3}}{\frac{8r^4}{22s}}$

19.________________

20. $\dfrac{2x-y^2}{x+\frac{y^2}{x}}$

20.________________

21. $\dfrac{r+\frac{3}{r}}{\frac{5}{r}+rt}$

21.________________

22. $\dfrac{\frac{x-2}{x+2}}{\frac{x}{x-2}}$

22.________________

23. $\dfrac{2s+\frac{3}{s}}{\frac{1}{s}-3s}$

23.________________

24. $\dfrac{\frac{15}{10k+10}}{\frac{5}{3k+3}}$

24.________________

25. $\dfrac{\frac{1}{h}-4}{\frac{1}{2}+2h}$

25.________________

Name: Date:
Instructor: Section:

26. $\dfrac{\dfrac{4}{x}-\dfrac{1}{2}}{\dfrac{5}{x}+\dfrac{1}{3}}$

26.________________

27. $\dfrac{\dfrac{1}{m-1}+4}{\dfrac{2}{m-1}-4}$

27.________________

28. $\dfrac{\dfrac{4}{x+4}}{\dfrac{3}{x^2-16}}$

28.________________

29. $\dfrac{\dfrac{6}{k+1}-\dfrac{5}{k-3}}{\dfrac{3}{k-3}+\dfrac{2}{k+2}}$

29.________________

30. $\dfrac{\dfrac{4}{t+2}+\dfrac{5}{t-1}}{\dfrac{3}{t+2}-\dfrac{1}{t-1}}$

30.________________

Name: Date:
Instructor: Section:

Chapter 6 RATIONAL EXPRESSIONS AND APPLICATIONS

6.6 Solving Equations with Rational Expressions

Learning Objectives

1 Distinguish between operations with rational expressions and equations with terms that are rational expressions.
2 Solve equations with rational expressions.
3 Solve a formula for a specified variable.

Key Terms

Use the vocabulary terms listed below to complete each statement in exercises 1-4.

simplified **solved** **adding or subtracting** **LCD**

1. Sums and differences are expressions to be ____________________.

2. When ____________________ rational expressions, keep the LCD throughout the simplification.

3. Equations are ____________________.

4. When solving an equation, multiply each side by the ____________________ so that denominators are eliminated.

Objective 1 Distinguish between operations with rational expressions and equations with terms that are rational expressions.

Identify each of the following as an operation or an equation. If it is an operation, perform it. If it is an equation, solve it.

1. $\frac{2x}{3}+\frac{2x}{5}=\frac{64}{15}$ **1.**________________

2. $\frac{3x}{5}-\frac{4x}{3}=\frac{22}{15}$ **2.**________________

3. $\frac{4x}{5}-\frac{5x}{10}$ **3.**________________

Name: Date:
Instructor: Section:

4. $\frac{2x}{5}+\frac{7x}{3}$

4.________________

Objective 2 Solve equations with rational expressions.

Solve each equation and check your answers.

5. $\frac{p}{p-2}=\frac{2}{p-2}+1$

5.________________

6. $\frac{4}{m}-\frac{2}{3m}=\frac{10}{9}$

6.________________

7. $\frac{4}{5x}+\frac{3}{2x}=\frac{23}{50}$

7.________________

8. $\frac{x-4}{5}=\frac{x+2}{3}$

8.________________

9. $\frac{4}{n+2}-\frac{2}{n}=\frac{1}{6}$

9.________________

10. $\frac{x}{3x+16}=\frac{4}{x}$

10.________________

11. $\frac{2p+3}{3}=\frac{4p+2}{15}$

11.________________

12. $\frac{4+x}{6}+\frac{x}{4}=\frac{19}{6}$

12.________________

13. $\frac{8}{2m+4}+\frac{2}{3m+6}=\frac{7}{9}$

13.________________

Name: Date:
Instructor: Section:

14. $\dfrac{2}{z-1}+\dfrac{3}{z+1}-\dfrac{17}{24}=0$

14.________________

15. $\dfrac{2}{m-3}+\dfrac{12}{9-m^2}=\dfrac{3}{m+3}$

15.________________

16. $\dfrac{9}{x^2-x-12}=\dfrac{3}{x-4}-\dfrac{x}{x+3}$

16.________________

17. $\dfrac{-16}{n^2-8n+12}=\dfrac{3}{n-2}+\dfrac{n}{n-6}$

17.________________

18. $\dfrac{1}{z^2+5z+6}+\dfrac{1}{12(z+2)}=\dfrac{-1}{z^2-2z-8}$

18.________________

Objective 3 Solve a formula for a specified variable.

Solve each formula for the specified variable.

19. $P=\dfrac{I}{rt}$ for t

19.________________

20. $F=\dfrac{k}{d-D}$ for D

20.________________

21. $S=\dfrac{a_1}{1-r}$ for r

21.________________

22. $h=\dfrac{2A}{B+b}$ for B

22.________________

Name: Date:
Instructor: Section:

23. $F = \dfrac{GmM}{d^2}$ for G

23.________________

24. $\dfrac{1}{R} = \dfrac{1}{R_1} + \dfrac{1}{R_2}$ for R

24.________________

25. $\dfrac{V_1P_1}{T_1} = \dfrac{V_2P_2}{T_2}$ for T_1

25.________________

26. $\dfrac{1}{f} = \dfrac{1}{d_0} + \dfrac{1}{d_1}$ for f

26.________________

27. $F = \dfrac{Gm_1m_2}{d^2}$ for G

27.________________

28. $S_n = \dfrac{n}{2}(a_1 + a_n)$ for a_1

28.________________

29. $A = \dfrac{1}{2}h(b_1 + b_2)$ for b_2

29.________________

30. $A = \dfrac{R_1R_2}{R_1 + R_r}$ for R_r

30.________________

Name: Date:
Instructor: Section:

Chapter 6 RATIONAL EXPRESSIONS AND APPLICATIONS

6.7 Applications of Rational Expressions

Learning Objectives
1 Solve problems about numbers.
2 Solve problems about distance, rate, and time.
3 Solve problems about work.

Key Terms

Use the vocabulary terms listed below to complete each statement in exercises 1-4.

read **check** **rate of work** **smaller**

1. If t is the amount of time needed to complete a job, then $\frac{1}{t}$ is the ____________________, or the amount of the job completed for one unit of time.

2. The last step when solving a problem is to ____________________ the solution.

3. The first step when solving a problem is to ____________________ the problem.

4. The rate traveling upstream is always ____________________ than the rate traveling downstream.

Objective 1 Solve problems about numbers.

Solve each problem.

1. One-fifth of a number is two less than one-third of the same number. What is the number? **1.**________________

2. If the same number is added to the numerator and denominator of the fraction $\frac{5}{9}$, the value of the resulting fraction is $\frac{2}{3}$. Find the number. **2.**________________

3. If two times a number is added to one-half of its reciprocal, the result is $\frac{13}{6}$. Find the number. **3.**________________

Name: Date:
Instructor: Section:

4. If a certain number is added to the numerator and twice that number is subtracted from the denominator of the fraction $\frac{3}{5}$, the result is equal to 5. Find the number.

4.________________

5. In a certain fraction, the numerator is 4 less than the denominator. If 5 is added to both the numerator and the denominator, the resulting fraction is equal to $\frac{7}{9}$. Find the original fraction.

5.________________

6. The sum of a number and its reciprocal is $\frac{13}{6}$. Find the number.

6.________________

7. If twice the reciprocal of a number is added to the number, the result is $\frac{9}{2}$. Find the number.

7.________________

8. If three times a number is subtracted from twice its reciprocal, the result is –1. Find the number.

8.________________

9. Sharon and Elaine worked as computer analysts. Last year, Sharon earned $\frac{3}{5}$ as much as Elaine. If they earned a total of $152,000, how much did each of them earn?

9.________________

10. Lauren takes $\frac{4}{5}$ the number of pills that David takes for the same illness. Together they use 45 pills. Find the number used by David.

10.________________

Name: Date:
Instructor: Section:

Objective 2 Solve problems about distance, rate, and time.

Solve each problem.

11. Mark can row 5 miles per hour in still water. It takes him as long to row 4 miles upstream as 16 miles downstream. How fast is the current?

11.________________

12. John flew from City A to City B at 200 miles per hour and from City B to City A at 180 miles per hour. The trip at the slower speed took $\frac{1}{2}$ hour longer. Find the distance between the two cities. (Assume there is no wind in either direction.)

12.________________

13. Yohannes traveled to his destination at an average speed of 70 miles per hour. Coming home, his average speed was 50 miles per hour and the trip took 2 hours longer. How far did he travel each way?

13.________________

14. Dipti flew her plane 600 miles against the wind in the same time it took her to fly 900 miles with the wind. If the speed of the wind was 30 miles per hour, what was the speed of the plane?

14.________________

15. Wendy drove a distance of 250 miles, at a speed that was 10 miles per hour faster than her speed on her return trip. If it took Wendy $\frac{5}{6}$ hour longer on the return trip, what was her speed on the return trip?

15.________________

Name: Date:
Instructor: Section:

16. A boat goes 6 miles per hour in still water. It takes as long to go 40 miles upstream as 80 miles downstream. Find the speed of the current.

16.________________

17. A ship goes 120 miles downriver in $2\frac{2}{3}$ hours less than it takes to go the same distance upriver. If the speed of the current is 6 miles per hour, find the speed of the ship.

17.________________

18. A plane traveling 450 miles per hour can go 1000 miles with the wind in $\frac{1}{2}$ hour less than when traveling against the wind. Find the speed of the wind.

18.________________

19. On Saturday, Pablo jogged 6 miles. On Monday, jogging at the same speed, it took him 30 minutes longer to cover 10 miles. How fast did Pablo jog?

19.________________

20. A plane made the trip from Redding to Los Angeles, a distance of 560 miles, in 1.5 hours less than it took to fly from Los Angeles to Portland, a distance of 1130 miles. Find the rate of the plane. (Assume there is no wind in either direction.)

20.________________

Name: Date:
Instructor: Section:

Objective 3 Solve problems about work.

Solve each problem.

21. Kelly can clean the house in 6 hours, but it takes Linda 4 hours. How long would it take them to clean the house if they worked together?

21.________________

22. Nina can wash the walls in a certain room in 2 hours and Mark can wash these walls in 5 hours. How long would it take them to complete the task if they work together?

22.________________

23. Phil can install the carpet in a room in 3 hours, but Lil needs 8 hours. How long will it take them to complete this task if they work together?

23.________________

24. One pipe can fill a swimming pool in 8 hours and another pipe can fill the pool in 12 hours. How long will it take to fill the pool if both pipes are open?

24.________________

25. Chuck can weed the garden in $\frac{1}{2}$ hour, but David takes 2 hours. How long does it take them to weed the garden if they work together?

25.________________

Name: Date:
Instructor: Section:

26. Jack can paint a certain room in $1\frac{1}{2}$ hours, but Joe needs 4 hours to paint the same room. How long does it take them to paint the room if they work together?

26.________________

27. Michael can type twice as fast as Sharon. Together they can type a certain job in 2 hours. How long would it take Michael to type the entire job by himself?

27.________________

28. Working together, Ethel and Al can balance the books for a certain company in 3 hours. Working alone, it would take Ethel $\frac{2}{3}$ as long as Al to balance the books. How long would it take Al to do the job alone?

28.________________

29. Judy and Tony can mow the lawn together in 4 hours. It takes Tony twice as long as Judy to do the job alone. How long would it take Judy working alone?

29.________________

30. Fred can seal an asphalt driveway in $\frac{1}{3}$ the time it takes John. Working together, it takes them $1\frac{1}{2}$ hours. How long would it have taken Fred working alone?

30.________________

Name: Date:
Instructor: Section:

Chapter 7 EQUATIONS OF LINES; FUNCTIONS

7.1 Review of Graphs and Slopes of Lines

Learning Objectives
1 Plot ordered pairs.
2 Graph lines and find intercepts.
3 Recognize equations of horizontal and vertical lines and lines passing through the origin.
4 Use the midpoint formula.
5 Find the slope of a line.
6 Graph a line, given its slope and a point on the line.
7 Use slopes to determine whether two lines are parallel, perpendicular, or neither.
8 Solve problems involving average rate of change.

Key Terms

Use the vocabulary terms listed below to complete each statement in exercises 1-12.

ordered pair **origin** **rectangular coordinate system** **coordinates**

quadrant **first-degree equation** **linear equation in two variables**

***x*-intercept** ***y*-intercept** **slope** **parallel lines** **perpendicular lines**

1. A(n) ____________________ is an equation that can be written in the form $Ax + By = C$, where *A*, *B*, and *C* are real numbers and *A* and *B* are both not 0.

2. A(n) ____________________ has no term with the variable to a power other than 1.

3. The ratio of the change in *y* to the change in *x* along a line is called the ____________________ of the line.

4. A(n) ____________________ is a pair of numbers written within parentheses in which the order of the numbers is important.

5. Two lines are ____________________ if their slopes are negative reciprocals.

6. The *x*-axis and *y*-axis placed at a right angle at their zero points form a ____________________, also called the Cartesian coordinate system.

7. A(n) point where a graph intersects the *y*-axis is called a(n) ____________________.

8. The point at which the *x*-axis and *y*-axis of a rectangular coordinate system intersect is called the ____________________.

9. Two nonvertical lines are ____________________ if their slopes are the same.

Name: Date:
Instructor: Section:

10. The numbers in an ordered pair are called the ____________________ of the corresponding point in the plane.

11. A(n) ____________________ is one of the four regions in the plane determined by a rectangular coordinate system.

12. A(n) point where a graph intersects the x-axis is called a(n) ____________________.

Objective 1 Plot ordered pairs.

Plot the ordered pair.

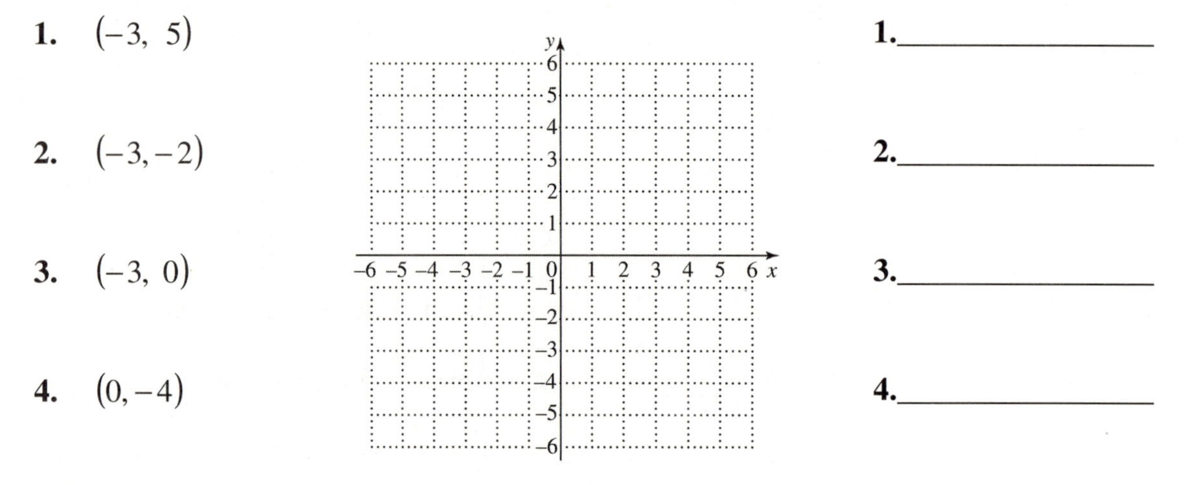

1. $(-3,\ 5)$ **1.**________________

2. $(-3,-2)$ **2.**________________

3. $(-3,\ 0)$ **3.**________________

4. $(0,-4)$ **4.**________________

Objective 2 Graph lines and find intercepts.

Find the x-intercept and y-intercept of the line.

5. $5x-4y=20$ **5.**________________

6. $4x-7y=-5$ **6.**________________

Name: Date:
Instructor: Section:

Graph the line.

7. $x - y = -7$ **7.**________________

8. $5x - 3y = 15$ **8.**________________

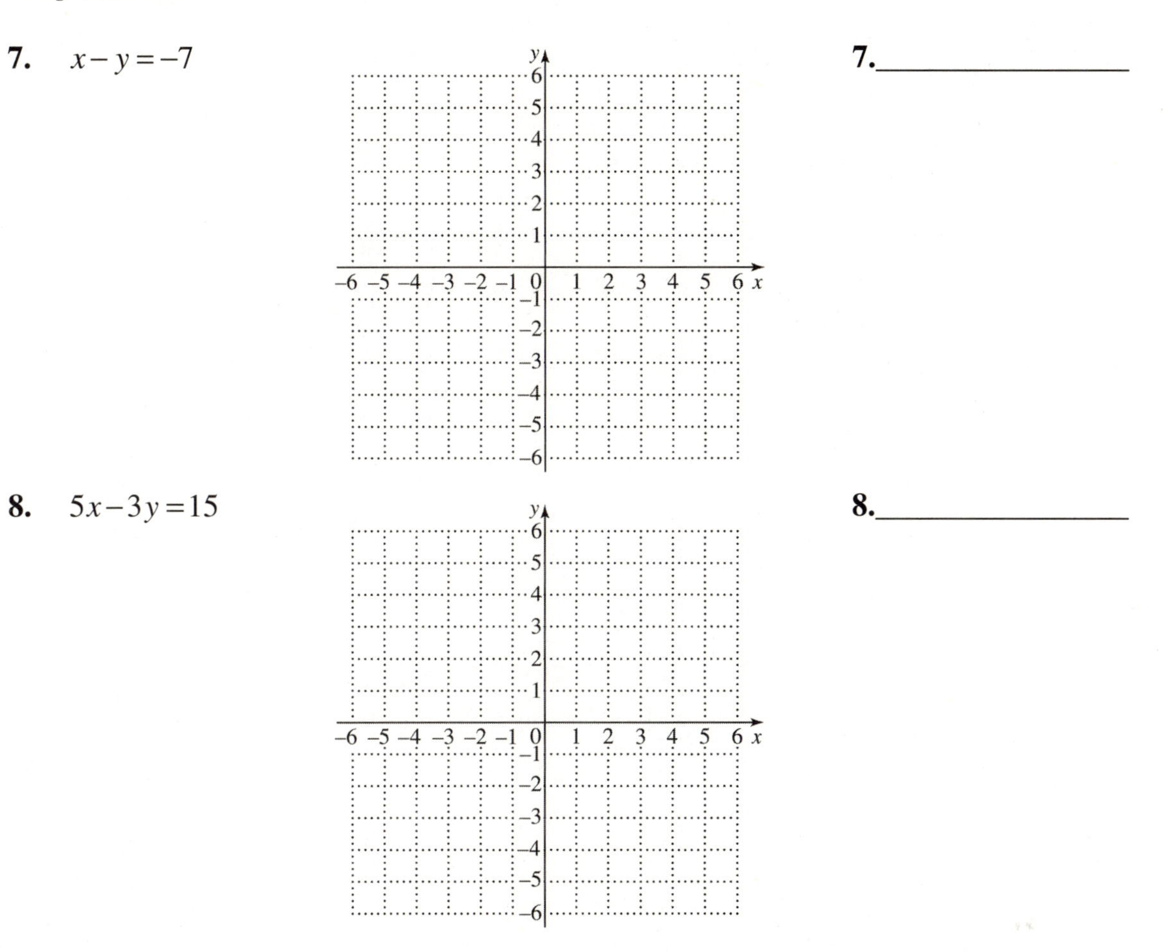

Objective 3 Recognize equations of horizontal and vertical lines and lines passing through the origin.

Graph the line.

9. $y = 0$ **9.**________________

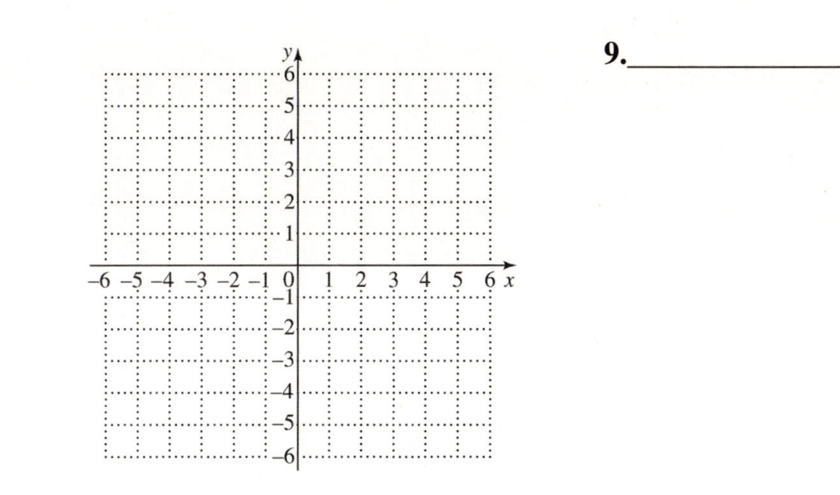

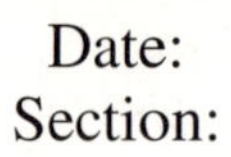

10. $x-4=0$ **10.**_______________

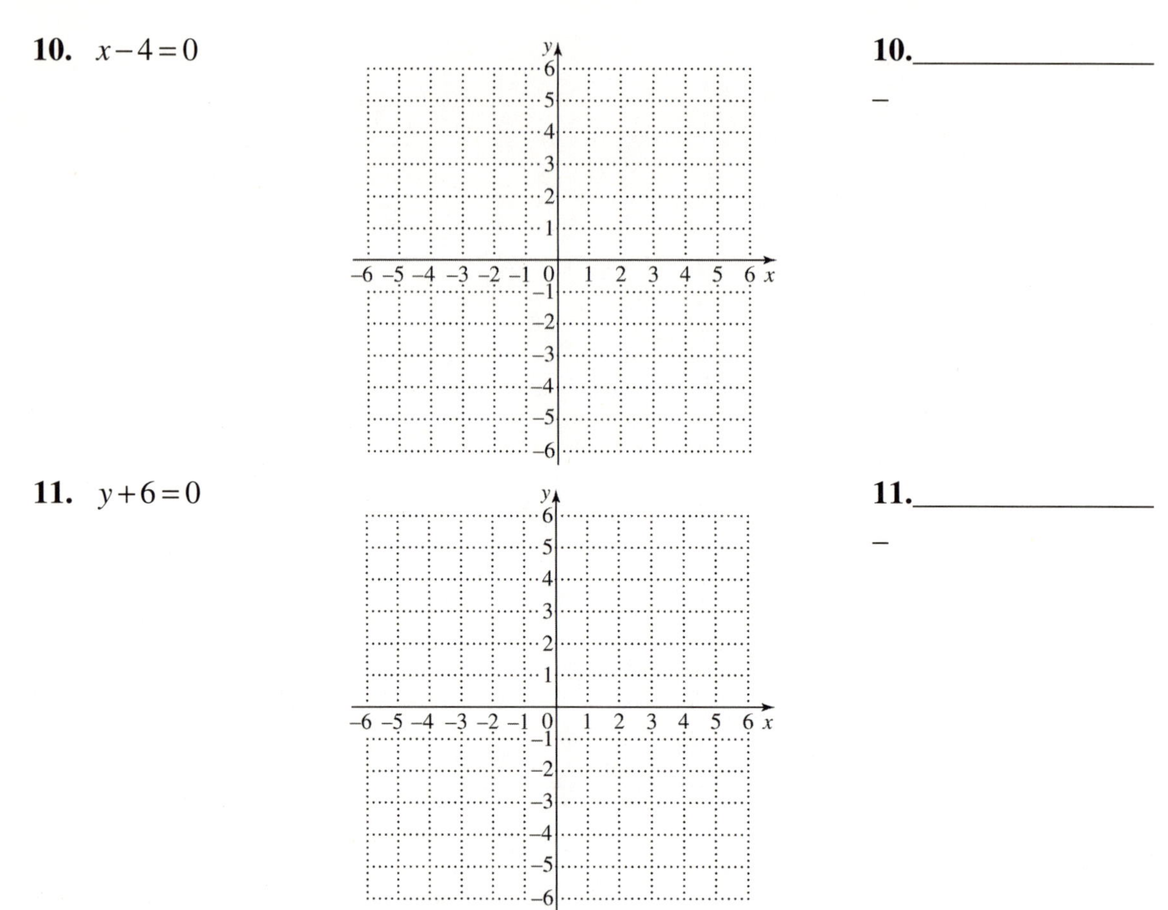

11. $y+6=0$ **11.**_______________

Objective 4 Use the midpoint formula.

Find the midpoint of each segment with the given endpoints.

12. $(-3,7)$ and $(5,3)$ **12.**_______________

13. $(4,-9)$ and $(-4,-3)$ **13.**_______________

14. $(6,5)$ and $(-1,-2)$ **14.**_______________

15. $(-12,7)$ and $(4,7)$ **15.**_______________

Name: Date:
Instructor: Section:

Objective 5 Find the slope of a line.

Find the slope of the line through the pair of points.

16. $(5,\ 7),\ (7,\ 9)$ **16.**________________

17. $(9,-4),\ (7,-7)$ **17.**________________

18. $(-1,-4),\ (-2,-3)$ **18.**________________

19. $(3,\ 6),\ (-2,\ 6)$ **19.**________________

Objective 6 Graph a line, given its slope and a point on the line.

Graph the line.

20. Slope: $\frac{2}{3}$ **20.**________________

Through: $(-1,\ 4)$

y 6 5 4 3 2 1 −6 −5 −4 −3 −2 −1 0 1 2 3 4 5 6 x −1 −2 −3 −4 −5 −6

21. Slope: $-\frac{1}{2}$ **21.**________________

Through: $(2,-3)$

y 6 5 4 3 2 1 −6 −5 −4 −3 −2 −1 0 1 2 3 4 5 6 x −1 −2 −3 −4 −5 −6

Name: Date:
Instructor: Section:

22. Slope: –1

Through: $(-1,\ 2)$

22.________________

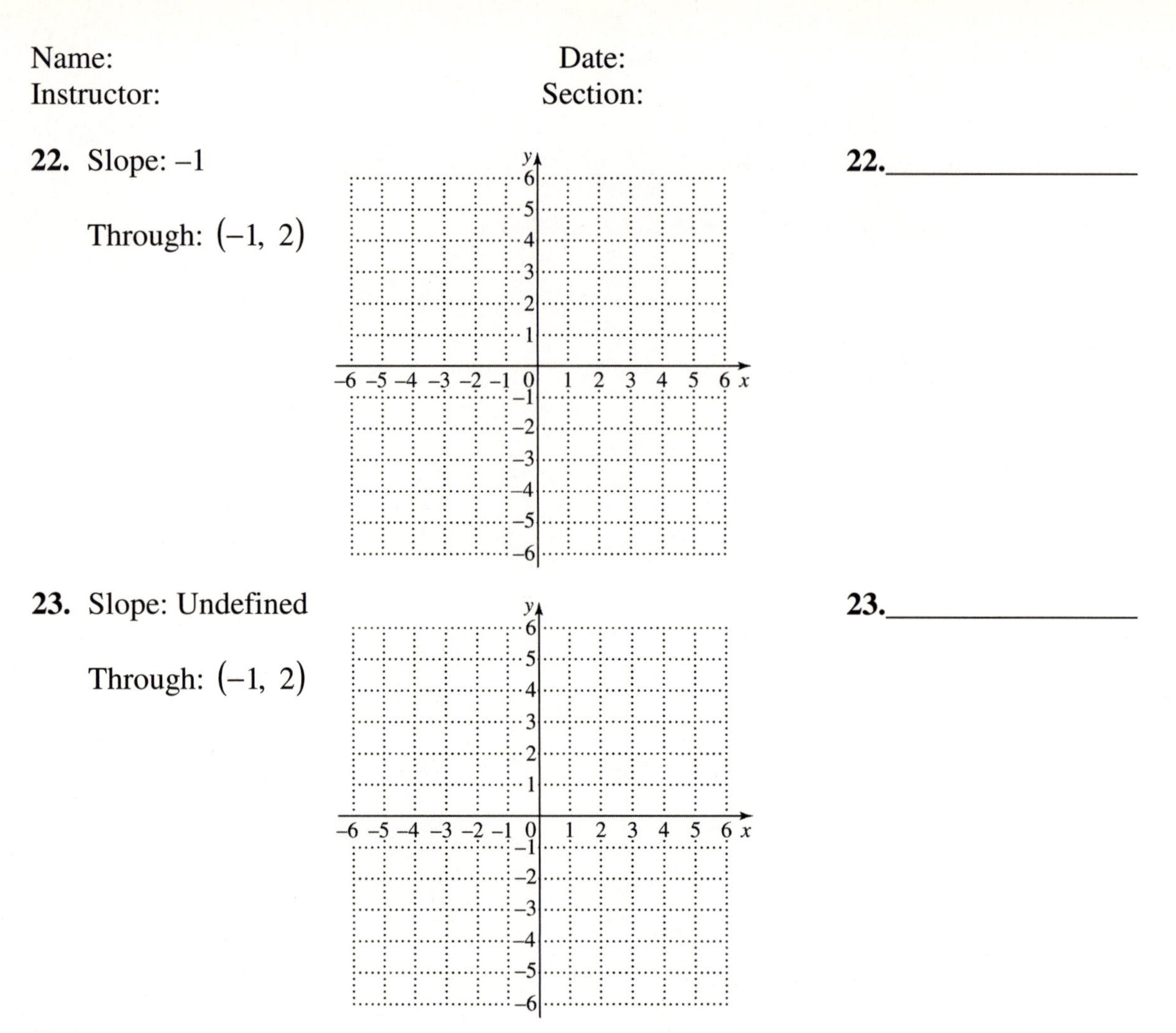

23. Slope: Undefined

Through: $(-1,\ 2)$

23.________________

Objective 7 Use slopes to determine whether two lines are parallel, perpendicular, or neither.

Decide whether the pair of lines is parallel, perpendicular, *or* neither.

24. $x+y=6,\ x+y=-2$

24.________________

25. $4x-3y=2,\ 3x-4y=2$

25.________________

26. The line through $(9,-2)$ and $(-1,\ 3)$ and the line through $(5,\ 7)$ and $(6,\ 5)$.

26.________________

Name: Date:
Instructor: Section:

27. The line through $(-7,\ 3)$ and $(2,-4)$ and the line through $(-8,\ 4)$ and $(-1,\ 13)$.

27.________________

Objective 8 Solve problems involving average rate of change.

Solve each problem.

28. Suppose the sales of a company are given by the linear equation $y = 1250x + 10{,}000$, where x is the number of years after 1980, and y is the sales in dollars. What is the average rate of change in sales per year?

28.________________

29. A plane had an altitude of 8500 feet at 4:02 P.M. and 12,700 feet at 4:39 P.M. What was the average rate of change in the altitude in feet per minute?

29.________________

30. A ramp is 10 feet high on the high end and 3 feet high on the low end. It covers a horizontal length of 29 feet. What is the average rate of change of the incline?

30.________________

Name: Date:
Instructor: Section:

Chapter 7 EQUATIONS OF LINES; FUNCTIONS

7.2 Review Equations of Lines; Linear Models

Learning Objectives
1. Write an equation of a line, given its slope and y-intercept.
2. Graph a line, using its slope and y-intercept.
3. Write an equation of a line, given its slope and a point on the line.
4. Write an equation of a line, given two points on the line.
5. Write an equation of a line parallel or perpendicular to a given line.
6. Write an equation of a line that models real data.

Key Terms

Use the vocabulary terms listed below to complete each statement in exercises 1-5.

slope-intercept form **point-slope form** **standard form** **horizontal line**

vertical line

1. A linear equation in two variables written in the form $Ax + By = C$, where A, B, and C are integers with no common factor (except 1) and $A \geq 0$, is in __________________.

2. A linear equation is written in __________________ if it is in the form $y = mx + b$, where m is the slope and $(0, b)$ is the y-intercept.

3. A __________________ has an equation with the form $x = a$.

4. A __________________ has an equation with the form $y = b$.

5. A linear equation is written in __________________ if it is in the form $y - y_1 = m(x - x_1)$, where m is the slope of the line and (x_1, y_1) is a point on the line.

Objective 1 Write an equation of a line, given its slope and y-intercept.

Write an equation in standard form for the line.

1. Slope: 2
y-intercept: $(0, -5)$

1.__________________

2. Slope: –5
y-intercept: $(0, 3)$

2.__________________

Name: Date:
Instructor: Section:

3. Slope: $\frac{3}{5}$ **3.**________________

y-intercept: $\left(0, \frac{2}{5}\right)$

4. Slope: 0 **4.**________________

y-intercept: $(0, 3)$

Objective 2 Graph a line, using its slope and *y*-intercept.

Graph each line using its slope and y-intercept.

5. $x + y = -4$ **5.**________________

6. $2x + y = 4$ **6.**________________

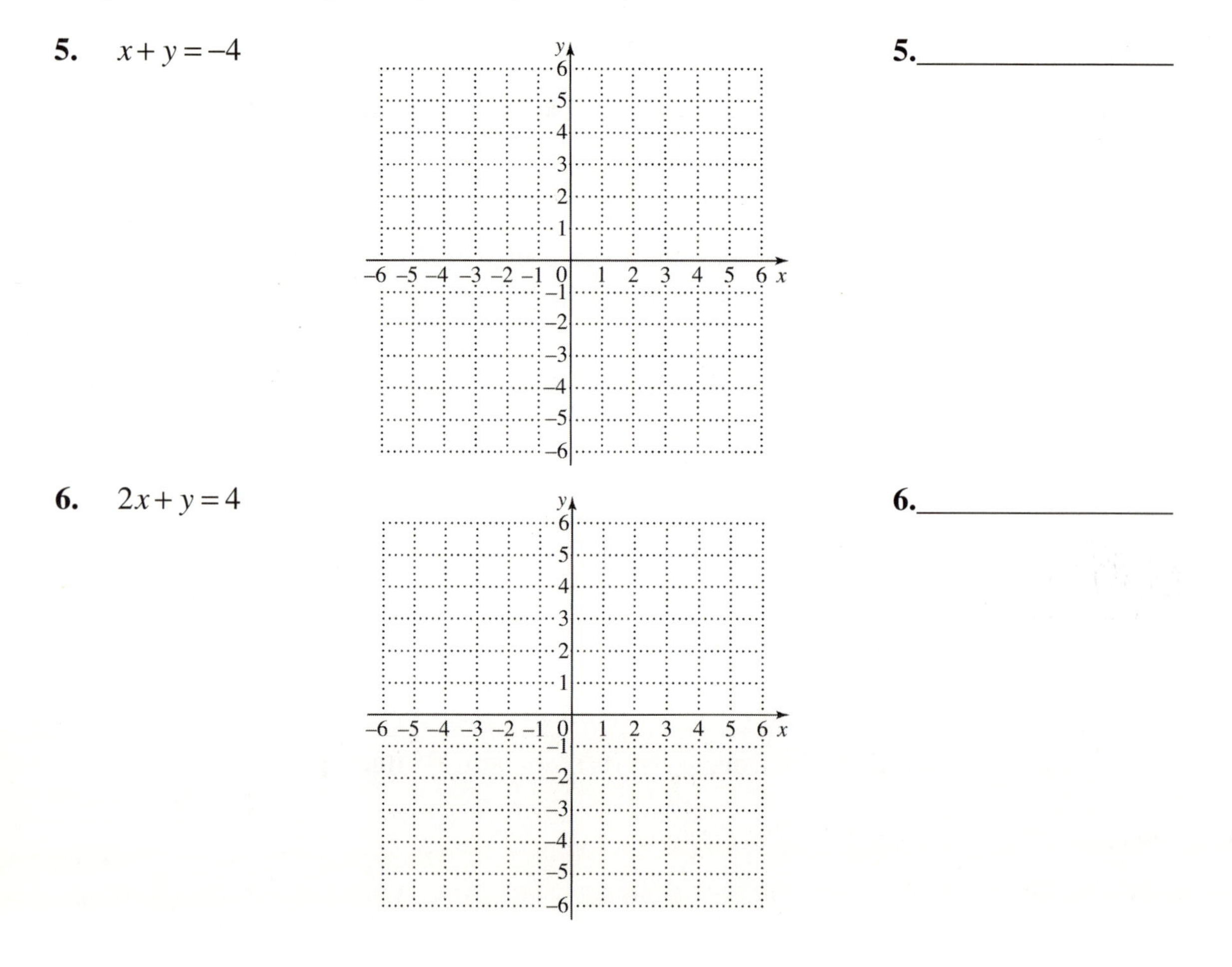

Name: Date:
Instructor: Section:

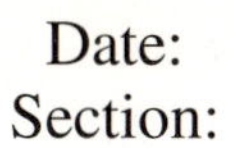

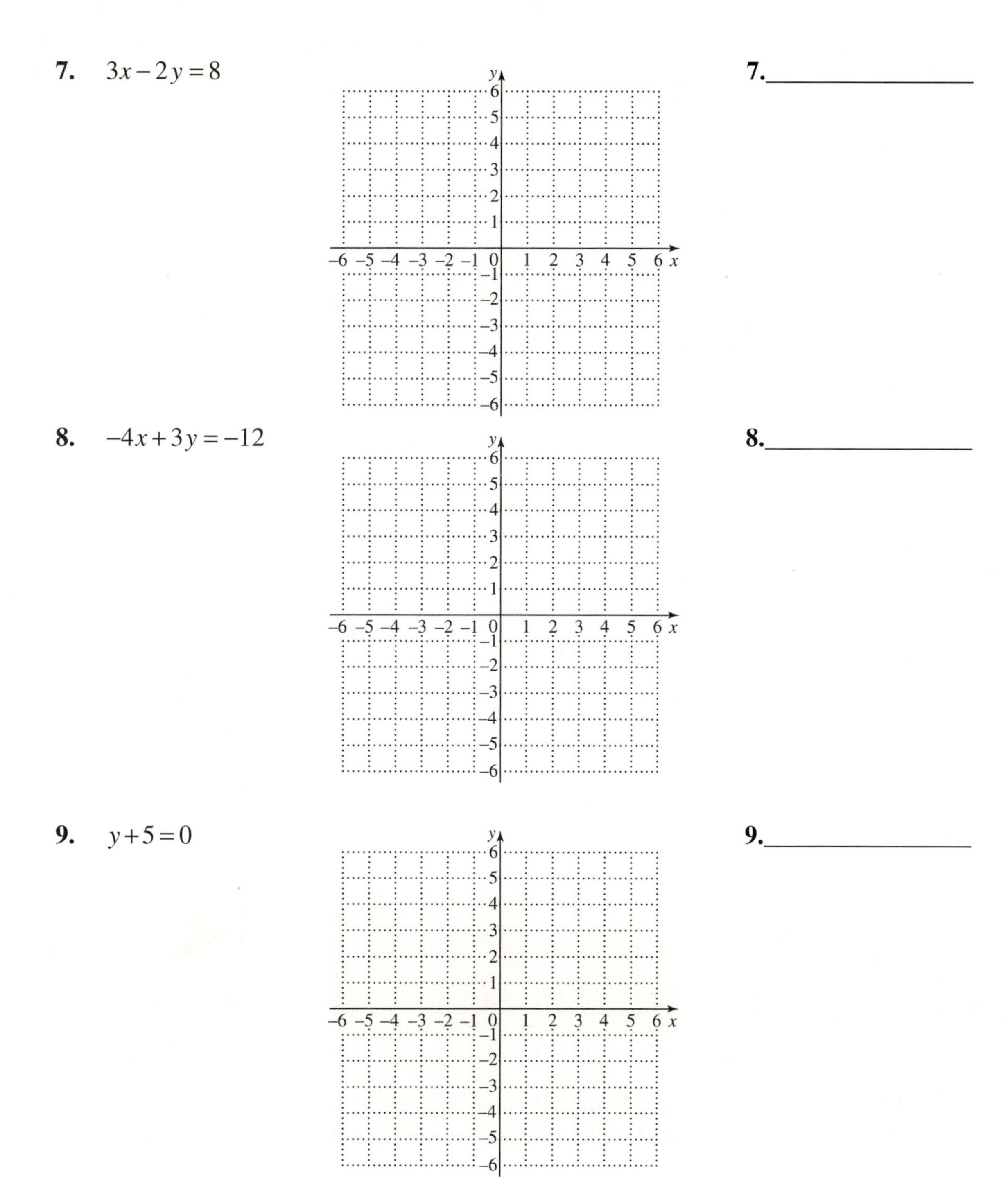

7. $3x - 2y = 8$

7. ________________

8. $-4x + 3y = -12$

8. ________________

9. $y + 5 = 0$

9. ________________

Name: Date:
Instructor: Section:

Objective 3 Write an equation of a line, given its slope and a point on the line.

Write an equation for the line. Write answer in the form $Ax + By = C$.

10. Slope: 2
Point: $(-1,\ 4)$

10.________________

11. Slope: 5
Point: $(3, -6)$

11.________________

12. Slope: –4
Point: $(-2,\ 5)$

12.________________

13. Slope: $-\frac{2}{3}$
Point: $(1, -5)$

13.________________

14. Slope: Undefined
Point: $(3,\ 0)$

14.________________

15. Slope: 0
Point: $(3, -4)$

15.________________

Objective 4 Write an equation of a line, given two points on the line.

Write an equation in standard form of the line passing through the given pair of points.

16. $(4,\ 9)$, $(3,\ 8)$

16.________________

Name: Date:
Instructor: Section:

17. $(7, 1), (6, 5)$ **17.**________________

18. $(2, -1), (5, -2)$ **18.**________________

19. $(-6, 2), (-4, 1)$ **19.**________________

20. $(-1, -4), (-2, -3)$ **20.**________________

21. $(3, -5), (-4, -5)$ **21.**________________

Objective 5 Write an equation of a line parallel or perpendicular to a given line.

Write an equation in standard form of the line.

22. Parallel to $x - y = 4$, through $(4, -7)$ **22.**________________

23. Parallel to $2x + 6y = 5$, through $(1, -2)$ **23.**________________

24. Parallel to $5x + y = 6$, through $(0, 4)$ **24.**________________

Name: Date:
Instructor: Section:

25. Perpendicular to $2x - y = 3$, through $(-1,\ 0)$ **25.**________________

26. Perpendicular to $5x + y = 8$, through $(2, -1)$ **26.**________________

27. Perpendicular to $3x = 2$, through $(-4,\ 5)$ **27.**________________

Objective 6 Write an equation of a line that models real data.

Suppose that you are in charge of your office Christmas party. You call the local caterer who informs you that the standard Christmas party package costs $4.95 per person.

28. Write a linear equation that represents the cost in dollars y, for catering the office party for x people. **28.**________________

29. If 71 people attend, what will be the total cost in dollars to cater the party? **29.**________________

30. The caterer is running a Christmas special. You can rent a punch bowl fountain for the low, low cost of only $19.50 for the evening. Write a new equation that defines the price you will pay with the punch bowl fountain included. **30.**________________

Name: Date:
Instructor: Section:

Chapter 7 EQUATIONS OF LINES; FUNCTIONS

7.3 Functions

Learning Objectives
1. Distinguish between independent and dependent variables.
2. Define and identify relations and functions.
3. Find the domain and range.
4. Identify functions defined by graphs and equations.
5. Use function notation.
6. Graph linear and constant functions.

Key Terms
Use the vocabulary terms listed below to complete each statement in exercises 1-10.

dependent variable **independent variable** **relation** **function**

domain **range** **vertical line test** **function notation**

linear function **constant function**

1. A ____________________ is a set of ordered pairs.

2. The ____________________ states that any vertical line drawn through the graph of a function must intersect the graph in at most one point.

3. A ____________________ is a set of ordered pairs in which each value of the first component *x* corresponds to exactly one value of the second component *y*.

4. In an equation relating *x* and *y*, if the value of the variable *y* depends on the variable *x*, then *y* is called the ____________________.

5. A function defined by an equation of the form $f(x) = ax + b$, for real numbers *a* and *b*, is a ____________________.

6. The set of all first components in the ordered pairs of a relation is called the ____________________.

7. A linear function of the form $f(x) = b$, where *b* is a constant, is called a ____________________.

8. ____________________ represents the value of the function at *x*, that is, the *y*-value that corresponds to *x*.

Name: Date:
Instructor: Section:

9. In an equation relating x and y, if the value of the variable y depends on the variable x, then x is called the ____________________.

10. The set of all second components in the ordered pairs of a relation is called the ____________________.

Objective 1 Distinguish between independent and dependent variables.

Identify the independent and dependent variable in each situation described.

1. the outside temperature; the cost to heat your home **1.**________________

2. water pressure exerted on a person; the depth of the person **2.**________________

Objective 2 Define and identify relations and functions.

Which of the following are functions?

3. $\{(1, 3), (1, 4), (2, -1), (3, 7)\}$ **3.**________________

4. $\{(6, -3), (4, -2), (2, -1), (0, 0)\}$ **4.**________________

5. $\{(-4, -1), (-3, -2), (-1, 0), (0, -1)\}$ **5.**________________

6. $\{(3, 4), (5, 2), (4, 3), (5, 3)\}$ **6.**________________

7. $x = \sqrt{y+1}$ **7.**________________

8. $x^2 + y^2 = 1$ **8.**________________

9. $xy = 5$ **9.**________________

Name: Date:
Instructor: Section:

Objective 3 Find the domain and range.

Decide whether the relation is a function, and give the domain and range of the relation.

10. $\{(-1, 1), (-2, 2), (0, 0)\}$ **10.**________________

11. $\{(3, 0), (2, 4), (1, 6), (-1, 3)\}$ **11.**________________

12. $\{(-2, -2), (-1, -1), (0, 0), (1, -1)\}$ **12.**________________

13. $\{(1, 3), (2, -1), (-1, 4), (1, 4)\}$ **13.**________________

14. $\{(2, -4), (1, -2), (-1, 2), (0, 3)\}$ **14.**________________

15. $\{(4, 2), (3, 2), (2, 2)\ (1, 2), (0, 2)\}$ **15.**________________

Objective 4 Identify functions defined by graphs and equations.

Determine whether each graph represents a function.

16. **16.**________________

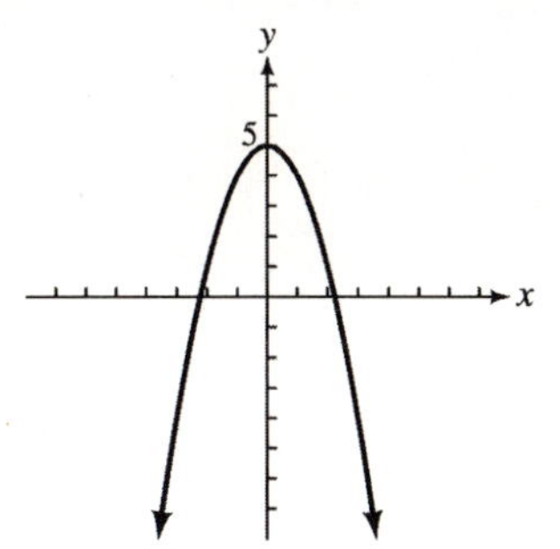

Name: Date:
Instructor: Section:

17.

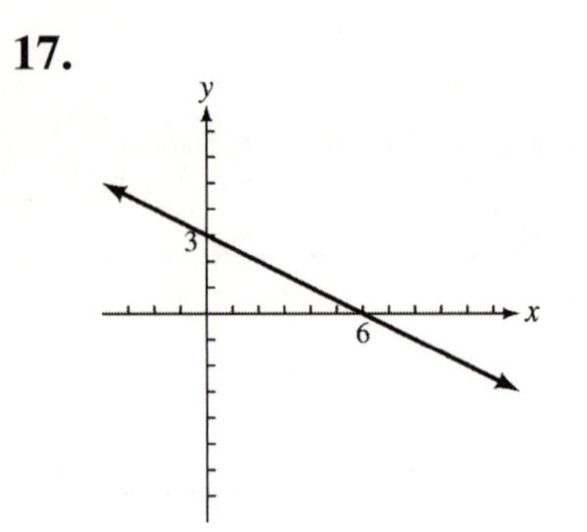

17.__________________

18.

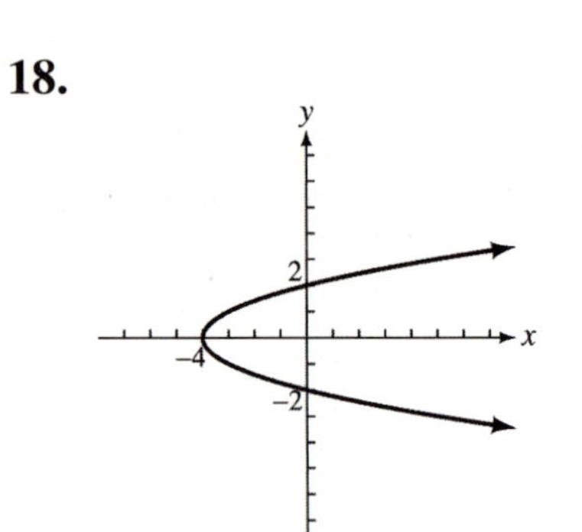

18.__________________

19.

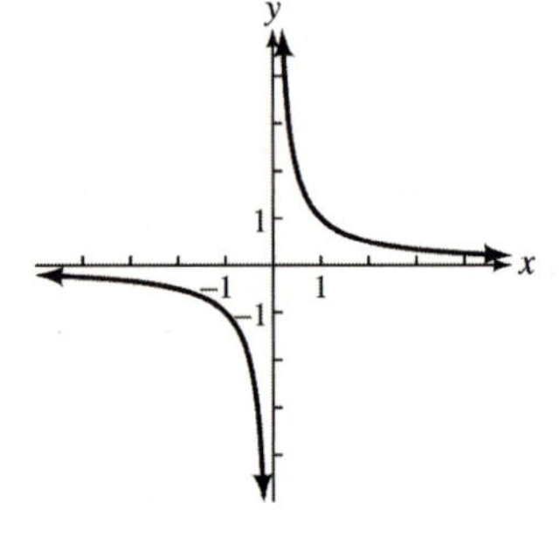

19.__________________

20.

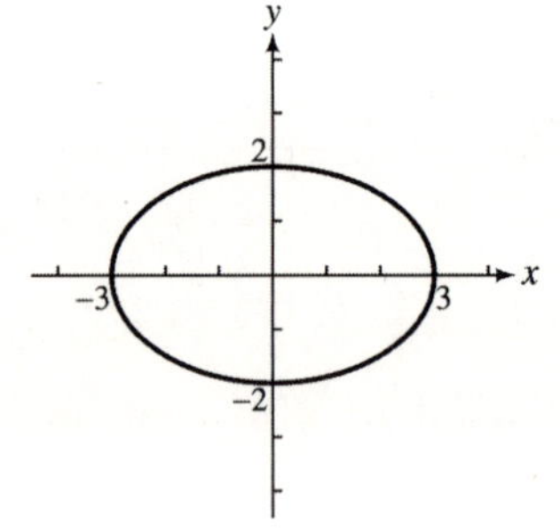

20.__________________

Name: Date:
Instructor: Section:

Decide whether each relation defines a function, and give the domain.

21. $y \le 2x$ **21.**________________

22. $y = \sqrt{x-3}$ **22.**________________

Objective 5 Use function notation.

Find $f(-2)$, $f(4)$, *and* $f(x+1)$.

23. $f(x) = 2x+5$ **23.**________________

24. $f(x) = 3x^2$ **24.**________________

25. $f(x) = \dfrac{4}{x^2+1}$ **25.**________________

Write the equation in the form $f(x) = mx+b$.

26. $x+4y=2$ **26.**________________

27. $\frac{1}{2}x+\frac{1}{3}y=-1$ **27.**________________

Name: Date:
Instructor: Section:

Objective 6 Graph linear and constant functions.

Graph each function. Give the domain and range.

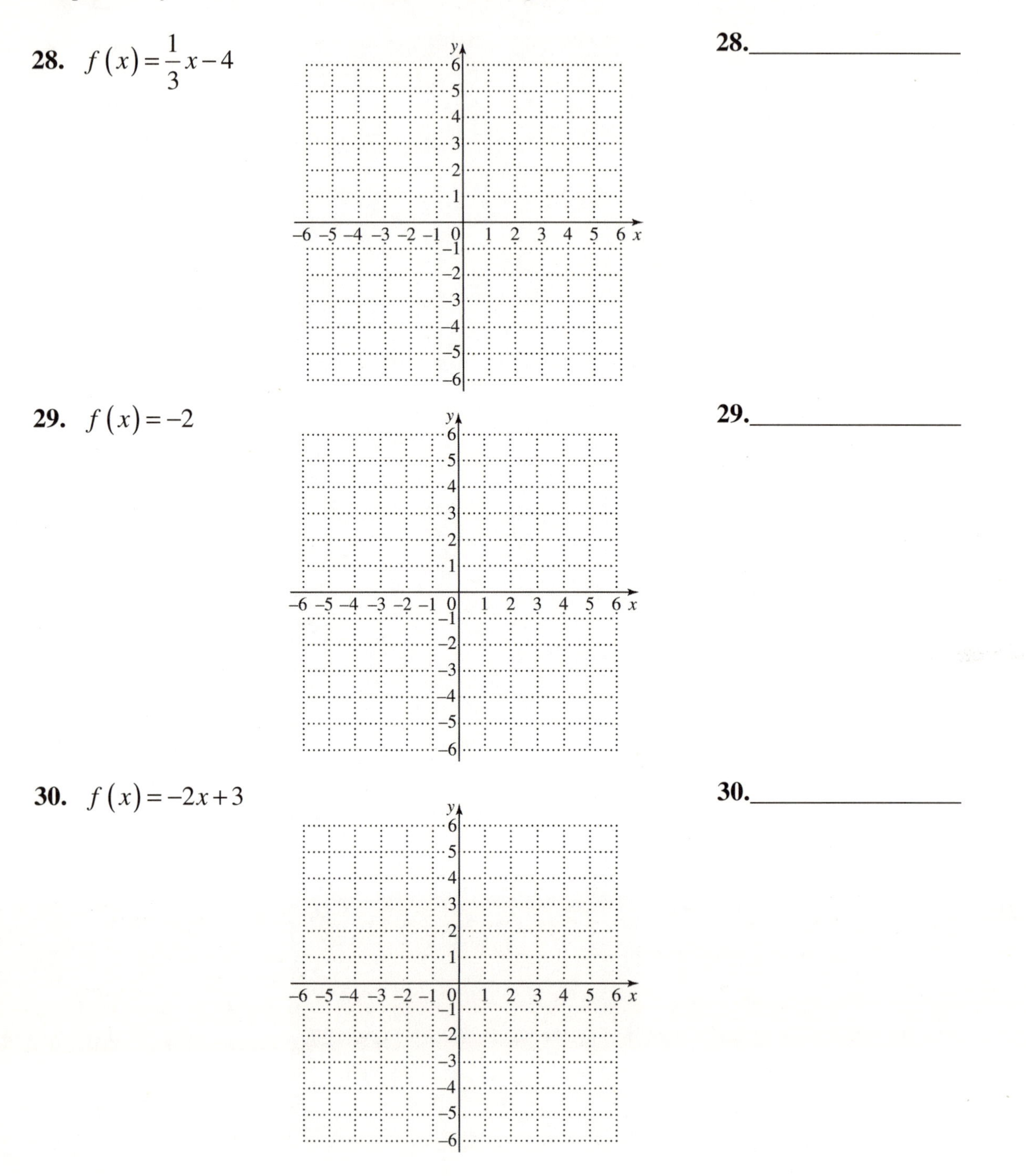

28. $f(x) = \frac{1}{3}x - 4$

28.________________

29. $f(x) = -2$

29.________________

30. $f(x) = -2x + 3$

30.________________

Name: Date:
Instructor: Section:

Chapter 7 EQUATIONS OF LINES; FUNCTIONS

7.4 Operations on Functions and Composition

Learning Objectives
1 Recognize polynomial functions.
2 Perform operations on polynomial functions.
3 Find the composition of functions.

Key Terms
Use the vocabulary terms listed below to complete each statement in exercises 1-3.

polynomial function **operations on functions** **composite function**

1. A function defined by a polynomial in one variable, consisting of one or more terms, is called a ___________________.

2. The sum, difference, product, and quotient of two functions, $f(x)$ and $g(x)$, are called ___________________.

3. A function in which some quantity depends on a variable that, in turn, depends on another variable is called a ___________________.

Objective 1 Recognize polynomial functions.

For each polynomial function, find $f(-1)$ and $f(2)$.

1. $f(x) = -3x + 7$ **1.**_______________

2. $f(x) = -\frac{1}{4}x + 1$ **2.**_______________

3. $f(x) = 5x^3 - 2x^2 + 9$ **3.**_______________

4. $f(x) = -x^4 - 2x^3 + 3x - 2$ **4.**_______________

Name: Date:
Instructor: Section:

Objective 2 Perform operations on polynomial functions.

Let $f(x)=2x^2+x-28$, $g(x)=2x-7$, *and* $h(x)=3x$. *Find each of the following.*

5. $(f+h)(x)$ **5.**________________

6. $(f-g)(x)$ **6.**________________

7. $(g-h)(x)$ **7.**________________

8. $(f-h)(2)$ **8.**________________

9. $(f+g)(-1)$ **9.**________________

10. $(g-h)(10)$ **10.**________________

11. $(g-f)(-3)$ **11.**________________

12. $(f+h)(0)$ **12.**________________

13. $(f-f)(x)$ **13.**________________

Name: Date:
Instructor: Section:

Let $f(x) = x^2 - 16$, $g(x) = 3x$, *and* $h(x) = x + 4$. *Find each of the following.*

14. $(fg)(x)$ **14.**________________

15. $(fh)(x)$ **15.**________________

16. $\left(\frac{f}{g}\right)(x)$ **16.**________________

17. $\left(\frac{f}{h}\right)(x)$ **17.**________________

18. $(fg)(-1)$ **18.**________________

19. $(gh)(4)$ **19.**________________

20. $\left(\frac{f}{g}\right)(3)$ **20.**________________

21. $\left(\frac{f}{h}\right)(-12)$ **21.**________________

Name: Date:
Instructor: Section:

22. $\left(\frac{f}{g}\right)(0)$ **22.**________________

Objective 3 Find the composition of functions.

Let $f(x)=x^2+2$, $g(x)=3x+1$, *and* $h(x)=x+4$. *Find each of the following.*

23. $(f\circ g)(5)$ **23.**________________

24. $(h\circ g)(5)$ **24.**________________

25. $(g\circ h)(-2)$ **25.**________________

26. $(f\circ h)(3)$ **26.**________________

27. $(f\circ g)(x)$ **27.**________________

28. $(g\circ f)(x)$ **28.**________________

29. $(f\circ h)(x)$ **29.**________________

30. $(f\circ h)\left(\frac{2}{3}\right)$ **30.**________________

Name: Date:
Instructor: Section:

Chapter 7 EQUATIONS OF LINES; FUNCTIONS

7.5 Variation

Learning Objectives
1 Write an equation expressing direct variation.
2 Find the constant of variation, and solve direct variation problems.
3 Solve inverse variation problems.
4 Solve joint variation problems.
5 Solve combined variation problems.

Key Terms
Use the vocabulary terms listed below to complete each statement in exercises 1-5.

direct variation **direct variation as a power** **inverse variation**

inverse variation as the *n*th power of *x* **joint variation**

1. The equation $y = kx$, where k is a real number, is called a(n) ____________________ equation.

2. The equation $y = \frac{k}{x^n}$, where k is a real number, is called a(n) ____________________ equation.

3. The equation $y = \frac{k}{x}$, where k is a real number, is called a(n) ____________________ equation.

4. The equation $y = kxz$, where k is a real number, is called a(n) ____________________ equation.

5. The equation $y = kx^n$, where k is a real number, is called a(n) ____________________ equation.

Objectives 1 and 2 Write an equation expressing direct variation. Find the constant of variation and solve direct variation problems.

Suppose y varies directly as x. Find the equation connecting y and x, given the following.

1. $y = 15$ when $x = 5$. **1.**________________

Name: Date:
Instructor: Section:

2. $y = 12$ when $x = 8$. **2.**________________

3. $y = 150$ when $x = 3$. **3.**________________

4. $y = 13.75$ when $x = 55$. **4.**________________

Solve the problem.

5. If r varies directly at t, and $r = 10$ when $t = 2$, find r when $t = 9$. **5.**________________

6. If x varies directly as y, and $x = 9$ when $y = 2$, find x when $y = 7$. **6.**________________

7. If a varies directly as b^2, and $a = 48$ when $b = 4$, find a when $b = 7$. **7.**________________

8. The circumference of a circle varies directly as the radius. A circle with a radius of 7 centimeters has a circumference of 43.96 centimeters. Find the circumference of the circle if the radius changes to 11 centimeters. **8.**________________

Objective 3 Solve inverse variation problems.

Suppose y varies inversely as x. Find the equation connecting y and x, given the following.

9. $y = 10$ when $x = 2$. **9.**________________

Name: Date:
Instructor: Section:

10. $y=2$ when $x=12$. **10.**________________

11. $y=5$ when $x=\frac{1}{6}$. **11.**________________

Solve the problem.

12. If y varies inversely as x, and $y=10$ when $x=3$, find y when $x=12$. **12.**________________

13. If p varies inversely as q^2, and $p=4$ when $q=\frac{1}{2}$, find p when $q=\frac{3}{2}$. **13.**________________

14. If z varies inversely as x^2, and $z=9$ when $x=\frac{2}{3}$, find z when $x=\frac{5}{4}$. **14.**________________

15. The current in a simple electrical circuit varies inversely as the resistance. If the current is 50 amperes (an ampere is a unit for measuring current) when the resistance is 10 ohms (an ohm is a unit for measuring resistance), find the current if the resistance is 5 ohms. **15.**________________

Name: Date:
Instructor: Section:

Objective 4 Solve joint variation problems.

Suppose y varies jointly as x and z. Find the equation connecting y, x, and z, given the following.

16. $y = 10$ when $x = 2$ and $z = 5$. **16.**________________

17. $y = 27$ when $x = 9$ and $z = 6$. **17.**________________

18. $y = 144$ when $x = 8$ and $z = 9$. **18.**________________

Solve the problem.

19. If r varies jointly as m and n^2, and $r = 72$ when $m = 4$ and $n = 6$, find r when $m = 3$ and $n = 4$. **19.**________________

20. If q varies jointly as p and r^2, and $q = 27$ when $p = 9$ and $r = 2$, find q when $p = 8$ and $r = 4$. **20.**________________

21. Suppose y varies jointly as x^2 and z^2, and $y = 72$ when $x = 2$ and $z = 3$. Find y when $x = 4$ and $z = 2$. **21.**________________

22. Suppose d varies jointly as f^2 and g^2, and $d = 384$ when $f = 3$ and $g = 8$. Find d when $f = 6$ and $g = 2$. **22.**________________

Name: Date:
Instructor: Section:

Objective 5 Solve combined variation problems.

Suppose y varies directly as x and inversely as z. Find the equation connecting y, x, and z, given the following.

23. $y=1$ when $x=2$ and $z=6$. **23.**________________

24. $y=6$ when $x=4$ and $z=8$. **24.**________________

25. $y=2$ when $x=6$ and $z=2$. **25.**________________

26. $y=10.5$ when $x=1.8$ and $z=.6$. **26.**________________

Solve the problem.

27. The time required to print a newsletter varies directly as the number of newsletters printed and inversely as the number of presses used. If 40,000 newsletters are printed in 1 hour when two presses are used, how long will it take to print 30,000 newsletters if three presses are used? **27.**________________

Name: Date:
Instructor: Section:

28. The time required to lay a sidewalk varies directly as its length and inversely as the number of people who are working on the job. If three people can lay a sidewalk 100 feet long in 15 hours, how long would it take two people to lay a sidewalk 40 feet long?

28.________________

29. When an object is moving in a circular path, the centripetal force varies directly as the square of the velocity and inversely as the radius of the circle. A stone that is whirled at the end of a string 50 centimeters long at 900 centimeters per second has a centripetal force of 3,240,000 dynes. Find the centripetal force if the stone is whirled at the end of a string 75 centimeters long at 1500 centimeters per second.

29.________________

30. The gravitational attraction between two objects varies directly as the product of their masses and inversely as the square of the distance between them. If the force of attraction between two spheres, each with a mass of 1 gram that are 1 centimeter apart, is 6.66×10^{-8} dynes, find the force of attraction between two spheres with masses of 2 grams and 4 grams that are 4 centimeters apart.

30.________________

Name: Date:
Instructor: Section:

Chapter 8 SYSTEMS OF LINEAR EQUATIONS

8.1 Solving Systems of Linear Equations by Graphing

Learning Objectives

1. Decide whether a given ordered pair is a solution of a system.
2. Solve linear systems by graphing.
3. Solve special systems by graphing.
4. Identify special systems without graphing.
5. Use a graphing calculator to solve a linear system.

Key Terms

Use the vocabulary terms listed below to complete each statement in exercises 1-7.

system of linear equations **solution set** **set-builder notation**

consistent system **inconsistent system** **independent equations**

dependent equations

1. A system of equations with a solution is called a(n) ____________________.

2. The ____________________ of a system of two equations is the set of ordered pairs (x, y) that make both equations true at the same time.

3. Equations of a system that have different graphs are called ____________________.

4. A(n) ____________________ of equations is a system with no solution.

5. Equations of a system that have the same graph are called ____________________.

6. A(n) ____________________ consists of two or more linear equations to be solved at the same time.

7. ____________________ is used to describe a set of numbers without actually having to list all the elements.

Name: Date:
Instructor: Section:

Objective 1 Decide whether a given ordered pair is a solution of a system.

Decide whether the ordered pair is a solution for the given system. Write solution *or* not a solution.

1. $\begin{aligned} x+2y&=7 \\ 2x-y&=4 \end{aligned}$ $(3,\ 2)$

1.________________

2. $\begin{aligned} 4x+5y&=11 \\ 3x-y&=-6 \end{aligned}$ $(-1,\ 3)$

2.________________

3. $\begin{aligned} 4x-5y&=26 \\ 3x+2y&=6 \end{aligned}$ $(4,-2)$

3.________________

4. $\begin{aligned} x&=4y \\ 3x&=y+33 \end{aligned}$ $(12,\ 3)$

4.________________

5. $\begin{aligned} 2x-9y&=27 \\ 5x+6y&=18 \end{aligned}$ $(0,-3)$

5.________________

6. $\begin{aligned} 7x+8y&=-14 \\ 5x-7y&=10 \end{aligned}$ $(-2,\ 0)$

6.________________

Name: Date:
Instructor: Section:

Objective 2 Solve linear systems by graphing.

Solve the system by graphing.

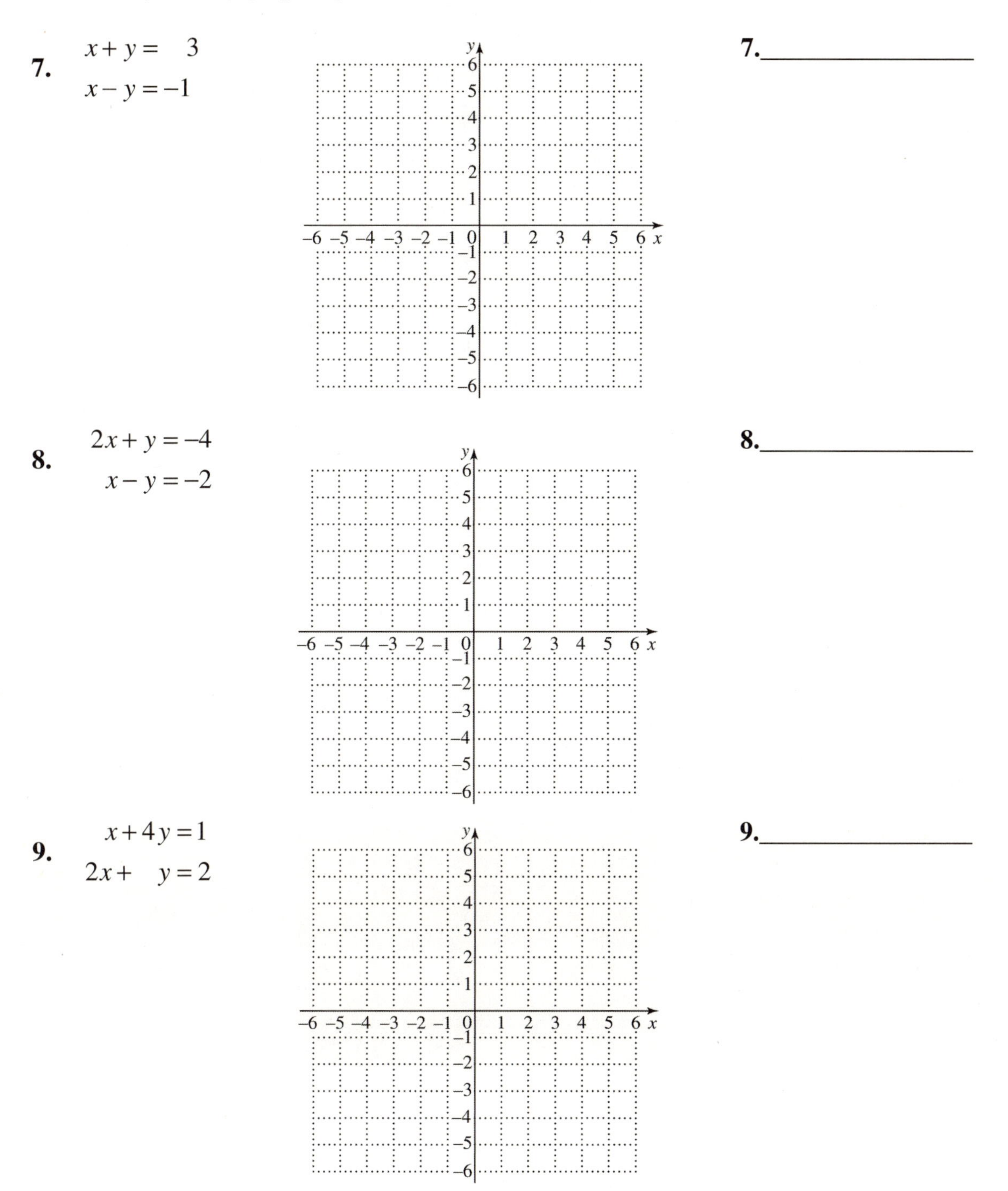

7. $x + y = 3$
$x - y = -1$

7.________________

8. $2x + y = -4$
$x - y = -2$

8.________________

9. $x + 4y = 1$
$2x + y = 2$

9.________________

Name: Date:
Instructor: Section:

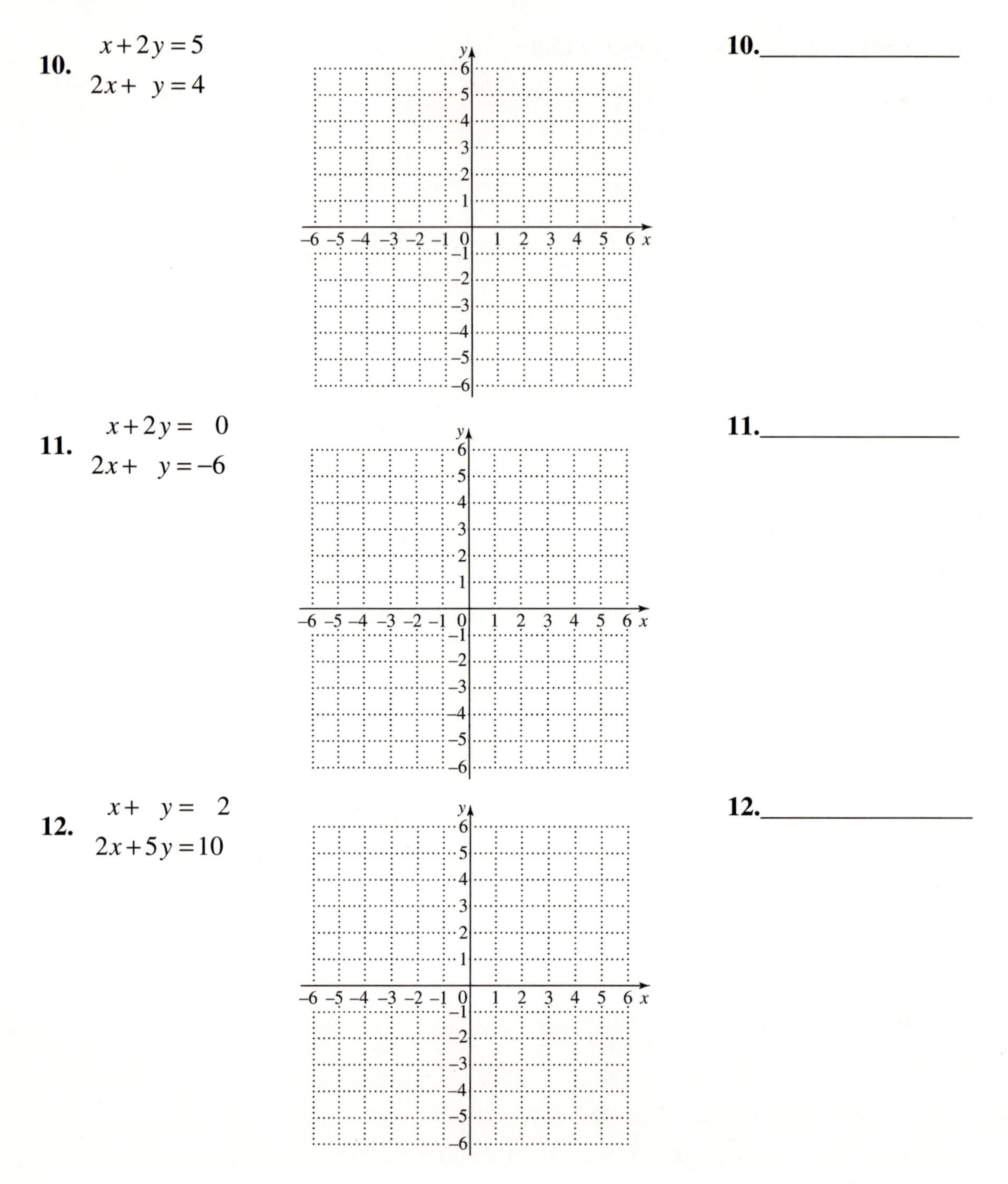

10. $x + 2y = 5$
$2x + y = 4$

10.________________

11. $x + 2y = 0$
$2x + y = -6$

11.________________

12. $x + y = 2$
$2x + 5y = 10$

12.________________

Name: Date:
Instructor: Section:

Objective 3 Solve special systems by graphing.

Solve the system by graphing both equations on the same axes. If the two equations produce parallel lines, write no solution. *If the two equations produce the same line, write* infinite number of solutions.

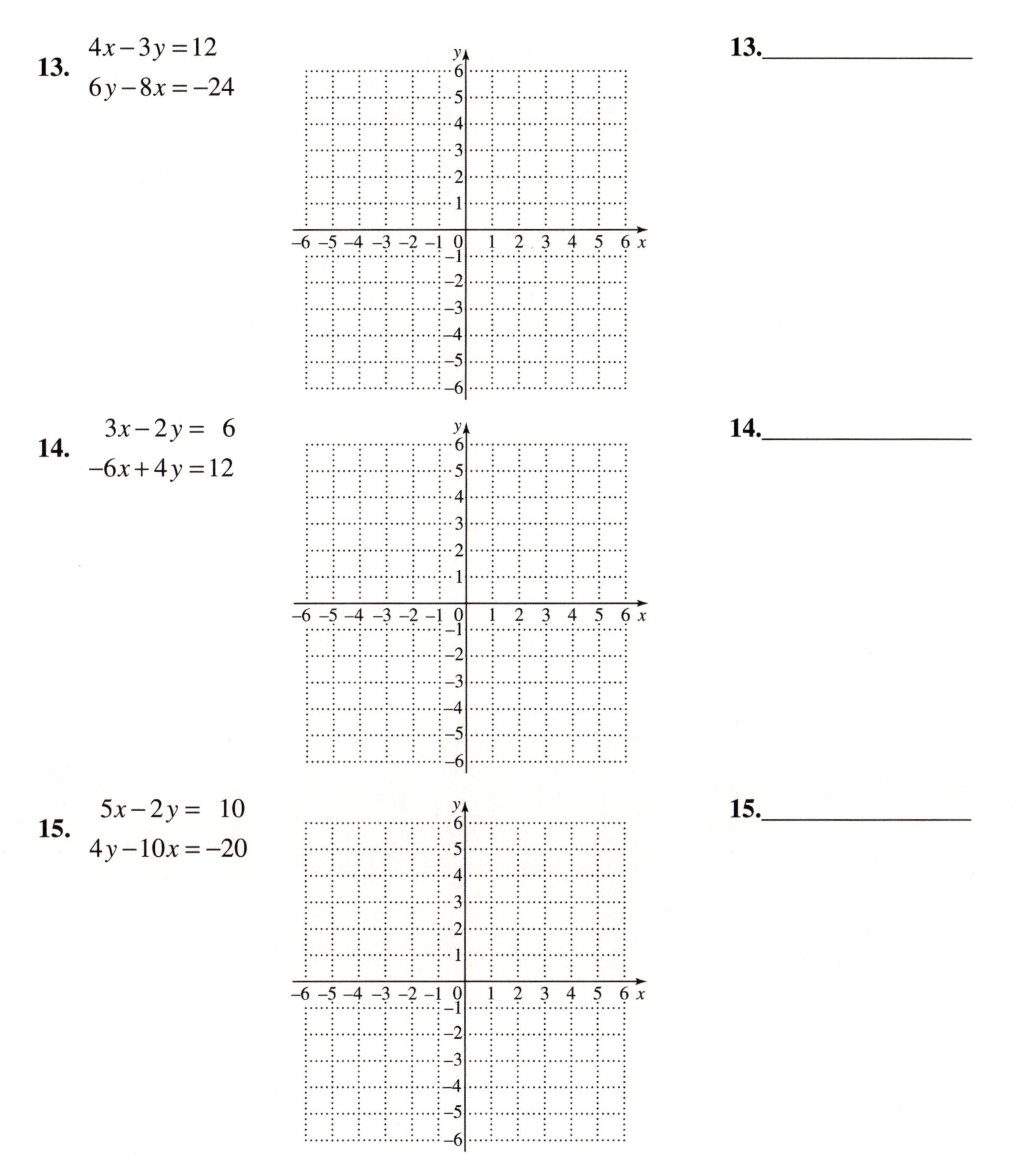

13. $4x-3y=12$
$6y-8x=-24$

13.________________

14. $3x-2y=6$
$-6x+4y=12$

14.________________

15. $5x-2y=10$
$4y-10x=-20$

15.________________

Name: Date:
Instructor: Section:

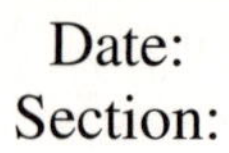

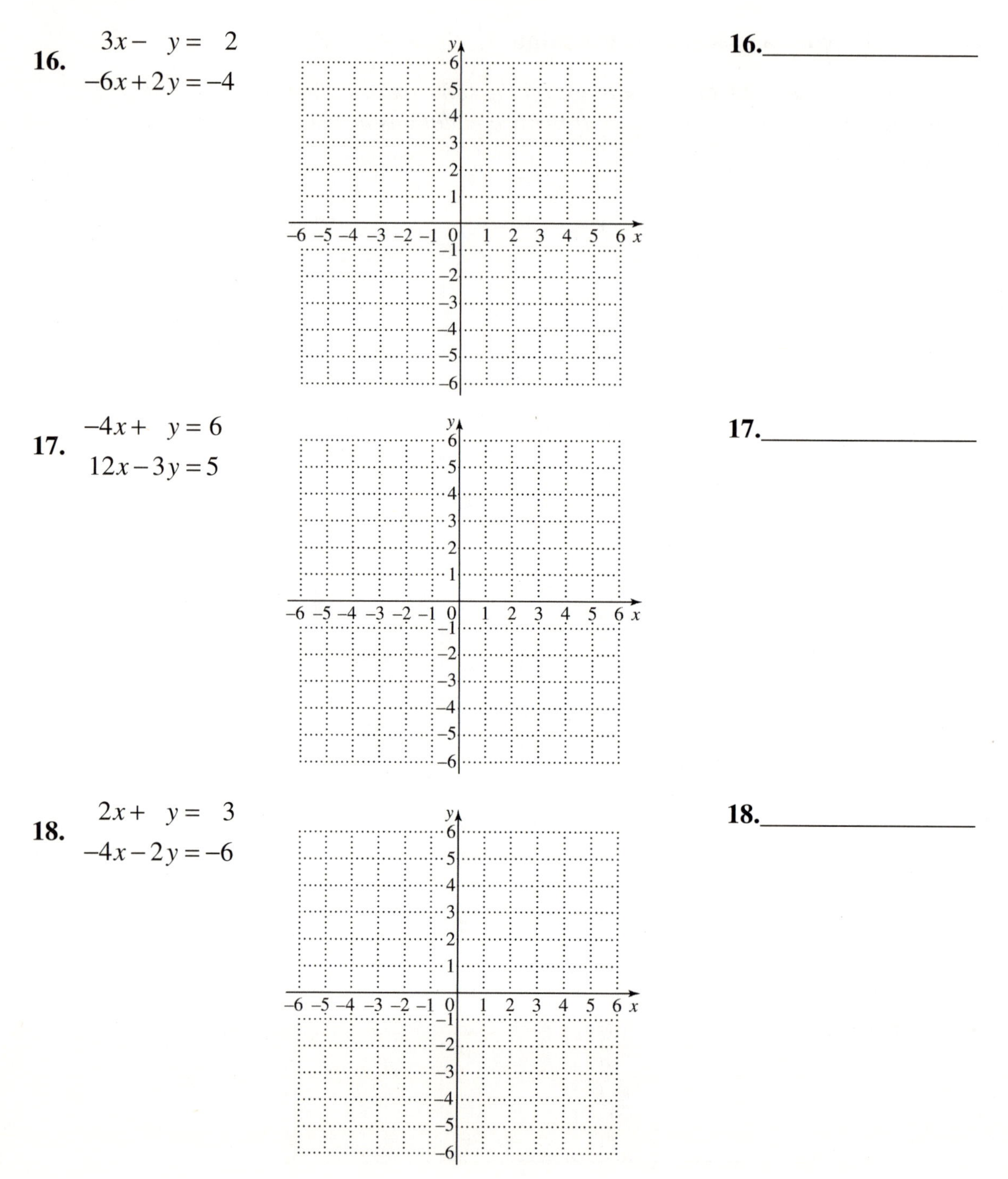

16. $\begin{aligned} 3x - y &= 2 \\ -6x + 2y &= -4 \end{aligned}$

16.________________

17. $\begin{aligned} -4x + y &= 6 \\ 12x - 3y &= 5 \end{aligned}$

17.________________

18. $\begin{aligned} 2x + y &= 3 \\ -4x - 2y &= -6 \end{aligned}$

18.________________

Name: Date:
Instructor: Section:

Objective 4 Identify special systems without graphing.

Without graphing, answer the following questions for each linear system.

(a) *Is the system* inconsistent, or *are the equations* dependent, *or* neither?

(b) *Is the graph a pair of* intersecting lines, *a pair of* parallel lines, *or* one line?

(c) *Does the system have* one solution, no solution, *or an* infinite number of solutions?

19. $x+y=2$
$x+y=5$

19.________________

20. $x-2y=5$
$2x-4y=10$

20.________________

21. $y=3x-2$
$3x-y=4$

21.________________

22. $2x-y=4$
$x+3y=2$

22.________________

23. $3x-9y=6$
$y=\frac{1}{3}x-\frac{2}{3}$

23.________________

24. $6x+2y=12$
$y=-3x+4$

24.________________

Name: Date:
Instructor: Section:

Objective 5 Use a graphing calculator to solve a linear system.

Use the graphing capabilities of your calculator to find the solutions to the given systems.

25. $y = -2x$
$2x - y = 16$

25.________________

26. $3x - y = 17$
$3y - x = 5$

26.________________

27. $3y = -x$
$9x + 8y = -76$

27.________________

28. $3x - 2y = 6$
$\frac{3}{2}x - 3 = y$

28.________________

29. $\frac{1}{3}x + y = 7$
$-\frac{1}{3}x + 4 = y$

29.________________

30. $1 + 5x = \frac{1}{2}y$
$1 + \frac{1}{4}y = 10x$

30.________________

Name: Date:
Instructor: Section:

Chapter 8 SYSTEMS OF LINEAR EQUATIONS

8.2 Solving Systems of Linear Equations by Substitution

Learning Objectives

1 Solve linear systems by substitution.
2 Solve special systems by substitution.
3 Solve linear systems with fractions.

Key Terms

Use the vocabulary terms listed below to complete each statement in exercises 1-3.

substitution method **substitute** **check**

1. The ___________________ is an algebraic method of solving a system of equations in which one equation is solved for one of the variables and the result is substituted in the other equation.

2. When solving a system by substitution, ___________________ one variable for an expression containing the other variable.

3. Always ___________________ the solution in both of the original equations. Then write the solution set.

Objective 1 Solve linear systems by substitution.

Solve the system by the substitution method.

1. $3x + y = -20$
$y = 2x$

1._______________

2. $4x - y = -7$
$y = -3x$

2._______________

3. $3x + 7y = 18$
$y = 2 - x$

3._______________

Name: Date:
Instructor: Section:

4. $\begin{aligned} 4x-3y&=15 \\ x&=y+4 \end{aligned}$ **4.**________________

5. $\begin{aligned} 3x+7y&=16 \\ y&=2x-5 \end{aligned}$ **5.**________________

6. $\begin{aligned} 2x-5y&=11 \\ 3x&=2y \end{aligned}$ **6.**________________

7. $\begin{aligned} 2x+5y&=26 \\ 4x&=3y \end{aligned}$ **7.**________________

8. $\begin{aligned} 3x+2y&=7 \\ 4x-3y&=-19 \end{aligned}$ **8.**________________

9. $\begin{aligned} 4x+5y&=13 \\ 3x-4y&=2 \end{aligned}$ **9.**________________

10. $\begin{aligned} 4x-3y&=-12 \\ x+3&=y \end{aligned}$ **10.**________________

Name: Date:
Instructor: Section:

Objective 2 Solve special systems by substitution.

Solve the system by the substitution method.

11. $2x + y = 3$
$y = 3 - 2x$

11.________________

12. $4x - 8y = 12$
$x - 2y = 3$

12.________________

13. $8x - 2y = 4$
$4x - y = 4$

13.________________

14. $6x + 2y = 12$
$3x + y = 4$

14.________________

15. $36x + 20y = 12$
$-27x - 15y = -9$

15.________________

16. $5x + 2y = -8$
$10x + 4y = 6$

16.________________

17. $48x - 56y = 32$
$21y - 18x = -12$

17.________________

Name: Date:
Instructor: Section:

18. $$\begin{aligned} 12x - 18y &= 25 \\ 4x - \ \ 6y &= 5 \end{aligned}$$

18.________________

19. $$\begin{aligned} 72x - 60y &= -12 \\ 25y - 30x &= \ \ 5 \end{aligned}$$

19.________________

20. $$\begin{aligned} x + \ \ y &= 8 \\ \frac{x}{2} + \frac{y}{2} &= 4 \end{aligned}$$

20.________________

Objective 3 Solve linear systems with fractions.

In the given systems, begin by clearing fractions and then solve the system by substitution.

21. $$\begin{aligned} \frac{x}{4} + \frac{y}{4} &= \ \ 1 \\ \frac{x}{2} + \frac{y}{2} &= -1 \end{aligned}$$

21.________________

22. $$\begin{aligned} \frac{7}{3}x + \ \ y &= 5 \\ 2x + \frac{3}{2}y &= 3 \end{aligned}$$

22.________________

23. $$\begin{aligned} \frac{5}{4}x - \ \ y &= 5 \\ -\frac{7}{8}x + \frac{5}{8}y &= 1 \end{aligned}$$

23.________________

Name: Date:
Instructor: Section:

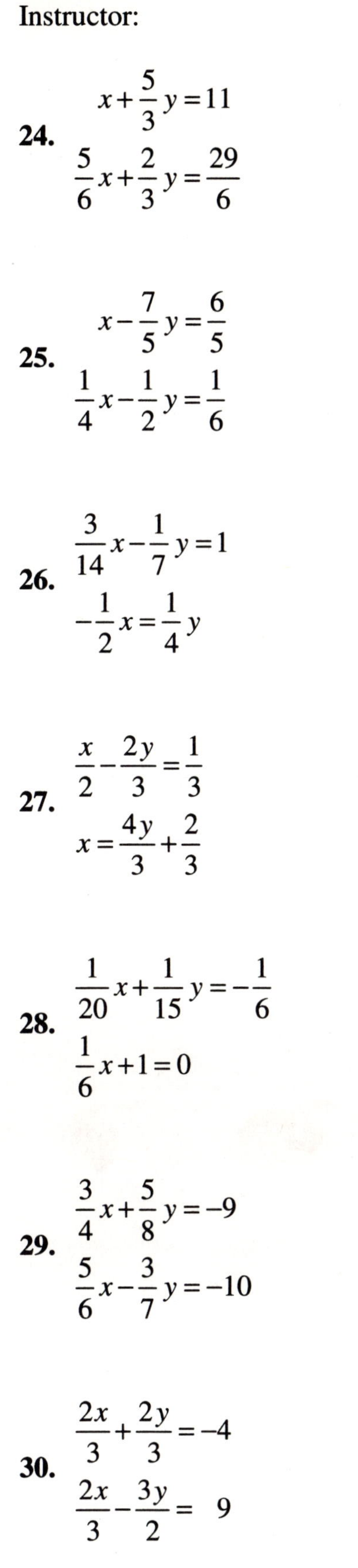

24.
$$x+\frac{5}{3}y=11$$
$$\frac{5}{6}x+\frac{2}{3}y=\frac{29}{6}$$

24.________________

25.
$$x-\frac{7}{5}y=\frac{6}{5}$$
$$\frac{1}{4}x-\frac{1}{2}y=\frac{1}{6}$$

25.________________

26.
$$\frac{3}{14}x-\frac{1}{7}y=1$$
$$-\frac{1}{2}x=\frac{1}{4}y$$

26.________________

27.
$$\frac{x}{2}-\frac{2y}{3}=\frac{1}{3}$$
$$x=\frac{4y}{3}+\frac{2}{3}$$

27.________________

28.
$$\frac{1}{20}x+\frac{1}{15}y=-\frac{1}{6}$$
$$\frac{1}{6}x+1=0$$

28.________________

29.
$$\frac{3}{4}x+\frac{5}{8}y=-9$$
$$\frac{5}{6}x-\frac{3}{7}y=-10$$

29.________________

30.
$$\frac{2x}{3}+\frac{2y}{3}=-4$$
$$\frac{2x}{3}-\frac{3y}{2}=9$$

30.________________

Name: Date:
Instructor: Section:

Chapter 8 SYSTEMS OF LINEAR EQUATIONS

8.3 Solving Systems of Linear Equations by Elimination

Learning Objectives
1 Solve linear systems by elimination.
2 Multiply when using the elimination method.
3 Use an alternative method to find the second value in a solution.
4 Use the elimination method to solve special systems.

Key Terms
Use the vocabulary terms listed below to complete each statement in exercises 1-4.

elimination method **opposites** **true** **false**

1. When solving a linear system by elimination, transform the equations as needed so that the coefficients of one pair of variables are ____________________.

2. When solving a linear system by elimination, a ____________________ statement indicates that the equations are inconsistent.

3. The ____________________ is an algebraic method used to solve a system of equations in which the equations of the system are combined so that one or more variables is eliminated.

4. When solving a linear system by elimination, a ____________________ statement indicates that the equations are dependent.

Objective 1 Solve linear systems by elimination.

Solve the system by the addition method. Check your answers.

1. $x + y = 4$
   $x - y = 2$

   1.________________

2. $3x - y = 7$
   $2x + y = 3$

   2.________________

3. $x - 8y = -8$
   $-x + y = -6$

   3.________________

Name: Date:
Instructor: Section:

4. $\begin{aligned} 2x - y &= 4 \\ 3x + y &= 11 \end{aligned}$

4.________________

5. $\begin{aligned} 4x + 2y &= 8 \\ 6x - 2y &= -13 \end{aligned}$

5.________________

6. $\begin{aligned} 3x - 4y &= 8 \\ x + 4y &= -2 \end{aligned}$

6.________________

7. $\begin{aligned} 8x + 2y &= 6 \\ 3x - 2y &= -6 \end{aligned}$

7.________________

Objective 2 Multiply when using the elimination method.

Solve the system by the addition method. Check your answers.

8. $\begin{aligned} x - 4y &= 12 \\ x + 6y &= -8 \end{aligned}$

8.________________

9. $\begin{aligned} 6x + y &= -2 \\ 3x - 4y &= 35 \end{aligned}$

9.________________

10. $\begin{aligned} 2x + y &= 3 \\ -3x + y &= -22 \end{aligned}$

10.________________

11. $\begin{aligned} 4x - 5y &= -17 \\ 3x + 2y &= -7 \end{aligned}$

11.________________

Name: Date:
Instructor: Section:

12. $$\begin{aligned} 4x-9y &= -5 \\ 3x+2y &= 5 \end{aligned}$$

12.________________

13. $$\begin{aligned} 3x+y &= 1 \\ 2x+3y &= -11 \end{aligned}$$

13.________________

14. $$\begin{aligned} 2x+2y &= 4 \\ 3x-5y &= -2 \end{aligned}$$

14.________________

Objective 3 Use an alternative method to find the second value in a solution.

Solve the system by using the elimination method twice. Check your answer.

15. $$\begin{aligned} 2x-y &= 6 \\ x+2y &= 8 \end{aligned}$$

15.________________

16. $$\begin{aligned} 3x+2y &= 6 \\ 9x+4y &= 24 \end{aligned}$$

16.________________

17. $$\begin{aligned} 4x-5y &= -22 \\ -16x+9y &= 110 \end{aligned}$$

17.________________

18. $$\begin{aligned} 6x-5y &= -22 \\ 9x+2y &= 5 \end{aligned}$$

18.________________

Name: Date:
Instructor: Section:

19. $\begin{aligned} x+5y &= 8 \\ x-3y &= -1 \end{aligned}$

19.________________

20. $\begin{aligned} -4y &= 2x-1 \\ y+2x &= 7 \end{aligned}$

20.________________

21. $\begin{aligned} 2x+3y &= 5 \\ 3x-2y &= -4 \end{aligned}$

21.________________

22. $\begin{aligned} 3x+5y &= 0 \\ 5x-4y &= 1 \end{aligned}$

22.________________

Objective 4 Use the elimination method to solve special systems.

Solve the system by the addition method.

23. $\begin{aligned} 3x-2y &= 6 \\ 6x-4y &= 7 \end{aligned}$

23.________________

24. $\begin{aligned} x+y &= 5 \\ -x-y &= -5 \end{aligned}$

24.________________

25. $\begin{aligned} 4x-8y &= 12 \\ x-2y &= 3 \end{aligned}$

25.________________

Name: Date:
Instructor: Section:

26. $\begin{aligned} 8x - 2y &= 4 \\ 4x - \ \ y &= 4 \end{aligned}$

26.________________

27. $\begin{aligned} 6x + 2y &= 12 \\ 3x + \ \ y &= -6 \end{aligned}$

27.________________

28. $\begin{aligned} 24x - 15y &= 30 \\ -16x + 10y &= -20 \end{aligned}$

28.________________

29. $\begin{aligned} 5x + 2y &= -8 \\ 10x + 4y &= 6 \end{aligned}$

29.________________

30. $\begin{aligned} 48x - 56y &= 32 \\ 21y - 18x &= -12 \end{aligned}$

30.________________

Name: Date:
Instructor: Section:

Chapter 8 SYSTEMS OF LINEAR EQUATIONS

8.4 Systems of Linear Equations in Three Variables

Learning Objectives

1 Understand the geometry of systems of three equations in three variables.
2 Solve linear systems (with three equations and three variables) by elimination.
3 Solve linear systems (with three equations and three variables) in which some of the equations have missing terms.
4 Solve special systems.

Key Terms

Use the vocabulary terms listed below to complete each statement in exercises 1-5.

ordered triple **common point** **line in common** **coincide**

no points common

1. The single solution to a system of equations in three variables can be visualized by three planes that meet at a single, ___________________.

2. Three planes may ___________________, so that the solution of a system is the set of all points on a plane.

3. A solution of an equation in three variables, written (x, y, z), is called a(n) ___________________.

4. Three planes may have a ___________________, so that the infinite set of points that satisfy the equation of the line is the solution of the system.

5. Three planes may have ___________________ to all three, so that there is no solution to the system.

Objective 2 Solve linear systems (with three equations and three variables) by elimination.

Solve the system of equations.

1.
$$\begin{aligned} x+y+z &= 2 \\ x-y+z &= -2 \\ x-y-z &= -4 \end{aligned}$$

1.________________

Name: Date:
Instructor: Section:

2.
$$\begin{aligned} x - y + z &= 2 \\ x + y - z &= 0 \\ x - y - z &= 4 \end{aligned}$$

2.________________

3.
$$\begin{aligned} 2x + y - z &= 9 \\ x + 2y + z &= 3 \\ 3x + 3y - z &= 14 \end{aligned}$$

3.________________

4.
$$\begin{aligned} 3x - y + 2z &= -6 \\ 2x + y + 2z &= -1 \\ 3x + y - z &= -10 \end{aligned}$$

4.________________

5.
$$\begin{aligned} 4x + 2y + 3z &= 11 \\ 2x + y - 4z &= -22 \\ 3x + 3y + z &= -1 \end{aligned}$$

5.________________

6.
$$\begin{aligned} x - 2y + 5z &= -7 \\ 2x + 3y - 4z &= 14 \\ 3x - 5y + z &= 7 \end{aligned}$$

6.________________

7.
$$\begin{aligned} 2x - 5y + 2z &= 30 \\ x + 4y + 5z &= -7 \\ \frac{1}{2}x - \frac{1}{4}y + z &= 4 \end{aligned}$$

7.________________

Name: Date:
Instructor: Section:

8.
$$\begin{aligned} 5x - 2y + z &= 28 \\ 3x + 5y - 2z &= -23 \\ \frac{2}{3}x + \frac{1}{3}y + z &= 1 \end{aligned}$$

8.________________

9.
$$\begin{aligned} \frac{1}{3}x + \frac{1}{6}y - \frac{2}{3}z &= -1 \\ \frac{3}{4}x + \frac{1}{3}y + \frac{1}{4}z &= -3 \\ \frac{1}{2}x + \frac{3}{2}y + \frac{3}{4}z &= 21 \end{aligned}$$

9.________________

10.
$$\begin{aligned} \frac{2}{3}x - \frac{1}{4}y + \frac{5}{8}z &= 0 \\ \frac{1}{5}x + \frac{2}{3}y - \frac{1}{4}z &= -7 \\ \frac{3}{5}x - \frac{4}{3}y + \frac{7}{8}z &= 5 \end{aligned}$$

10.________________

Objective 3 Solve linear systems (with three equations and three variables) in which some of the equations have missing terms.

Solve the system of equations.

11.
$$\begin{aligned} x \quad - z &= -3 \\ y + z &= 4 \\ x - y \quad &= 3 \end{aligned}$$

11.________________

Name: Date:
Instructor: Section:

12.
$$\begin{aligned} x+2y \quad &= 1 \\ y-z &= -6 \\ x \quad +z &= 8 \end{aligned}$$

12.________________

13.
$$\begin{aligned} x+5y \quad &= -23 \\ 4y-3z &= -29 \\ 2x \quad +5z &= 19 \end{aligned}$$

13.________________

14.
$$\begin{aligned} 3x \quad -4z &= -23 \\ y+5z &= 24 \\ x-3y \quad &= 2 \end{aligned}$$

14.________________

15.
$$\begin{aligned} 4x-5y \quad &= -13 \\ 3x \quad + z &= 9 \\ 2y+5z &= 10 \end{aligned}$$

15.________________

16.
$$\begin{aligned} 7x \quad + z &= -1 \\ 3y-2z &= 8 \\ 5x+y \quad &= 2 \end{aligned}$$

16.________________

Name: Date:
Instructor: Section:

17.

$$\begin{aligned} 4x \qquad\quad - z &= -6 \\ \frac{3}{5}y + \frac{1}{2}z &= 0 \\ \frac{1}{3}x \qquad + \frac{2}{3}z &= -5 \end{aligned}$$

17.________________

18.

$$\begin{aligned} 2x + 5y \qquad &= 18 \\ 3y + 2z &= 4 \\ \frac{1}{4}x - y \qquad &= -1 \end{aligned}$$

18.________________

19.

$$\begin{aligned} 5x \qquad - 2z &= 8 \\ 4y + 3z &= -9 \\ \frac{1}{2}x + \frac{2}{3}y \qquad &= -1 \end{aligned}$$

19.________________

20.

$$\begin{aligned} x + 2y \qquad &= -2 \\ \frac{1}{2}y + z &= -1 \\ \frac{2}{3}x - \frac{3}{4}y \qquad &= 7 \end{aligned}$$

20.________________

Name: Date:
Instructor: Section:

Objective 4 Solve special systems.

Solve the system of equations.

21.
$$\begin{aligned} x - y + z &= 7 \\ 2x + 5y - 4z &= 2 \\ -x + y - z &= 4 \end{aligned}$$

21.________________

22.
$$\begin{aligned} 8x - 7y + 2z &= 1 \\ 3x + 4y - z &= 6 \\ -8x + 7y - 2z &= 5 \end{aligned}$$

22.________________

23.
$$\begin{aligned} 3x - 2y + 4z &= 5 \\ -3x + 2y - 4z &= -5 \\ \frac{3}{2}x - y + 2z &= \frac{5}{2} \end{aligned}$$

23.________________

24.
$$\begin{aligned} -x + 5y - 2z &= 3 \\ 2x - 10y + 4z &= -6 \\ -3x + 15y - 6z &= 9 \end{aligned}$$

24.________________

25.
$$\begin{aligned} 8x - 4y + 2z &= 0 \\ 3x + y - 4z &= 0 \\ 5x + y + 2z &= 0 \end{aligned}$$

25.________________

Name: Date:
Instructor: Section:

26.
$$\begin{aligned} x-3y+4z&=0\\ 2x+y-z&=0\\ -x+y-5z&=0 \end{aligned}$$

26.________________

27.
$$\begin{aligned} 3x-2y+5z&=6\\ x+4y-z&=1\\ \frac{3}{2}x-y+\frac{5}{2}z&=-3 \end{aligned}$$

27.________________

28.
$$\begin{aligned} 2x+7y-8z&=3\\ 5x-y-z&=1\\ x+\frac{7}{2}y-4z&=3 \end{aligned}$$

28.________________

29.
$$\begin{aligned} x-5y+2z&=0\\ -x+5y-2z&=0\\ \frac{1}{2}x-\frac{5}{2}y+z&=0 \end{aligned}$$

29.________________

30.
$$\begin{aligned} 3x-2y+5z&=0\\ 6x-4y+10z&=0\\ \frac{3}{2}x-y+\frac{5}{2}z&=0 \end{aligned}$$

30.________________

Name: Date:
Instructor: Section:

Chapter 8 SYSTEMS OF LINEAR EQUATIONS

8.5 Applications of Systems of Linear Equations

Learning Objectives

1. Solve geometry problems by using two variables.
2. Solve money problems by using two variables.
3. Solve mixture problems by using two variables.
4. Solve distance-rate-time problems by using two variables.
5. Solve problems with three variables by using a system of three equations.

Key Terms

Use the vocabulary terms listed below to complete each statement in exercises 1-3.

read **assign variables** **state the answer**

1. After finding unknown values, __________________ to the problem and check to see if the result is reasonable.

2. The first step when solving an applied problem is to __________________ the problem carefully until you understand what is given and what is to be found.

3. After reading and understanding the problem, __________________ to represent the unknown values, using diagrams or tables as needed.

Objective 1 Solve geometry problems by using two variables.

For the word problem, select variables to represent the two unknowns, write two equations using the two variables, and solve the resulting system.

1. The length of a rectangle is 5 feet more than the width. The perimeter of the rectangle is 58 feet. Find the width of the rectangle.

1.________________

2. The length of a rectangle is 7 centimeters more than the width. The perimeter of the rectangle is 134 centimeters. Find the width of the rectangle.

2.________________

Name: Date:
Instructor: Section:

3. The perimeter of a rectangle is 70 inches. If the width were doubled, the width would be 10 inches more than the length. Find the width of the rectangle.

3.________________

4. A triangle has one side 18 centimeters long. The two unknown sides are of equal length. The perimeter is 44 centimeters. Find the length of the unknown sides.

4.________________

5. The perimeter of a rectangle is 96 inches. If the width were tripled, the width would be 36 inches more than the length. Find the width of the rectangle.

5.________________

6. The width of a rectangle is 42 yards less then the length. The perimeter of the rectangle is 220 yards. Find the length of the rectangle.

6.________________

Objective 2 Solve money problems by using two variables.

For the word problem, select variables to represent the two unknowns, write two equations using the two variables, and solve the resulting system.

7. Norma Foust has some \$5-bills and some \$10-bills. The total value of the money is \$260, with a total of 32 bills. How many of each are there?

7.________________

8. Pablo Gomez has some \$10-bills and some \$20-bills. The total value of the money is \$650, with a total of 40 bills. How many of each are there?

8.________________

Name: Date:
Instructor: Section:

9. The cost of 7 small boxes and 15 large boxes is $26.70. The cost of 10 small boxes and 3 large boxes is $10.50. Find the cost of a small box and the cost of a large box.

9._______________

10. The cost of 42 small oranges and 12 large oranges is $18.60. The cost of 24 small oranges and 12 large oranges is $13.20. Find the cost of a small orange and the cost of a large orange.

10._______________

11. The Garden Center ordered 6 ounces of marigold seed and 8 ounces of carnation seed paying $214.54. They later ordered another 12 ounces of marigold seed and 18 ounces of carnation seed, paying $464.28. Find the price per ounce for each type of seed.

11._______________

12. Hugh bought 105 pounds of cattle feed and 62 pounds of rabbit feed, paying $109.75. He later bought 70 pounds of cattle feed and 35 pounds of rabbit feed for $70. Find the cost per pound for each type of feed.

12._______________

Objective 3 Solve mixture problems by using two variables.

For the word problem, select variables to represent the two unknowns, write two equations using the two variables, and solve the resulting system.

13. How many ounces each of 20% acid and 50% acid must be mixed together to get 120 ounces of a 30% acid?

13._______________

Name: Date:
Instructor: Section:

14. How many ounces each of 40% acid and 80% acid must be mixed together to get 160 ounces of a 70% acid?

14.________________

15. A radiator holds 10 liters. How much pure antifreeze must be added to a mixture that is 10% antifreeze to make enough of a 20% mixture to fill the radiator?

15.________________

16. A radiator holds 15 liters. How much pure antifreeze must be added to a mixture that is 10% antifreeze to make enough of a 25% mixture to fill the radiator?

16.________________

17. A candy mix is to be made by mixing candy worth $12 per kilogram with candy worth $15 per kilogram to get 120 kilograms of a mixture worth $13 per kilogram. How many kilograms of each should be used?

17.________________

18. A coffee mix is to be made by mixing coffee worth $10 per kilogram with coffee worth $18 per kilogram to get 80 kilograms of a mixture worth $12 per kilogram. How many kilograms of each should be used?

18.________________

Name: Date:
Instructor: Section:

Objective 4 Solve distance-rate-time problems by using two variables.

For the word problem, select variables to represent the two unknowns, write two equations using the two variables, and solve the resulting system.

19. Two cars start at the same point and travel in opposite directions. One car travels at 65 miles per hour and the other at 60 miles per hour. How far has each traveled when they are 375 miles apart?

19.________________

20. A plane flying with the wind flew 600 miles in 5 hours. Against the wind, the plane required 6 hours to fly the same distance. Find the rate of the wind.

20.________________

21. Two trains start at the same point going in the same direction on parallel tracks. One train travels at 70 miles per hour and the other at 42 miles per hour. In how many hours will they be 154 miles apart?

21.________________

22. Dave and his sister Ann jog to school daily. Dave jogs at 9 miles per hour, and Ann jogs at 5 miles per hour. When Dave reaches school, Ann is 1/2 mile from the school. How far do Dave and Ann live from their school?

22.________________

23. A motorboat traveling with the current went 72 miles in 3 hours. Against the current, the boat could only go 48 miles in the same amount of time. Find the rate of the boat in still water.

23.________________

Name: Date:
Instructor: Section:

24. A plane can travel 600 miles per hour with the wind and 530 miles per hour against the wind. Find the speed of the plane in still air.

24._______________

Objective 5 Solve problems with three variables by using a system of three equations.

For the problem, select variables to represent the three unknowns, write three equations using the three variables, and solve the resulting system.

25. Three numbers have a sum of 31. The middle number is 1 more than the smallest number. The sum of the smaller two numbers is 7 more than the largest number. Find the three numbers.

25._______________

26. The sum of three numbers is 99. The difference of the smaller two is 3. The sum of the smallest and twice the largest is 108. Find the three numbers.

26._______________

27. The sum of the measures of the angles of any triangle is 180°. In a certain triangle, the first angle measures 20° less than the second angle, and the second angle measures 10° more than the third. Find the three angles.

27._______________

28. In a certain triangle, the sum of the measures of the smallest and largest angles is 50° more than the measure of the medium angle. The medium angle measures 25° more than the smallest angle. Find the three angles.

28._______________

Name: Date:
Instructor: Section:

29. A triangle has a perimeter of 197 centimeters. The sum of the smallest and medium sides is 49 more than the longest side. The difference of the longest and shortest sides is 22. Find the lengths of the three sides.

29._______________

30. Lee has some $5, $10, and $20-bills. He has a total of 51 bills, worth $795. The number of $5-bills is 25 less than the number of $20-bills. Find the number of each type of bill he has.

30._______________

Name: Date:
Instructor: Section:

Chapter 8 SYSTEMS OF LINEAR EQUATIONS

8.6 Solving Systems of Linear Equations by Matrix Methods

Learning Objectives
1 Define a matrix.
2 Write the augmented matrix of a system.
3 Use row operations to solve a system with two equations.
4 Use row operations to solve a system with three equations.
5 Use row operations to solve special systems.

Key Terms
Use the vocabulary terms listed below to complete each statement in exercises 1-6.

elements **dimensions** **square matrix** **augmented matrix**

row operations **row echelon form**

1. ____________________ are operations on a matrix that produce equivalent matrices leading to systems that have the same solutions as the original system of equations.

2. The numbers in a matrix are called the ____________________ of the matrix.

3. If a matrix is written with 1s on the diagonal from upper left to lower right and 0s below the 1s, it is said to be in ____________________.

4. A(n) ____________________ is a matrix that has a vertical bar that separates the columns of the matrix into two groups.

5. A(n) ____________________ is a matrix that has the same number of rows as columns.

6. The number of rows followed by the number of columns gives the ____________________ of a matrix.

Objective 1 Define a matrix.

1. Give an example of a 3×2 matrix and a 1×4 matrix. **1.**__________________

Name: Date:
Instructor: Section:

Give the number of rows and columns in the matrix.

2. $\begin{bmatrix} -2 & 5 & 3 \\ 1 & 0 & 1 \end{bmatrix}$ **2.**____________

3. $\begin{bmatrix} 5 & -8 \\ 1 & -9 \end{bmatrix}$ **3.**____________

4. $\begin{bmatrix} -4 & 7 & 6 & 5 \\ 0 & 0 & 9 & 8 \\ 3 & 2 & 5 & 1 \end{bmatrix}$ **4.**____________

5. $\begin{bmatrix} 5 \\ -1 \end{bmatrix}$ **5.**____________

6. $\begin{bmatrix} -8 & 9 & 7 \end{bmatrix}$ **6.**____________

Objective 2 Write the augmented matrix of a system.

Write the augmented matrix for the system. Do not try to solve the system.

7. $\begin{aligned} 5x - y &= 6 \\ 2x + 3y &= 4 \end{aligned}$ **7.**____________

8. $\begin{aligned} x - y + 4 &= 0 \\ 3x + y - 1 &= 0 \end{aligned}$ **8.**____________

9. $\begin{aligned} x + 6 &= 0 \\ y - 2 &= 0 \end{aligned}$ **9.**____________

10. $\begin{aligned} x - y + z &= 4 \\ 2x + y - 3z &= 1 \\ 5x - 8y + z &= 2 \end{aligned}$ **10.**____________

11. $\begin{aligned} 4x - 7y + z - 1 &= 0 \\ 2x + 3y - 5z + 2 &= 0 \\ 6x - y + 8z - 5 &= 0 \end{aligned}$ **11.**____________

Name: Date:
Instructor: Section:

12.
$$\begin{aligned} x - \quad y &= 2 \\ y + 3z &= 5 \\ x - \quad z &= 1 \end{aligned}$$

12.________________

Objective 3 Use row operations to solve a system with two equations.

Use row operations to solve the system.

13.
$$\begin{aligned} x + y &= 5 \\ x - y &= 3 \end{aligned}$$

13.________________

14.
$$\begin{aligned} x + 4y &= 1 \\ 2x + \quad y &= 2 \end{aligned}$$

14.________________

15.
$$\begin{aligned} x - 2y &= 1 \\ x + 4y &= 7 \end{aligned}$$

15.________________

16.
$$\begin{aligned} x + 2y &= \quad 0 \\ 2x + \quad y &= -6 \end{aligned}$$

16.________________

17.
$$\begin{aligned} 3x + 4y &= 5 \\ 9x - 8y &= 0 \end{aligned}$$

17.________________

18.
$$\begin{aligned} x + \quad 5y &= -1 \\ 3x - 10y &= \quad 2 \end{aligned}$$

18.________________

Name: Date:
Instructor: Section:

Objective 4 Use row operations to solve a system with three equations.

Use row operations to solve the system.

19.
$$\begin{aligned} x+y+z &= 2 \\ x-y-z &= 0 \\ x+y-z &= -2 \end{aligned}$$

19.________________

20.
$$\begin{aligned} x-y+z &= -1 \\ x-y+2z &= 0 \\ x+y-z &= -3 \end{aligned}$$

20.________________

21.
$$\begin{aligned} x+2y-z &= 0 \\ 2x-y+z &= 4 \\ 3x+y-z &= 1 \end{aligned}$$

21.________________

22.
$$\begin{aligned} x+y+2z &= 7 \\ 3x-y+z &= 10 \\ 2x+y-3z &= -6 \end{aligned}$$

22.________________

23.
$$\begin{aligned} x-y &= 1 \\ 2y+3z &= 8 \\ 2x+z &= 6 \end{aligned}$$

23.________________

Name: Date:
Instructor: Section:

24.
$$\begin{aligned} 2x + y \quad &= 1 \\ 2y + 7z &= -3 \\ 3x \quad - z &= 8 \end{aligned}$$

24.________________

Objective 5 Use row operations to solve special systems.

Use row operations to solve the system.

25.
$$\begin{aligned} 5x - 10y &= 8 \\ -x + 2y &= 3 \end{aligned}$$

25.________________

26.
$$\begin{aligned} -3x + 4y &= 6 \\ 9x - 12y &= -18 \end{aligned}$$

26.________________

27.
$$\begin{aligned} \frac{2}{3}x - \frac{4}{5}y &= 7 \\ x - \frac{6}{5}y &= \frac{21}{2} \end{aligned}$$

27.________________

28.
$$\begin{aligned} .2x - .3y &= 6.1 \\ -4x + 6y &= -122 \end{aligned}$$

28.________________

29.
$$\begin{aligned} 3x + 6y + 9z &= 5 \\ x + 2y + 3z &= 4 \\ 2x + 4y + 6z &= 8 \end{aligned}$$

29.________________

30.
$$\begin{aligned} x + 3y + z &= 5 \\ -3x - 9y - 3z &= -15 \\ 4x + 12y + 4z &= 20 \end{aligned}$$

30.________________

Name: Date:
Instructor: Section:

Chapter 9 INEQUALITIES AND ABSOLUTE VALUE

9.1 Set Operations and Compound Inequalities

Learning Objectives
1 Find the intersection of two sets.
2 Solve compound inequalities with the word *and*.
3 Find the union of two sets.
4 Solve compound inequalities with the word *or*.

Key Terms
Use the vocabulary terms listed below to complete each statement in exercises 1-3.

intersection of two sets **compound inequality** **union of two sets**

1. The ____________________ is the set of elements that belongs to both sets.

2. A(n) ____________________ consists of two inequalities linked by a connective word such as *and* or *or.*

3. The ____________________ is the set of elements that belong to either set.

Objective 1 Find the intersection of two sets.

Find the intersection.

1. $\{0, 1, 2, 3\} \cap \{2, 3, 4, 5\}$ **1.**________________

2. $\{-6, -5, -4\} \cap \{-3, -2, -1\}$ **2.**________________

3. $\{7, 8, 9, 10\} \cap \varnothing$ **3.**________________

Name: Date:
Instructor: Section:

Let $A = \{0, 1, 2, 3, 4, 5\}$, $B = \{2, 4, 6, 8, 10\}$, $C = \{1, 3, 5, 7, 9\}$, $D = \{0, 2, 4\}$, *and* $E = \{0\}$. *Find the indicated intersection.*

4. $A \cap B$ **4.**________________

5. $A \cap C$ **5.**________________

6. $A \cap E$ **6.**________________

7. $B \cap C$ **7.**________________

Objective 2 Solve compound inequalities with the word *and*.

For the compound inequality, give the solution set in both interval and graph forms.

8. $r < 3$ *and* $r > 0$ **8.**________________

9. $m \leq 4$ *and* $m \leq 7$ **9.**________________

10. $x - 3 \leq 6$ *and* $x + 2 \geq 7$ **10.**________________

11. $2q < -2$ *and* $q + 3 > 1$ **11.**________________

12. $2z + 1 < 3$ *and* $3z - 3 > 3$ **12.**________________

13. $3k - 4 \leq 8$ *and* $4k - 2 \leq 16$ **13.**________________

Name: Date:
Instructor: Section:

14. $q < -1$ *and* $q \geq 2$ 14.________________

15. $r \geq 2$ *and* $r \leq -2$ 15.________________

Objective 3 Find the union of two sets.

Find the union.

16. $\{0, 1, 2, 3\} \cup \{2, 3, 4, 5\}$ 16.________________

17. $\{-6, -5, -4\} \cup \{-3, -2, -1\}$ 17.________________

18. $\{2, 6, 8\} \cup \{2, 6, 8\}$ 18.________________

Let $A = \{1, 2, 3, 4, 5, 6\}$, $B = \{0, 2, 4, 6, 8, 10\}$, $C = \{1, 3, 5, 7, 9\}$, $D = \{1, 2, 3\}$, *and* $E = \{0\}$. *Find the indicated union.*

19. $A \cup B$ 19.________________

20. $A \cup E$ 20.________________

21. $B \cup C$ 21.________________

22. $B \cup D$ 22.________________

Name: Date:
Instructor: Section:

Objective 4 Solve compound inequalities with the word *or*.

For the compound inequality, give the solution set in both interval and graph forms.

23. $m > 4$ *or* $m < -1$ **23.**_______________

24. $y \leq 1$ *or* $y \geq 6$ **24.**_______________

25. $k \leq 3$ *or* $k \geq 6$ **25.**_______________

26. $r \geq -1$ *or* $r \geq 4$ **26.**_______________

27. $p \geq -1$ *or* $p \leq 6$ **27.**_______________

28. $q + 3 > 7$ *or* $q + 1 \leq -3$ **28.**_______________

29. $s - 5 > 0$ *or* $s + 7 < 6$ **29.**_______________

30. $4x < x - 5$ *or* $6x > 2x + 3$ **30.**_______________

Name: Date:
Instructor: Section:

Chapter 9 INEQUALITIES AND ABSOLUTE VALUE

9.2 Solving Systems of Linear Equations by Substitution

Learning Objectives

1. Use the distance definition of absolute value.
2. Solve equations of the form $|ax+b|=k$, for $k>0$.
3. Solve inequalities of the form $|ax+b|<k$ and of the form $|ax+b|>k$, for $k>0$.
4. Solve inequalities that involve rewriting.
5. Solve equations of the form $|ax+b|=|cx+d|$.
6. Solve special cases of absolute value equations and inequalities.

Key Terms

Use the vocabulary terms listed below to complete each statement in exercises 1-3.

absolute value equation **absolute value inequality** **negative**

1. An absolute value expression will not have a solution if, when isolated, it equals a ____________________ number.

2. An ____________________ is an equation that involves the absolute value of a variable expression.

3. An ____________________ is an inequality that involves the absolute value of a variable expression.

Objective 1 Use the distance definition of absolute value.

Graph the solution set of the equation or inequality.

1. $|m|=7$ **1.**________________

2. $|k|<8$ **2.**________________

3. $|x|\geq 6$ **3.**________________

4. $|p|\geq -2$ **4.**________________

Name: Date:
Instructor: Section:

5. $|t| \le 0$ **5.**________________

Objective 2 Solve equations of the form $|ax+b| = k$, for $k > 0$.

Solve.

6. $|m+4| = 8$ **6.**________________

7. $|3k-1| = 6$ **7.**________________

8. $|m+6| = 2$ **8.**________________

9. $|2r+3| = 0$ **9.**________________

10. $\left|5-\frac{4}{3}x\right| = 9$ **10.**________________

Objective 3 Solve inequalities of the form $|ax+b| < k$ and of the form $|ax+b| > k$, for $k > 0$.

Solve the absolute value inequality. Graph the solution set.

11. $|x-2| > 8$ **11.**________________

12. $|n+5| < 8$ **12.**________________

Name: Date:
Instructor: Section:

13. $|2x+3|<7$ 13.________________

14. $|3q-5|+2\geq 6$ 14.________________

15. $|4y-1|-3\leq -1$ 15.________________

Objective 4 Solve absolute value equations that involve rewriting.

Solve.

16. $|z|-6=3$ 16.________________

17. $|y|-5=-7$ 17.________________

18. $|2q+4|-3=-2$ 18.________________

19. $|2w-1|+7=12$ 19.________________

20. $|3y-4|+2=8$ 20.________________

Name: Date:
Instructor: Section:

Objective 5 Solve equations of the form $|ax+b|=|cx+d|$.

Solve.

21. $|a-4|=|a-3|$ **21.**________________

22. $|5-z|=|2z+3|$ **22.**________________

23. $|y+5|=|3y+1|$ **23.**________________

24. $|2p-4|=|7-p|$ **24.**________________

25. $\left|y-\frac{1}{4}\right|=\left|\frac{1}{2}y+1\right|$ **25.**________________

Objective 6 Solve special cases of absolute value equations and inequalties.

Solve.

26. $|2x-4|=-6$ **26.**________________

27. $|p|=0$ **27.**________________

28. $|k+5|\le -2$ **28.**________________

Name: Date:
Instructor: Section:

29. $|4+t|<0$

29.________________

30. $|m-2|\geq -1$

30.________________

Name: Date:
Instructor: Section:

Chapter 9 INEQUALITIES AND ABSOLUTE VALUE

9.3 Linear Inequalities in Two Variables

Learning Objectives

1. Graph linear inequalities in two variables.
2. Graph the intersection of two linear inequalities.
3. Graph the union of two inequalities.
4. Use a graphing calculator to solve linear inequalities in one variable.

Key Terms

Use the vocabulary terms listed below to complete each statement in exercises 1-6.

linear inequality in two variables **boundary line** **test point**

1. A ____________________ is used to determine which region of the graph of an inequality to shade.

2. A ____________________ divides a plane into two regions.

3. A ____________________ can be written in the form $Ax + By < C$ or $Ax + By > C$ (or with $\leq$ or $\geq$), where A, B, and C are real numbers and A and B are both not 0.

Objective 1 Graph linear inequalities in two variables.

Graph the linear inequality.

1. $x + y \geq 2$

1.________________

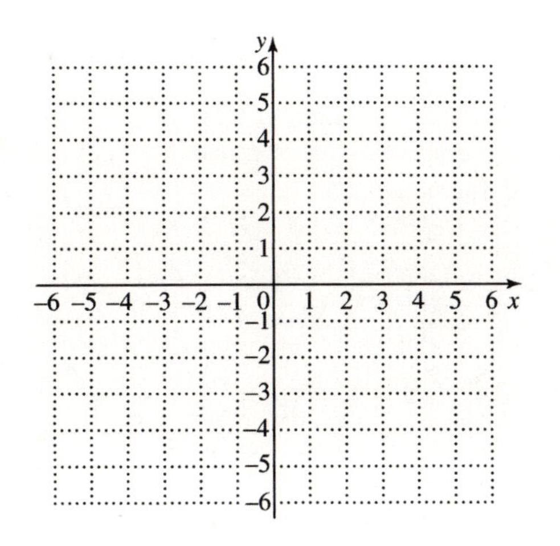

Name: Date:
Instructor: Section:

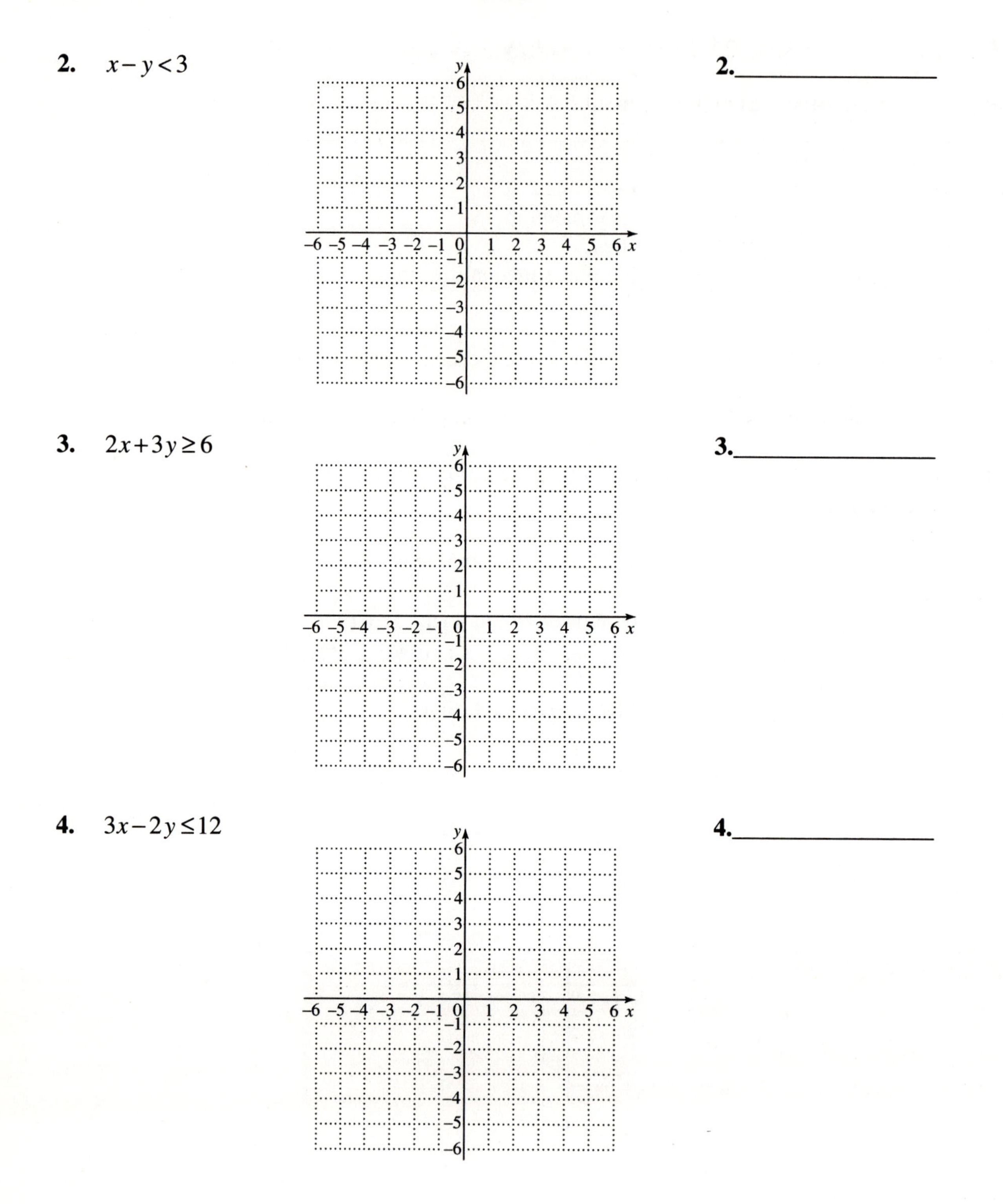

2. $x - y < 3$ **2.**________________

3. $2x + 3y \geq 6$ **3.**________________

4. $3x - 2y \leq 12$ **4.**________________

Name: Date:
Instructor: Section:

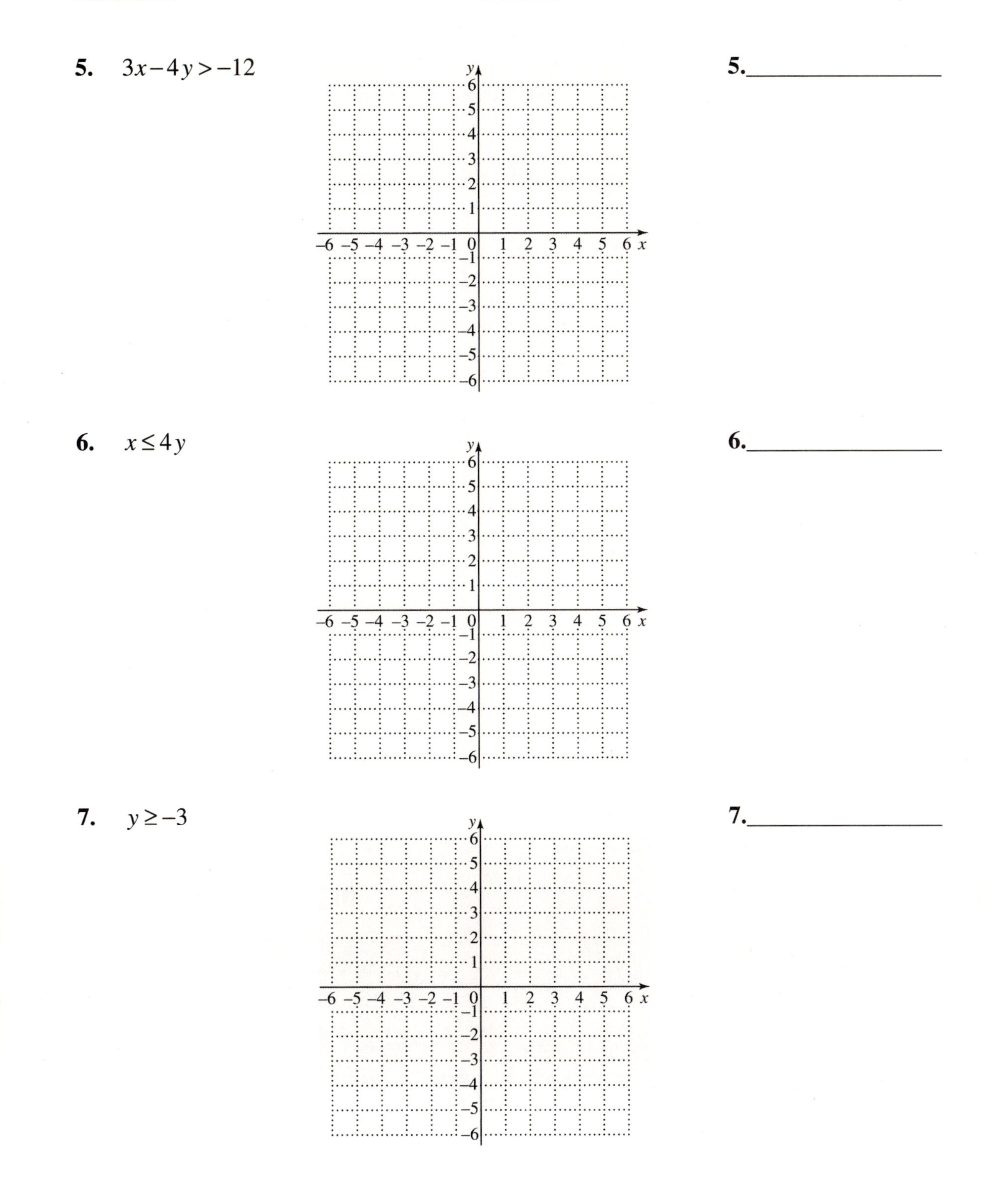

5. $3x-4y>-12$

5.________________

6. $x\leq 4y$

6.________________

7. $y\geq -3$

7.________________

Name: Date:
Instructor: Section:

Objective 2 Graph the intersection of two linear inequalities.

Graph the intersection of the solutions of the pair of inequalities.

8. $x+y\leq 4$ and $x-y>-2$ **8.**________________

9. $2x+y<6$ and $x-3y>-6$ **9.**________________

10. $3x+4y\leq 12$ and $2x-y\leq 4$ **10.**________________

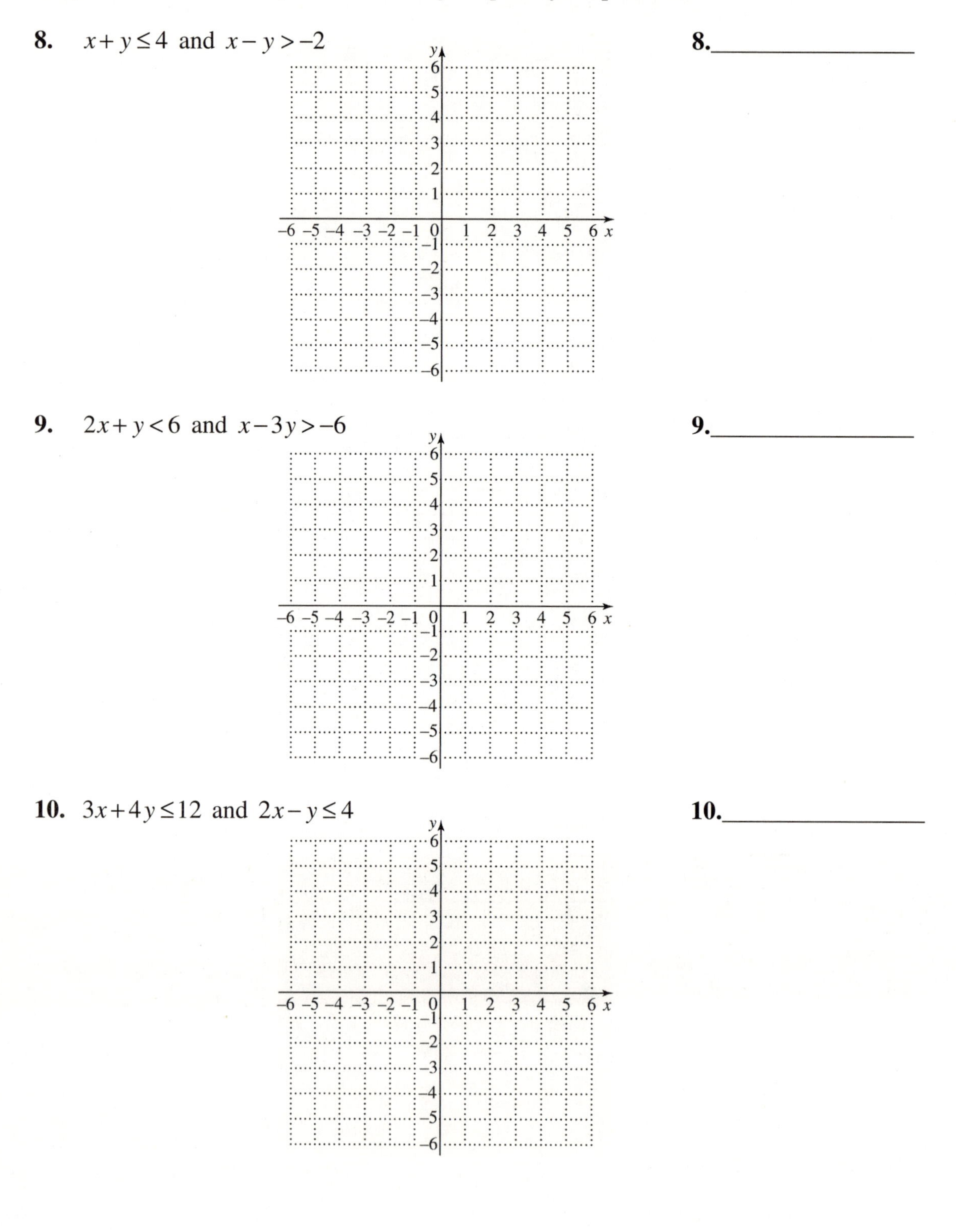

Name: Date:
Instructor: Section:

11. $2x + y \geq 6$ and $y \geq 4$ **11.**________________

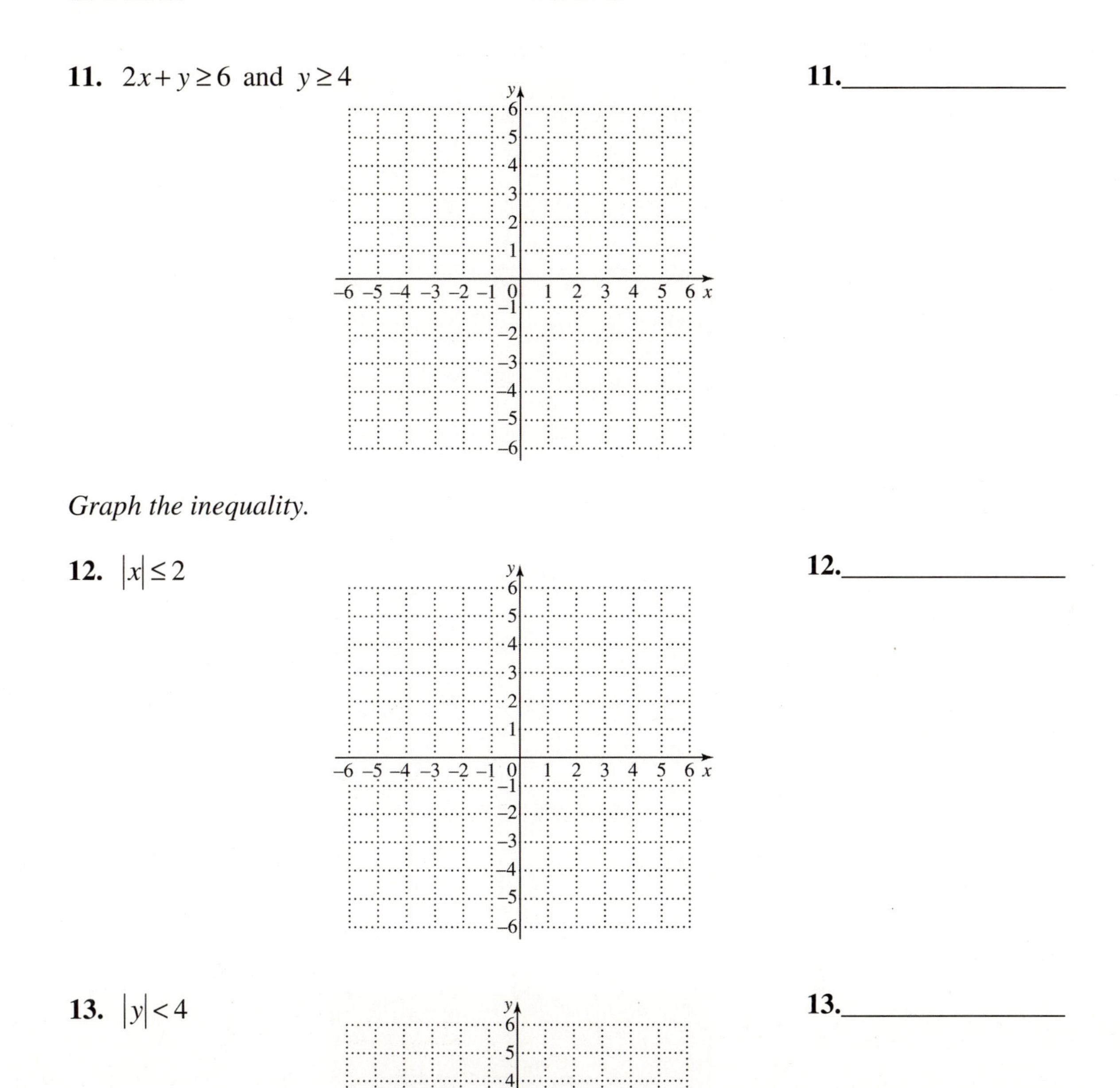

Graph the inequality.

12. $|x| \leq 2$ **12.**________________

13. $|y| < 4$ **13.**________________

Name: Date:
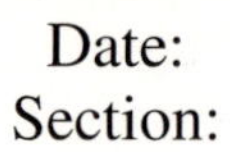
Instructor: Section:

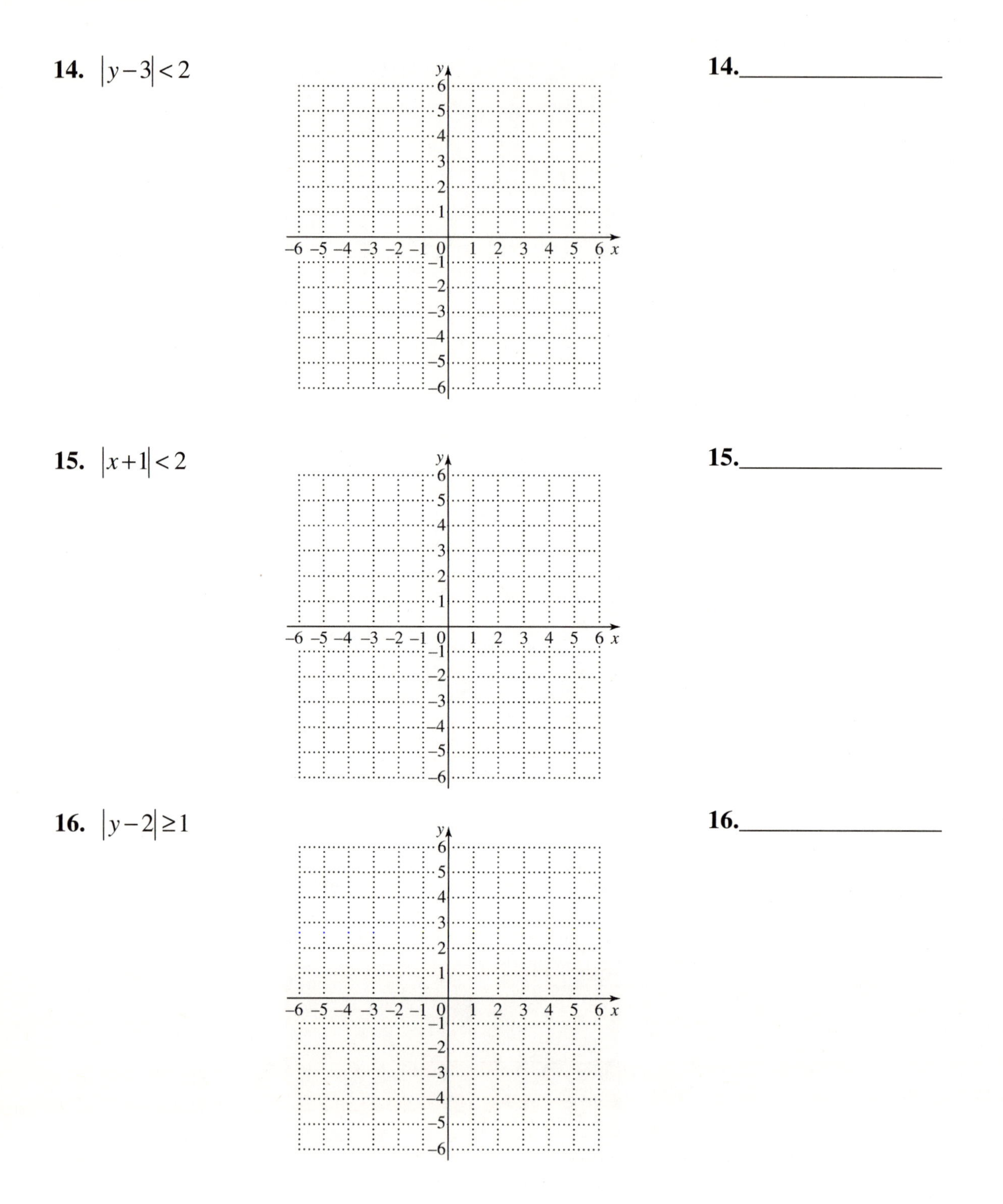

14. $|y-3|<2$

14.________________

15. $|x+1|<2$

15.________________

16. $|y-2|\geq 1$

16.________________

Name: Date:
Instructor: Section:

Objective 3 Graph the union of two linear inequalities.

Graph the union of the solutions of the pair of inequalities.

17. $4x-2y\geq-4$ or $x\geq1$

17.________________

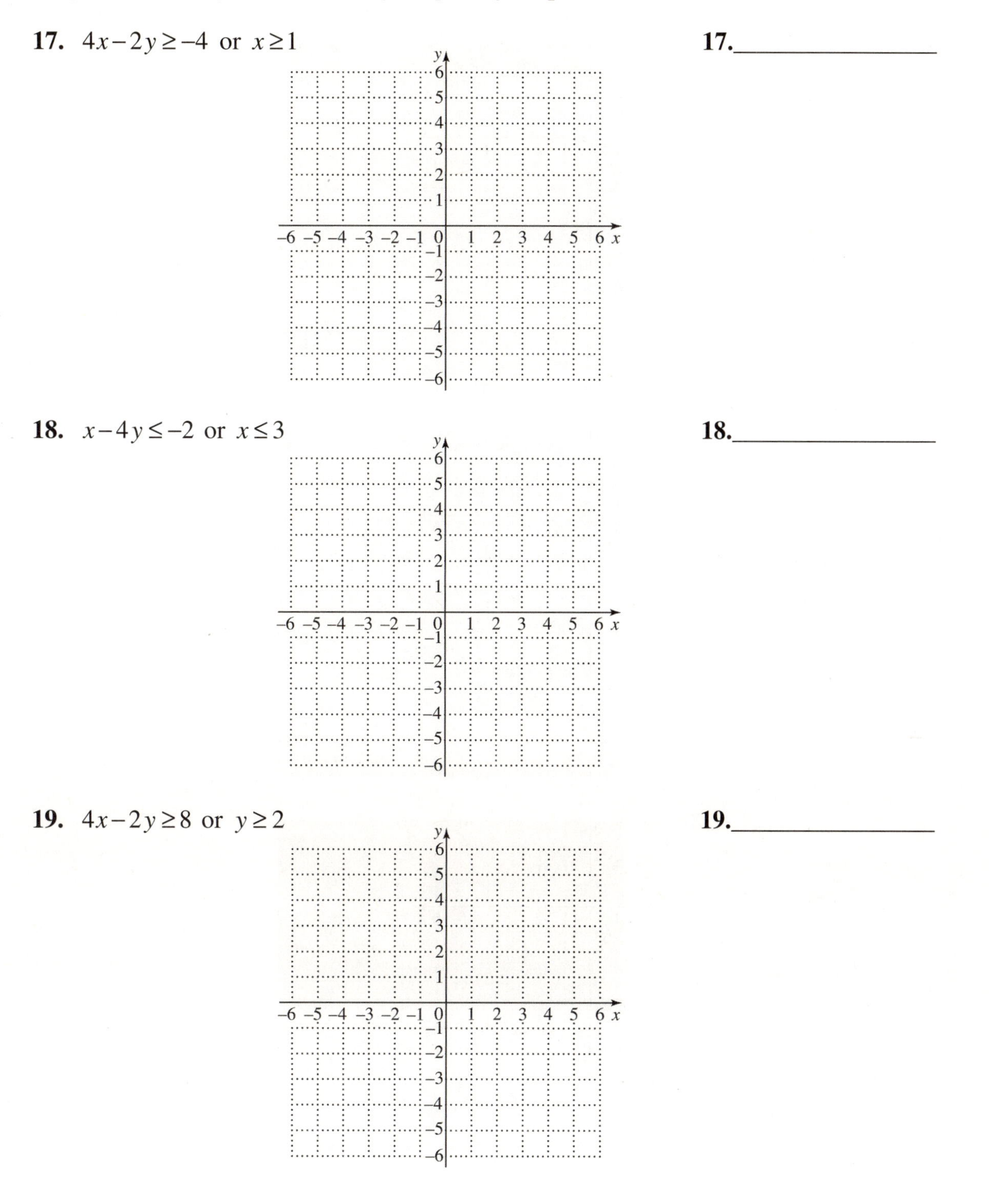

18. $x-4y\leq-2$ or $x\leq3$

18.________________

19. $4x-2y\geq8$ or $y\geq2$

19.________________

Name: Date:
Instructor: Section:

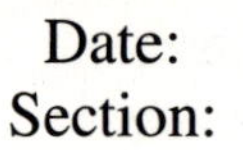

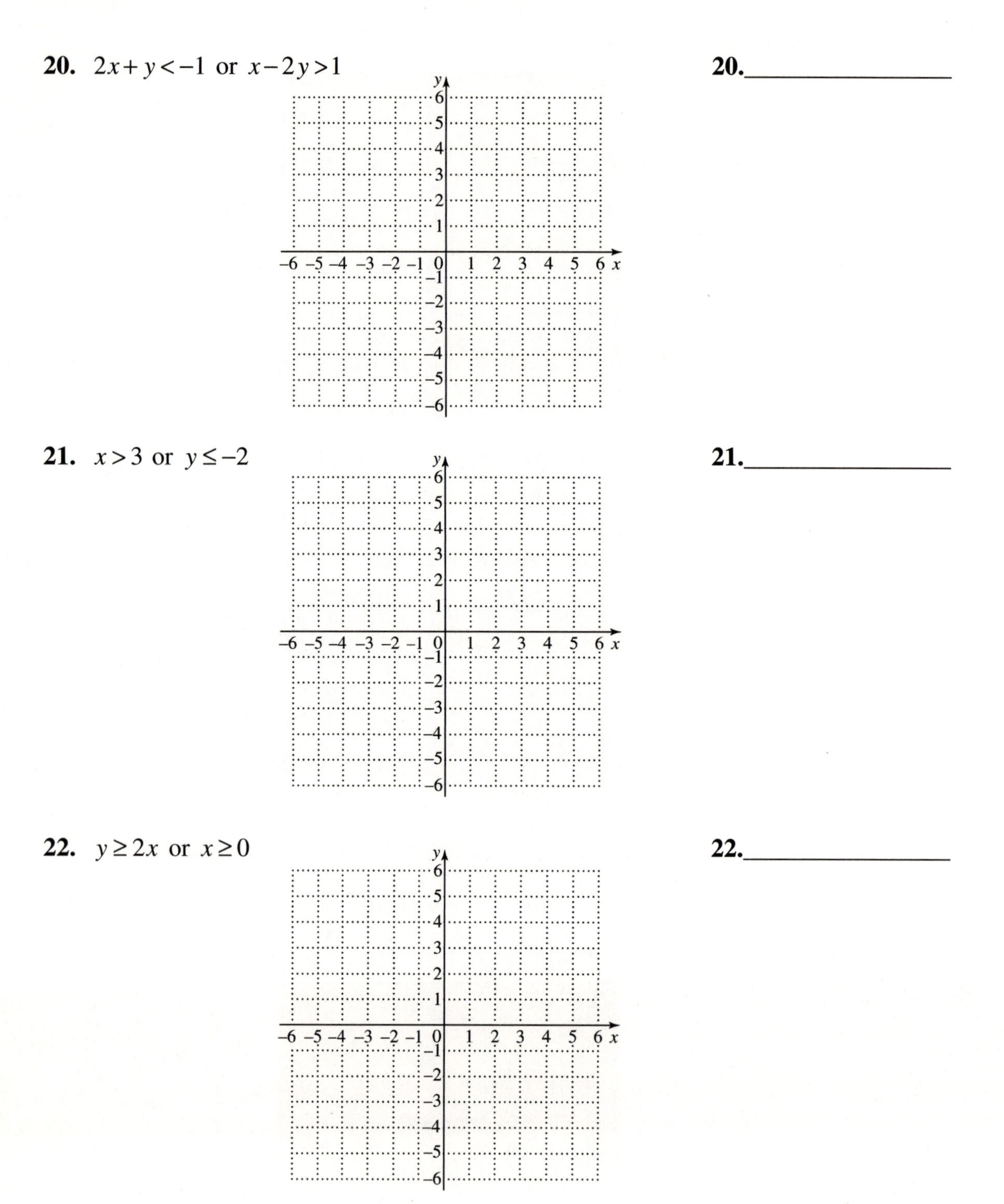

20. $2x + y < -1$ or $x - 2y > 1$

20.________________

21. $x > 3$ or $y \le -2$

21.________________

22. $y \ge 2x$ or $x \ge 0$

22.________________

Name: Date:
Instructor: Section:

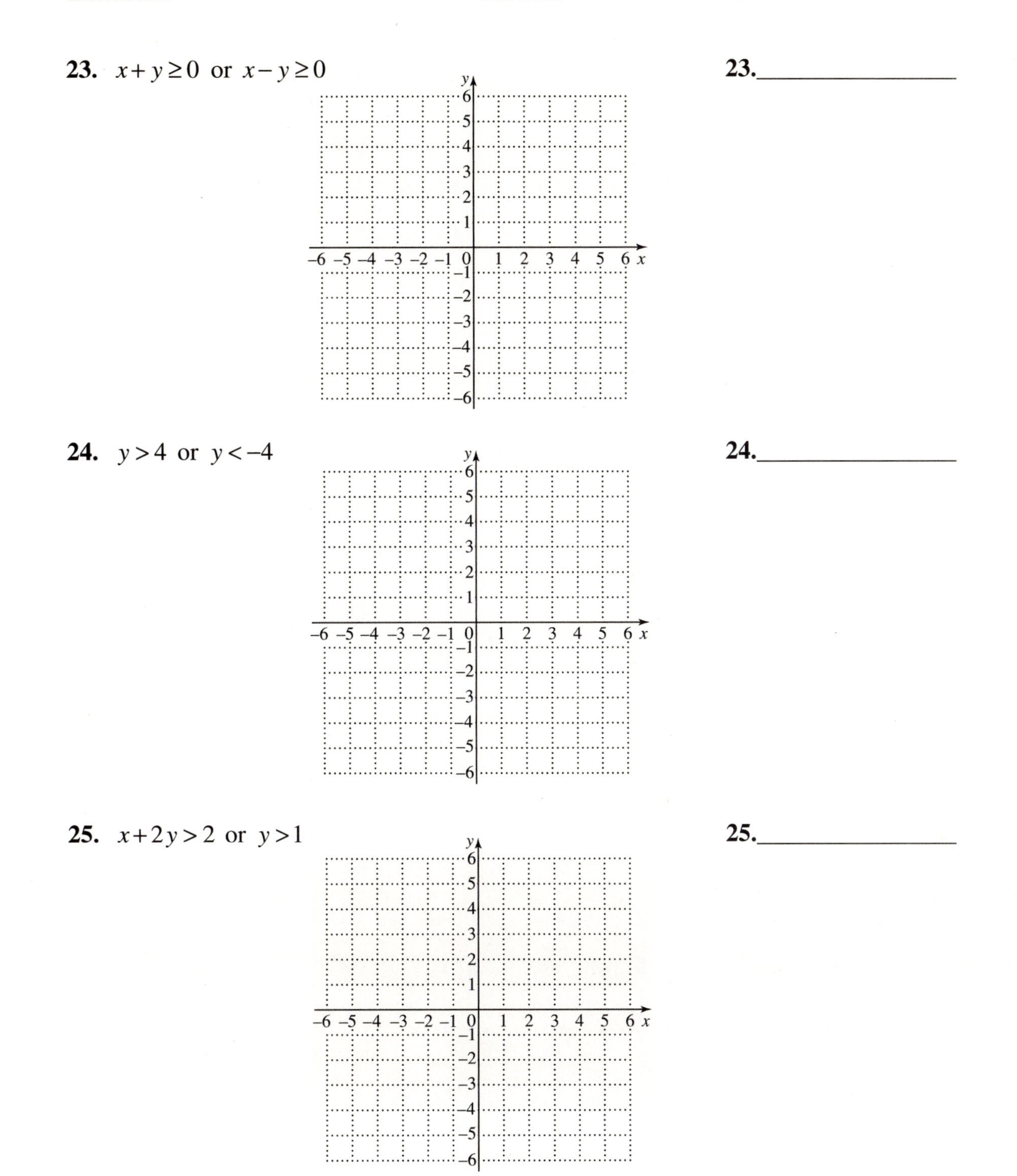

23. $x+y\geq 0$ or $x-y\geq 0$

23.________________

24. $y>4$ or $y<-4$

24.________________

25. $x+2y>2$ or $y>1$

25.________________

Name: Date:
Instructor: Section:

Objective 4 Use a graphing calculator to solve linear inequalities.

Use a graphing calculator to solve the inequality to the nearest hundredth.

26. $\frac{7}{2}x - 22 < 3$ **26.**________________

27. $18 - 7x > 0$ **27.**________________

28. $21.3 - 2x \geq 17x + 1$ **28.**________________

Use a graphing calculator to graph the inequality.

29. $y \leq 1.9x - 3.6$ **29.**________________

30. $y \geq 21.6x + 46.9$ **30.**________________

Name: Date:
Instructor: Section:

Chapter 10 ROOTS, RADICALS, AND ROOT FUNCTIONS

10.1 Radical Expressions and Graphs

Learning Objectives

1. Find square roots.
2. Decide whether a given root is rational, irrational, or not a real number.
3. Find cube, fourth, and other roots.
4. Graph functions defined by radical expressions.
5. Find *n*th roots of *n*th powers.
6. Use a calculator to find roots.

Key Terms

Use the vocabulary terms listed below to complete each statement in exercises 1-7.

principal roots **radical sign** **radicand** **radical expression**

perfect square ***n*th root** **index**

1. The symbol $\sqrt{}$ is called a(n) ___________________.

2. A(n) ___________________ is a number with a rational square root.

3. In a radical of the form $\sqrt[n]{a}$, n is called the ___________________ or order.

4. For even indexes, the symbols $\sqrt{}, \sqrt[4]{}, \sqrt[6]{}, \ldots, \sqrt[n]{}$, are used for nonnegative roots, which are called ___________________.

5. A(n) ___________________ is an algebraic expression that contains radicals.

6. The ___________________ of a, written $\sqrt[n]{a}$, is a number whose nth power equals a.

7. The number or expression under the radical sign is called the ___________________.

Objective 1 Find square roots.

Find all square roots of the number.

1. 169 **1.**________________

2. 324 **2.**________________

Name: Date:
Instructor: Section:

3. 1225 **3.**________________

4. $\frac{121}{144}$ **4.**________________

Find the square root.

5. $\sqrt{100}$ **5.**________________

6. $-\sqrt{900}$ **6.**________________

7. $-\sqrt{529}$ **7.**________________

8. $\sqrt{\frac{256}{361}}$ **8.**________________

Objective 2 Decide whether a given root is rational, irrational, or not a real number.

Tell whether the square root is rational, irrational, *or* not a real number.

9. $-\sqrt{16}$ **9.**________________

10. $\sqrt{95}$ **10.**________________

11. $\sqrt{2.5}$ **11.**________________

12. $\sqrt{-0.9}$ **12.**________________

Objective 3 Find cube, fourth, and other roots.

Find each root that is a real number.

13. $\sqrt[3]{27}$ **13.**________________

Name: Date:
Instructor: Section:

14. $-\sqrt[5]{1}$ **14.**________________

15. $\sqrt[3]{-216}$ **15.**________________

16. $-\sqrt[4]{625}$ **16.**________________

17. $-\sqrt[5]{-32}$ **17.**________________

18. $-\sqrt[4]{-1}$ **18.**________________

Objective 4 Graph functions defined by radical expressions.

Graph each function by creating a table of values. Give the domain and the range.

19. $f(x)=\sqrt{x+6}$ **19.**________________

20. $f(x)=\sqrt[3]{x}-4$ **20.**________________

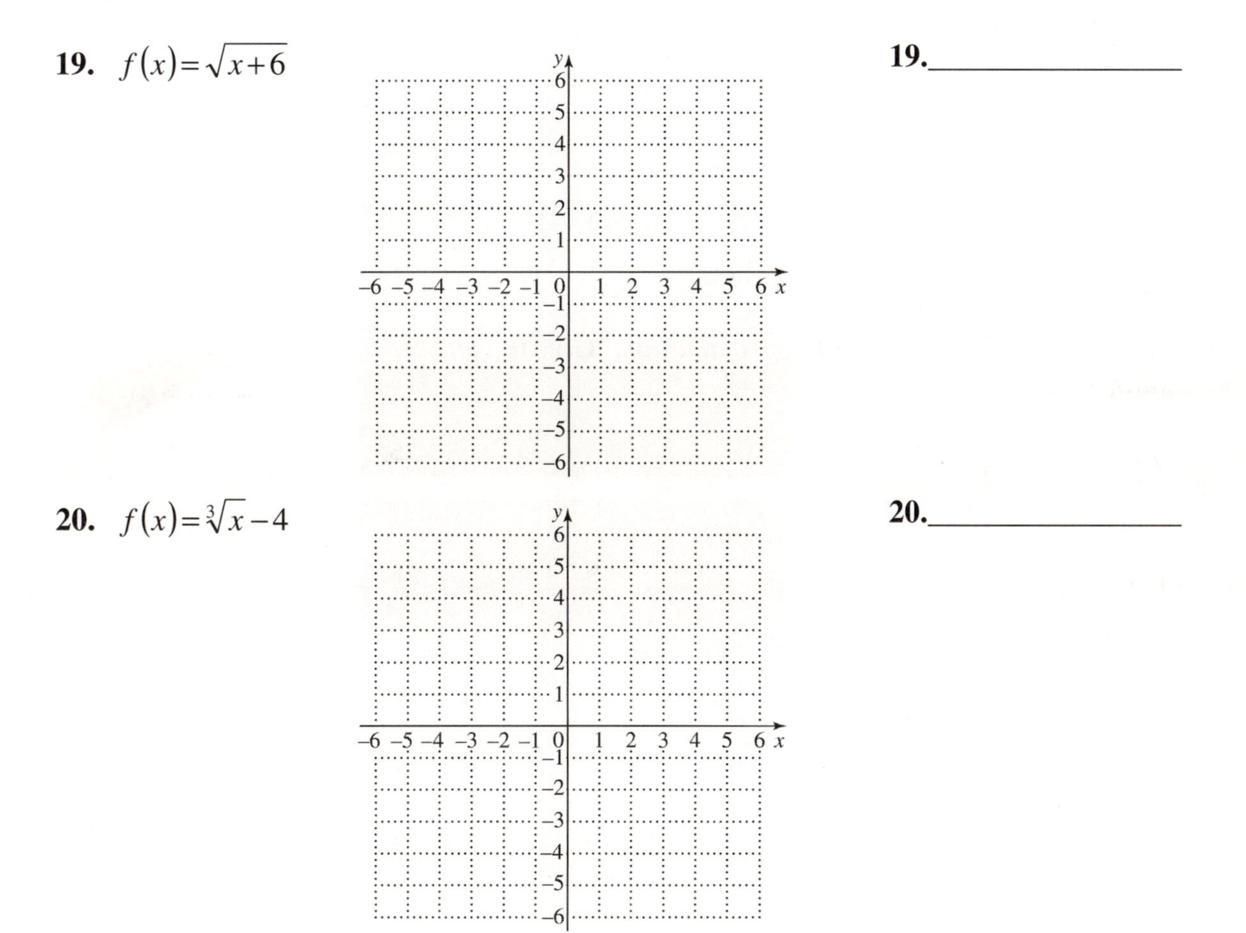

Name: Date:
Instructor: Section:

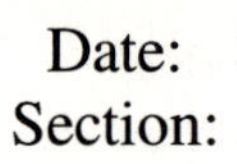

21. $f(x) = 3 - \sqrt[3]{x}$ **21.**_______________

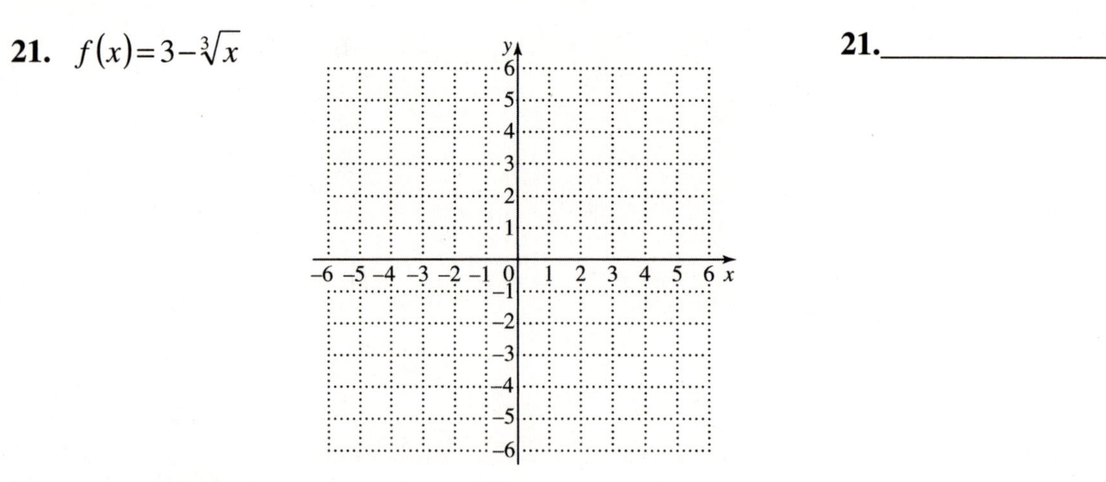

Objective 5 Find *n*th roots of *n*th powers.

Simplify each root.

22. $\sqrt{(-9)^2}$ **22.**_______________

23. $-\sqrt{y^6}$ **23.**_______________

24. $\sqrt[6]{a^{18}}$ **24.**_______________

Objective 6 Use a calculator to find roots.

Use a calculator to find a decimal approximation. Give the answer to the nearest thousandth.

25. $\sqrt{76}$ **25.**_______________

26. $\sqrt{640}$ **26.**_______________

27. $-\sqrt{990}$ **27.**_______________

Name: Date:
Instructor: Section:

28. $\sqrt[3]{701}$ **28.**________________

29. $\sqrt[4]{128.7}$ **29.**________________

30. $629^{-\frac{2}{5}}$ **30.**________________

Name: Date:
Instructor: Section:

Chapter 10 ROOTS, RADICALS, AND ROOT FUNCTIONS

10.2 Rational exponents

Learning Objectives

1. Use exponential notation for *n*th roots.
2. Define and use expressions of the form $a^{m/n}$.
3. Convert between radicals and rational exponents.
4. Use the rules for exponents with rational exponents.

Key Terms

Use the vocabulary terms listed below to complete each statement in exercises 1-4.

$a^{1/n}$ $a^{m/n}$ $a^{-m/n}$ **radical form of** $a^{m/n}$

1. The equivalent form of $\sqrt[n]{a}$ **is** ____________________.

2. The equivalent form of $\frac{1}{a^{m/n}}$ **is** ____________________.

3. The ____________________ is written as $\sqrt[n]{a^m} = \left(\sqrt[n]{a}\right)^m$.

4. The equivalent form of $\left(a^{1/n}\right)^m$ **or** $\left(a^m\right)^{1/n}$ **is** ____________________.

Objective 1 Use exponential notation for *n*th roots.

Simplify.

1. $32^{\frac{1}{5}}$ 1.________________

2. $-8^{\frac{1}{3}}$ 2.________________

3. $625^{\frac{1}{4}}$ 3.________________

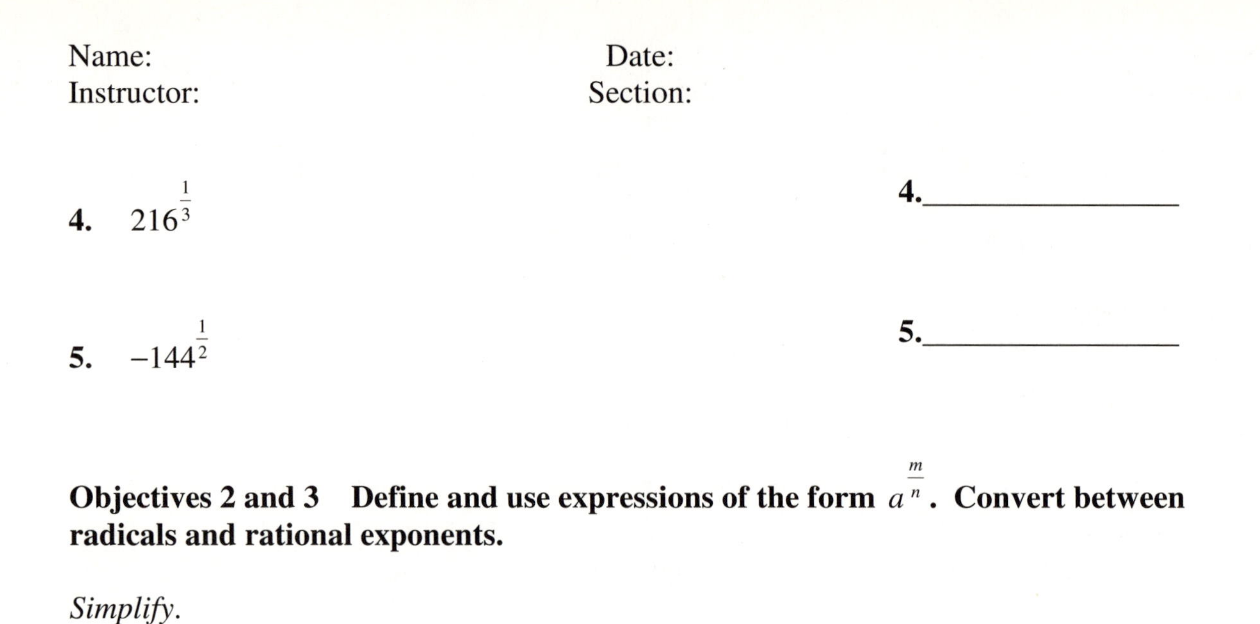

Name: Date:
Instructor: Section:

4. $216^{\frac{1}{3}}$ **4.**________________

5. $-144^{\frac{1}{2}}$ **5.**________________

Objectives 2 and 3 Define and use expressions of the form $a^{\frac{m}{n}}$. Convert between radicals and rational exponents.

Simplify.

6. $27^{\frac{2}{3}}$ **6.**________________

7. $-81^{\frac{5}{4}}$ **7.**________________

8. $216^{\frac{-2}{3}}$ **8.**________________

9. $-625^{\frac{3}{4}}$ **9.**________________

10. $729^{\frac{5}{6}}$ **10.**________________

Find each root that is a real number. Use a calculator as needed.

11. $\sqrt[3]{64}$ **11.**________________

Name: Date:
Instructor: Section:

12. $\sqrt[5]{-a^{15}}$ **12.**________________

13. $-\sqrt{\frac{121}{25}}$ **13.**________________

Write with radicals. Assume that all variables represent positive real numbers.

14. $8b^{\frac{3}{5}}$ **14.**________________

15. $(7x^2y)^{\frac{2}{3}}$ **15.**________________

16. $(p^4+q^2)^{\frac{3}{4}}$ **16.**________________

Write with rational exponents. Assume that all variables represent positive real numbers.

17. $12\sqrt[4]{x^5}$ **17.**________________

18. $\sqrt{(6ab)^3}$ **18.**________________

19. $\frac{1}{\sqrt[3]{n^4}}$ **19.**________________

Name: Date:
Instructor: Section:

Objective 4 Use the rules for exponents with rational exponents.

Use the rules of exponents to simplify. Assume that all variables represent positive real numbers. Write the answer with only positive exponents.

20. $13^{\frac{4}{5}} \cdot 13^{\frac{6}{5}}$

20.________________

21. $\dfrac{8^{\frac{3}{4}}}{8^{-\frac{1}{4}}}$

21.________________

22. $\dfrac{a^{\frac{2}{3}} \cdot a^{-\frac{1}{3}}}{\left(a^{-\frac{1}{6}}\right)^3}$

22.________________

23. $\left(a^4\right)^{\frac{1}{2}} \cdot \left(a^6\right)^{\frac{1}{3}}$

23.________________

24. $\left(\dfrac{c^6}{x^3}\right)^{\frac{2}{3}}$

24.________________

Name: Date:
Instructor: Section:

Multiply. Assume that all variables represent positive real numbers.

25. $y^{\frac{3}{4}}\left(y^{\frac{1}{4}}-3y^{\frac{9}{4}}\right)$

25.________________

26. $7z^{\frac{2}{5}}\left(z^{-\frac{2}{5}}+5z^{-\frac{1}{5}}\right)$

26.________________

27. $d^{-\frac{8}{7}}\left(d^{2}+4d^{3}\right)$

27.________________

Factor, using the given common factor. Assume that all variables represent positive real numbers.

28. $2y^{\frac{1}{2}}-3y^{-\frac{1}{2}};\ y^{-\frac{1}{2}}$

28.________________

29. $x^{3}-2x^{\frac{3}{2}};\ x^{\frac{3}{2}}$

29.________________

30. $7s^{-\frac{2}{3}}-3s^{-\frac{1}{3}};\ s^{-\frac{1}{3}}$

30.________________

Name: Date:
Instructor: Section:

Chapter 10 ROOTS, RADICALS, AND ROOT FUNCTIONS

10.3 Simplifying radical expressions

Learning Objectives
1 Use the product rule for radicals.
2 Use the quotient rule for radicals.
3 Simplify radicals.
4 Simplify products and quotients of radicals with different indexes.
5 Use the Pythagorean formula.
6 Use the distance formula.

Key Terms
Use the vocabulary terms listed below to complete each statement in exercises 1-4.

product rule for radicals **quotient rule for radicals** **Pythagorean formula**

distance formula

1. The ____________________ states that, in a right triangle, the sum of the squares of the legs equals the square of the hypotenuse.

2. The rule $\sqrt[n]{a} \cdot \sqrt[n]{b} = \sqrt[n]{ab}$ is called the ____________________.

3. The rule $\sqrt[n]{\frac{a}{b}} = \frac{\sqrt[n]{a}}{\sqrt[n]{b}}$ is called the ____________________.

4. The formula $d = \sqrt{(x_2 - x_1)^2 + (y_2 - y_1)^2}$ is called the ____________________.

Objective 1 Use the product rule for radicals.

Use the product rule to simplify. Assume that all variables represent nonnegative real numbers.

1. $\sqrt{14} \cdot \sqrt{5}$ 1.________________

2. $\sqrt{7x} \cdot \sqrt{6t}$ 2.________________

3. $\sqrt{\frac{11}{r}} \cdot \sqrt{\frac{3}{p}}$ 3.________________

Name: Date:
Instructor: Section:

4. $\sqrt[6]{9t} \cdot \sqrt[6]{3t^2}$ 4.________________

5. $\sqrt{3} \cdot \sqrt[3]{7}$ 5.________________

6. $\sqrt[4]{8} \cdot \sqrt[4]{15}$ 6.________________

Objective 2 Use the quotient rule for radicals.

Use the quotient rule to simplify. Assume that all variables represent positive real numbers.

7. $\sqrt{\dfrac{25}{16}}$ 7.________________

8. $\sqrt{\dfrac{t^4}{81}}$ 8.________________

9. $\sqrt[3]{\dfrac{a^6}{125}}$ 9.________________

10. $\sqrt{\dfrac{r}{121}}$ 10.________________

11. $\sqrt[4]{\dfrac{p}{16}}$ 11.________________

12. $\sqrt[3]{\dfrac{343}{125}}$ 12.________________

Name: Date:
Instructor: Section:

Objective 3 Simplify radicals.

Simplify. Assume that all variables represent nonnegative real numbers.

13. $\sqrt{27}$ **13.**________________

14. $\sqrt{200}$ **14.**________________

15. $\sqrt[3]{24}$ **15.**________________

16. $\sqrt[3]{54x^{11}}$ **16.**________________

17. $\sqrt[4]{16x^{12}y^{10}}$ **17.**________________

18. $\sqrt{49r^{9}}$ **18.**________________

Objective 4 Simplify products and quotients of radicals with different indexes.

Simplify. Assume that all variables represent positive real numbers.

19. $\sqrt[12]{x^{16}}$ **19.**________________

20. $\sqrt[24]{5^{4}}$ **20.**________________

21. $\sqrt[24]{x^{18}y^{12}}$ **21.**________________

Name: Date:
Instructor: Section:

22. $\sqrt[6]{z^4y^2}$ **22.**________________

Objective 5 Use the Pythagorean Formula.

Find the missing length in each right triangle. Simplify the answer if necessary.

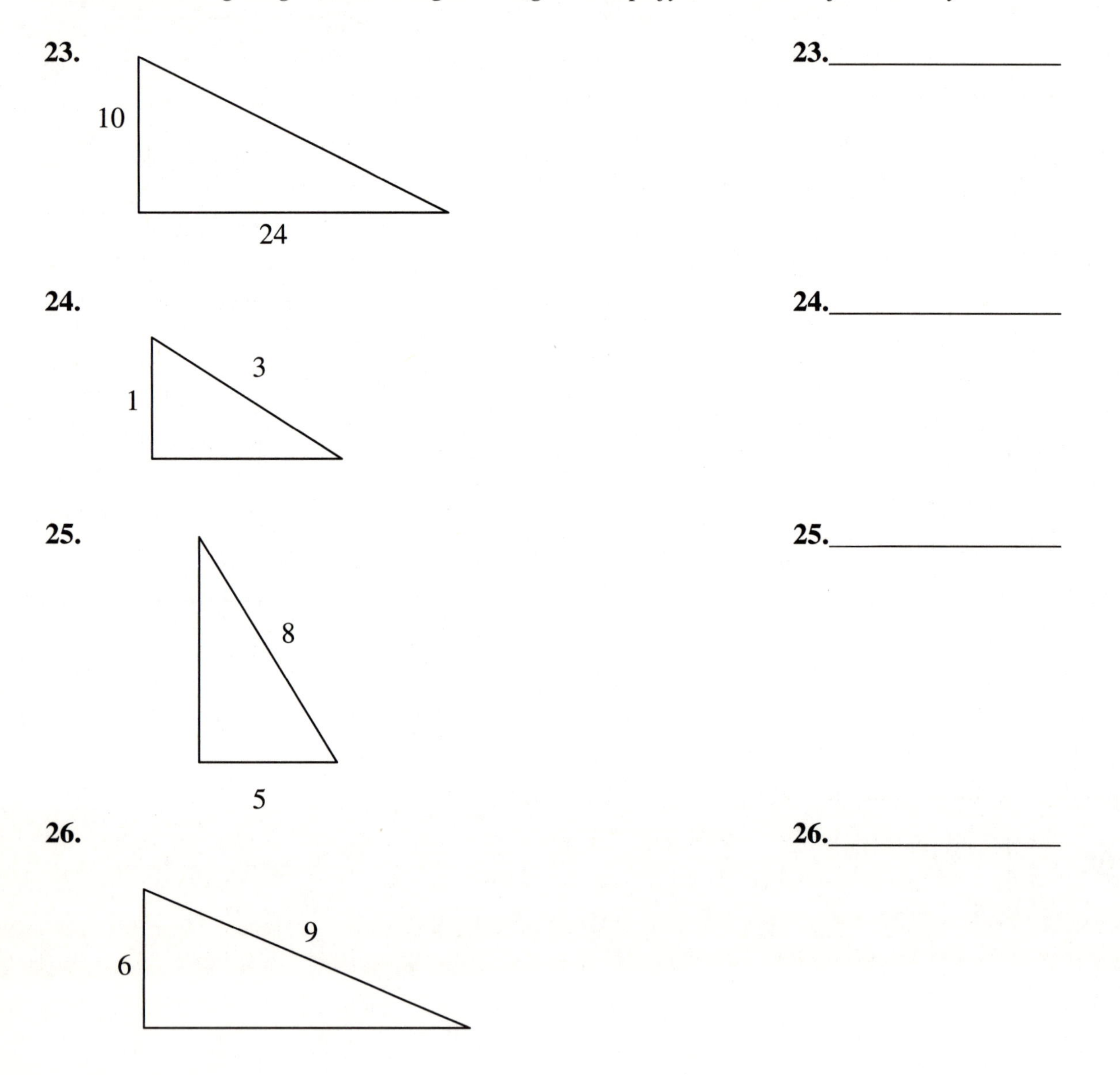

Name: Date:
Instructor: Section:

Objective 6 Use the distance formula.

Find the distance between the given pairs of points.

27. $(4, 9)$ and $(1, 5)$ **27.**________________

28. $(-4, 5)$ and $(1, -3)$ **28.**________________

29. $(2, -3)$ and $(-5, 4)$ **29.**________________

30. $(21, 4)$ and $(111, 60)$ **30.**________________

Name: Date:
Instructor: Section:

Chapter 10 ROOTS, RADICALS, AND ROOT FUNCTIONS

10.4 Adding and subtracting radical expressions

Learning Objectives
1 Simplify radical expressions involving addition and subtraction.

Key Terms
Use the vocabulary terms listed below to complete each statement in exercises 1-2.

same radicand **sum of the roots**

1. Only radical expressions with the same index and the __________________ may be combined.

2. The root of a sum does not equal the __________________.

Objective 1 Simplify radical expressions involving addition and subtraction.

Simplify. Assume that all variables represent positive real numbers.

1. $2\sqrt{7}+3\sqrt{7}$ **1.**________________

2. $3\sqrt{13}-\sqrt{13}+5\sqrt{52}$ **2.**________________

3. $3\sqrt{48}+5\sqrt{27}$ **3.**________________

4. $2\sqrt{18}-5\sqrt{32}+7\sqrt{162}$ **4.**________________

5. $7\sqrt{3}-6\sqrt{21}$ **5.**________________

6. $6\sqrt[3]{135}+3\sqrt[3]{40}$ **6.**________________

Name: Date:
Instructor: Section:

7. $5\sqrt[3]{24}+4\sqrt[3]{81}$ **7.**________________

8. $-3\sqrt[4]{243}-2\sqrt[4]{48}$ **8.**________________

9. $5\sqrt{13}+4\sqrt{13}-6\sqrt{13}$ **9.**________________

10. $3\sqrt{6}-8\sqrt{6}-5\sqrt{24}$ **10.**________________

11. $\sqrt{48y}+\sqrt{12y}+\sqrt{27y}$ **11.**________________

12. $\sqrt{98}-2\sqrt{8}+\sqrt{32}$ **12.**________________

13. $\sqrt[3]{81}+\sqrt[3]{24}+\sqrt[3]{192}$ **13.**________________

14. $\sqrt[3]{625}+\sqrt[3]{135}-\sqrt[3]{40}$ **14.**________________

15. $\sqrt{100x}-\sqrt{9x}+\sqrt{25x}$ **15.**________________

16. $-2\sqrt[3]{81}+\sqrt[3]{24}$ **16.**________________

Name: Date:
Instructor: Section:

17. $2\sqrt{162}-\sqrt{72}+6\sqrt{128}$

17.________________

18. $4\sqrt{27}-2\sqrt{48}+\sqrt{147}$

18.________________

19. $4\sqrt{6}-5\sqrt{7}$

19.________________

20. $2\sqrt{8}+4\sqrt{50}+3\sqrt{18}$

20.________________

Simplify.

21. $\dfrac{\sqrt{16}}{\sqrt{9}}-\dfrac{\sqrt{3}}{3}$

21.________________

22. $\dfrac{\sqrt{32}}{3}+\dfrac{\sqrt{8}}{\sqrt{18}}$

22.________________

23. $\sqrt{\dfrac{125}{5}}-\sqrt{\dfrac{135}{15}}$

23.________________

24. $\dfrac{\sqrt{243}}{4}-\sqrt{\dfrac{3}{16}}$

24.________________

Name: Date:
Instructor: Section:

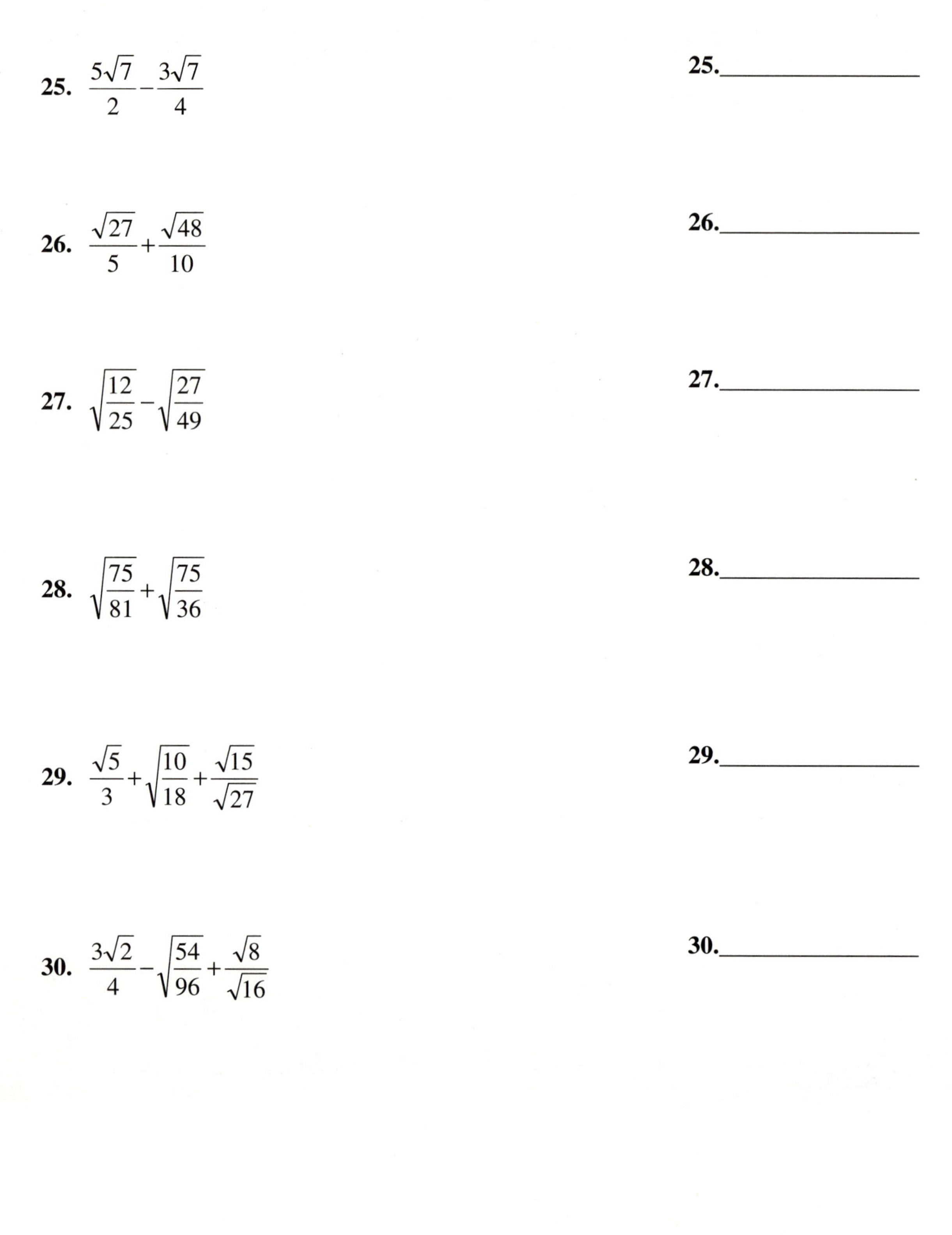

25. $\dfrac{5\sqrt{7}}{2}-\dfrac{3\sqrt{7}}{4}$

25.________________

26. $\dfrac{\sqrt{27}}{5}+\dfrac{\sqrt{48}}{10}$

26.________________

27. $\sqrt{\dfrac{12}{25}}-\sqrt{\dfrac{27}{49}}$

27.________________

28. $\sqrt{\dfrac{75}{81}}+\sqrt{\dfrac{75}{36}}$

28.________________

29. $\dfrac{\sqrt{5}}{3}+\sqrt{\dfrac{10}{18}}+\dfrac{\sqrt{15}}{\sqrt{27}}$

29.________________

30. $\dfrac{3\sqrt{2}}{4}-\sqrt{\dfrac{54}{96}}+\dfrac{\sqrt{8}}{\sqrt{16}}$

30.________________

Name: Date:
Instructor: Section:

Chapter 10 ROOTS, RADICALS, AND ROOT FUNCTIONS

10.5 Multiplying and dividing radical expressions

Learning Objectives
1 Multiply radicals.
2 Rationalize denominators with one radical term.
3 Rationalize denominators with binomials involving radicals.
4 Write radical quotients in lowest terms.

Key Terms
Use the vocabulary terms listed below to complete each statement in exercises 1-3.

FOIL method **rationalizing the denominator** **conjugates**

1. The expressions $x+y$ and $x-y$ are called ____________________.

2. The ____________________ is used when multiplying binomial expressions involving radicals.

3. The process of removing radicals from a denominator so that the denominator contains only rational numbers is called ____________________.

Objective 1 Multiply radicals.

Find the product and simplify.

1. $(3+\sqrt{2})(2+\sqrt{7})$

2. $(\sqrt{10}+\sqrt{3})(\sqrt{6}-\sqrt{11})$

3. $(3+\sqrt[3]{5})(3-\sqrt[3]{5})$

4. $(2\sqrt{x}-3)(3\sqrt{x}-2)$

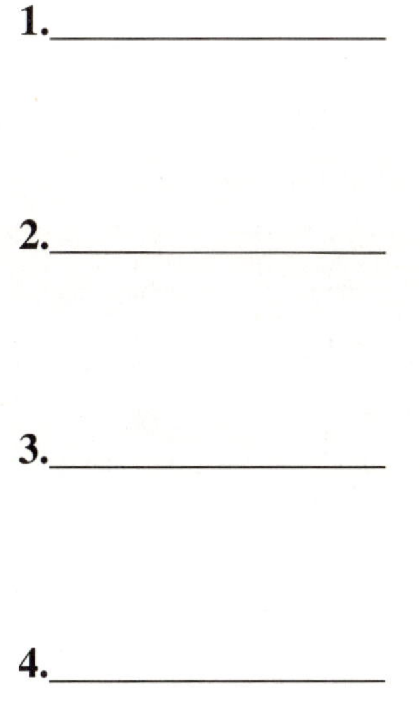

Name: Date:
Instructor: Section:

Objective 2 Rationalize denominators with one radical term.

Rationalize the denominator. Simplify if necessary.

5. $\dfrac{6}{\sqrt{5}}$ **5.**________________

6. $\dfrac{\sqrt{15}}{\sqrt{2}}$ **6.**________________

7. $\dfrac{7}{\sqrt{75}}$ **7.**________________

8. $\dfrac{4}{\sqrt{6}}$ **8.**________________

9. $\dfrac{5}{2\sqrt{5}}$ **9.**________________

10. $\dfrac{3}{\sqrt{98}}$ **10.**________________

Simplify. Assume that all variables represent positive real numbers.

11. $\sqrt{\dfrac{27}{48}}$ **11.**________________

12. $\sqrt{\dfrac{50}{r}}$ **12.**________________

13. $\sqrt{\dfrac{8}{m}}$ **13.**________________

Name: Date:
Instructor: Section:

14. $\sqrt{\frac{5}{8}}$

14.________________

15. $\sqrt{\frac{18}{7}}$

15.________________

16. $\sqrt{\frac{5}{8x^2}}$

16.________________

Simplify. Assume that all variables represent positive real numbers.

17. $\sqrt[3]{\frac{1}{9}}$

17.________________

18. $\sqrt[3]{\frac{8}{100}}$

18.________________

19. $\sqrt[3]{\frac{3}{8}}$

19.________________

20. $\sqrt[3]{\frac{7}{36}}$

20.________________

21. $\sqrt[3]{\frac{t^6}{x^7}}$

21.________________

Name: Date:
Instructor: Section:

Objective 3 Rationalize denominators with binomials involving radicals.

Rationalize the denominator.

22. $\dfrac{5}{7-\sqrt{3}}$ **22.**________________

23. $\dfrac{-6}{\sqrt{7}+3}$ **23.**________________

24. $\dfrac{26}{\sqrt{11}+\sqrt{2}}$ **24.**________________

25. $\dfrac{1}{4+\sqrt{5}}$ **25.**________________

26. $\dfrac{\sqrt{3}}{\sqrt{5}-\sqrt{2}}$ **26.**________________

Objective 4 Write radical quotients in lowest terms.

Write in lowest terms. Assume that all variables represent positive real numbers.

27. $\dfrac{12-3\sqrt{2}}{3}$ **27.**________________

28. $\dfrac{5+2\sqrt{75}}{25}$ **28.**________________

Name: Date:
Instructor: Section:

29. $\dfrac{2x-\sqrt{8x^2}}{4x}$

29.________________

30. $\dfrac{16-12\sqrt{72}}{24}$

30.________________

Name: Date:
Instructor: Section:

Chapter 10 ROOTS, RADICALS, AND ROOT FUNCTIONS

10.6 Solving equations with radicals

Learning Objectives
1 Solve radical equations by using the power rule.
2 Solve radical equations that require additional steps.
3 Solve radical equations with indexes greater than 2.
4 Solve radical equations by using a graphing calculator.
5 Use the power rule to solve a formula for a specified variable.

Key Terms
Use the vocabulary terms listed below to complete each statement in exercises 1-3.

power rule for solving an equation with radicals **extraneous solutions**

isolate the radical

1. The ___________________ states that, if both sides of an equation are raised to the same power, all solutions of the original equation are also solutions to the new equation.

2. Solutions that do not satisfy the original equation are called ___________________.

3. The first step when solving an equation with radicals is to ___________________.

Objective 1 Solve radical equations by using the power rule.

Solve.

1. $\sqrt{t}=5$ **1.**________________

2. $\sqrt{9c+9}=9$ **2.**________________

3. $\sqrt{2m+6}=6$ **3.**________________

4. $\sqrt{3n-8}=5$ **4.**________________

5. $\sqrt{2q-1}=9$ **5.**________________

Name: Date:
Instructor: Section:

6. $\sqrt{3w+4}=7$ 6.________________

7. $\sqrt{x+2}=3$ 7.________________

8. $\sqrt{a}+6=-2$ 8.________________

9. $\sqrt{t+1}-4=0$ 9.________________

10. $\sqrt{5r-4}-9=0$ 10.________________

Objective 2 Solve radical equations that require additional steps.

Solve.

11. $\sqrt{z^2+5z+2}=z+2$ 11.________________

12. $\sqrt{9a^2+6a-23}=3a+5$ 12.________________

13. $\sqrt{4x^2+3x+7}=2x+1$ 13.________________

14. $\sqrt{9x^2-5x+12}=1-3x$ 14.________________

Name: Date:
Instructor: Section:

15. $\sqrt{4x^2-5x+43}=2x+3$ 15.________________

16. $\sqrt{4a^2+27a+48}=2a+7$ 16.________________

17. $\sqrt{p^2+6p+12}=-p$ 17.________________

18. $2w+1-\sqrt{4w^2+3w+2}=0$ 18.________________

19. $\sqrt{3k+7}+\sqrt{k+1}=2$ 19.________________

20. $\sqrt{10d+6}-\sqrt{4d+4}=2$ 20.________________

Objective 3 Solve radical equations with indexes greater than 2.

Solve.

21. $\sqrt[3]{7x-5}=\sqrt[3]{3x+7}$ 21.________________

22. $\sqrt[5]{2t+1}-1=0$ 22.________________

23. $\sqrt[5]{b-1}-2=0$ 23.________________

Name: Date:
Instructor: Section:

24. $\sqrt[4]{2m+1} = \sqrt[4]{m+22}$

24.________________

Objective 4 Solve radical equations by using a graphing calculator.

Use a graphing calculator to solve the equation.

25. $\sqrt{x+4} = 8 - x$

25.________________

26. $3\sqrt{4-x} = x$

26.________________

27. $\sqrt{1-5x} = -2x - 2$

27.________________

Objective 5 Use the power rule to solve a formula for a specified variable.

Solve each formula for the indicated variable.

28. $Z = \sqrt{\frac{L}{C}}$, for L

28.________________

29. $f = \frac{1}{2\pi\sqrt{LC}}$, for C

29.________________

30. $N = \frac{1}{2\pi}\sqrt{\frac{a}{r}}$, for r

30.________________

Name: Date:
Instructor: Section:

Chapter 10 ROOTS, RADICALS, AND ROOT FUNCTIONS

10.7 Complex numbers

Learning Objectives

1. Simplify numbers of the form $\sqrt{-b}$, where $b > 0$.
2. Recognize complex numbers.
3. Add and subtract complex numbers.
4. Multiply complex numbers.
5. Divide complex numbers.
6. Find powers of i.

Key Terms

Use the vocabulary terms listed below to complete each statement in exercises 1-7.

imaginary unit $i\sqrt{b}$ **complex number** **real part** **imaginary part**

pure imaginary number **standard form**

1. The ____________________ of a complex number $a+bi$ is a.
2. A(n) ____________________ is any number that can be written in the form $a+bi$, where a and b are real numbers.
3. A complex number $a+bi$ with $a = 0$ and $b \neq 0$ is called a(n) ____________________.
4. The symbol i is called the ____________________.
5. The ____________________ of a complex number $a+bi$ is b.
6. For any positive real number b, the expression $\sqrt{-b}$ equals ____________________.
7. The ____________________ of a complex number is $a+bi$.

Objective 1 Simplify numbers of the form $\sqrt{-b}$, where $b > 0$.

Simplify.

1. $\sqrt{-49}$ **1.**__________________

2. $\sqrt{-6}$ **2.**__________________

Name: Date:
Instructor: Section:

3. $-\sqrt{-63}$ 3.________________

4. $\sqrt{-288}$ 4.________________

5. $-\sqrt{-72}$ 5.________________

Multiply or divide as indicated.

6. $\sqrt{-3}\cdot\sqrt{-15}$ 6.________________

7. $\sqrt{-14}\cdot\sqrt{3}$ 7.________________

8. $\sqrt{-6}\cdot\sqrt{-3}\cdot\sqrt{2}$ 8.________________

9. $\dfrac{\sqrt{-125}}{\sqrt{-5}}$ 9.________________

10. $\dfrac{\sqrt{-56}\cdot\sqrt{-6}}{\sqrt{16}}$ 10.________________

Objective 2 Recognize complex numbers.

The real numbers are a subset of the complex numbers. Classify the complex number as real, nonreal complex, *or* pure imaginary..

11. $7-3i$ 11.________________

12. $\sqrt{5}$ 12.________________

13. $i\sqrt{7}$ 13.________________

Name: Date:
Instructor: Section:

Objective 3 Add and subtract complex numbers.

Add or subtract as indicated.

14. $(5+7i)+(-2+4i)$ **14.**________________

15. $4+(3+6i)$ **15.**________________

16. $(-7-2i)-(-3-3i)$ **16.**________________

17. $(\sqrt{3}-2\sqrt{2}i)+(2\sqrt{3}-2\sqrt{2}i)$ **17.**________________

18. $[(8+4i)-(5-3i)]+(4-2i)$ **18.**________________

Objective 4 Multiply complex numbers.

Multiply.

19. $(2+5i)(3-i)$ **19.**________________

20. $(5-3i)(5+3i)$ **20.**________________

21. $(\sqrt{2}-i\sqrt{3})^2$ **21.**________________

Objective 5 Divide complex numbers.

Find the quotient.

22. $\frac{1+i}{2-i}$ **22.**________________

Name: Date:
Instructor: Section:

23. $\dfrac{5+2i}{9-4i}$

23.________________

24. $\dfrac{7}{5+2i}$

24.________________

25. $\dfrac{1+i}{(2-i)(2+i)}$

25.________________

Objective 6 Find powers of *i*.

Find powers of i.

26. i^{11}

26.________________

27. i^{100}

27.________________

28. i^{-62}

28.________________

29. i^{236}

29.________________

30. i^{-115}

30.________________

Name: Date:
Instructor: Section:

Chapter 11 QUADRATIC EQUATIONS, INEQUALITIES, AND FUNCTIONS

11.1 Solving Quadratic Equations by the Square Root Property

Learning Objectives

1. Review the zero-factor property.
2. Solve equations of the form $x^2 = k$, where $k > 0$.
3. Solve equations of the form $(ax+b)^2 = k$, where $k > 0$.
4. Solve quadratic equations with solutions that are not real numbers.

Key Terms

Use the vocabulary terms listed below to complete each statement in exercises 1-4.

quadratic equation **standard form** **zero-factor property**

square root property

1. A quadratic equation written in the form $ax^2 + bx + c = 0$, where *a*, *b*, and *c* are real numbers, with $a \neq 0$, is written in ____________________.

2. A ____________________ is an equation that can be written in the form $ax^2 + bx + c = 0$, where *a*, *b*, and *c* are real numbers, with $a \neq 0$.

3. The ____________________ states that if $x^2 = k$, then $x = \sqrt{k}$ or $x = -\sqrt{k}$.

4. The ____________________ states that if two numbers have a product of 0, then at least one of the numbers must be 0.

Objective 1 Review the zero-factor property.

Use the zero factor property to solve the equation.

1. $r^2 + r - 72 = 0$ **1.**________________

2. $p^2 + p - 20 = 0$ **2.**________________

3. $6z^2 + 19z + 10 = 0$ **3.**________________

Name: Date:
Instructor: Section:

4. $8x^2+2x-15=0$ **4.**________________

5. $12a^2+11a-5=0$ **5.**________________

6. $5p^2+3p=0$ **6.**________________

7. $16x^2-25=0$ **7.**________________

Objective 2 Solve equations of the form $x^2=k$, where $k>0$.

Use the square root property to solve the equation.

8. $p^2=81$ **8.**________________

9. $r^2-1=15$ **9.**________________

10. $x^2+9=16$ **10.**________________

11. $4-y^2=0$ **11.**________________

12. $t^2=11$ **12.**________________

13. $3x^2+11=38$ **13.**________________

Name: Date:
Instructor: Section:

14. $12-4y^2=8$ **14.**____________

Objective 3 Solve equations of the form $(ax+b)^2=k$, where $k>0$.

Use the square root property to solve the equation.

15. $(x-5)^2=16$ **15.**____________

16. $(y+7)^2=25$ **16.**____________

17. $(4t+3)^2=5$ **17.**____________

18. $(r-2)^2=9$ **18.**____________

19. $(3t+4)^2=49$ **19.**____________

20. $(a+5)^2=7$ **20.**____________

21. $(4b+5)^2=12$ **21.**____________

Objective 4 Solve quadratic equations with solutions that are not real numbers.

Solve the equation.

22. $x^2+1=0$ **22.**____________

Name: Date:
Instructor: Section:

23. $9+y^2=0$ **23.**________________

24. $16=-q^2$ **24.**________________

25. $k^2+25=0$ **25.**________________

26. $m^2=-36$ **26.**________________

27. $4h^2+9=0$ **27.**________________

28. $49+16w^2=0$ **28.**________________

29. $(7c-1)^2+5=0$ **29.**________________

30. $(2x-1)^2+16=0$ **30.**________________

Name: Date:
Instructor: Section:

Chapter 11 QUADRATIC EQUATIONS, INEQUALITIES, AND FUNCTIONS

11.2 Solving Quadratic Equations by Completing the Square

Learning Objectives

1 Solve quadratic equations by completing the square when the coefficient of the squared term is 1.

2 Solve quadratic equations by completing the square when the coefficient of the squared term is not 1.

3. Simplify an equation before solving.

Key Terms

Use the vocabulary terms listed below to complete each statement in exercises 1-3.

completing the square **square root property** ***a***

1. The process of adding to a binomial the number that makes it a perfect square trinomial is called ____________________.

2. The first step when solving quadratic equations of the form $ax^2+bx+c=0$, where $a \neq 1$, is to divide both sides of the equation by ____________________.

3. After completing the square, use the ____________________ to solve the resulting equation.

Objective 1 Solve quadratic equations by completing the square when the coefficient of the squared term is 1.

Solve the equation by completing the square.

1. $x^2+4x+3=0$

1.

2. $x^2-9x+8=0$

2.________________

3. $p^2+3p-10=0$

3.________________

Name: Date:
Instructor: Section:

4. $c^2 - 8c - 9 = 0$ **4.**________________

5. $x^2 - 12x + 27 = 0$ **5.**________________

6. $x^2 + 6x + 5 = 0$ **6.**________________

7. $x^2 - 2x - 3 = 0$ **7.**________________

8. $x^2 - 13x + 40 = 0$ **8.**________________

9. $t^2 - 14t + 45 = 0$ **9.**________________

10. $z^2 - 6z - 72 = 0$ **10.**________________

Objective 2 Solve quadratic equations by completing the square when the coefficient of the squared term is not 1.

Solve.

11. $3x^2 - 5x = 0$ **11.**________________

12. $3t^2 + 4t - 1 = 0$ **12.**________________

Name: Date:
Instructor: Section:

13. $2x^2 - 4x = 1$

13.________________

14. $2t^2 - 5t + 2 = 0$

14.________________

15. $6y^2 + 13y + 6 = 0$

15.________________

16. $3t^2 + 8t - 3 = 0$

16.________________

17. $4x^2 - 13x - 12 = 0$

17.________________

18. $5x^2 - 11x + 2 = 0$

18.________________

19. $3m^2 + 10m + 8 = 0$

19.________________

20. $2t^2 + t - 15 = 0$

20.________________

Name: Date:
Instructor: Section:

Objective 3 Simplify an equation before solving.

Solve.

21. $m(m+7)+12=0$ **21.**________________

22. $x(x-9)=-20$ **22.**________________

23. $(x-5)(x+4)=63$ **23.**________________

24. $6(x^2-1)=-5x$ **24.**________________

25. $(x+2)(x+4)=63$ **25.**________________

26. $x^2+9=6x$ **26.**________________

27. $4(t^2-3)=-13t$ **27.**________________

Name: Date:
Instructor: Section:

28. $10(n^2-1)=-21n$

28.________________

29. $(y+1)(y-3)=12$

29.________________

30. $x^2+1=-2x$

30.________________

Name: Date:
Instructor: Section:

Chapter 11 QUADRATIC EQUATIONS, INEQUALITIES, AND FUNCTIONS

11.3 Solving Quadratic Equations by the Quadratic Formula

Learning Objectives
1 Derive the quadratic formula.
2 Solve quadratic equations by using the quadratic formula.
3 Use the discriminant to determine the number and type of solutions.

Key Terms
Use the vocabulary terms listed below to complete each statement in exercises 1-4.

quadratic formula **discriminant** **two irrational solutions**

one rational solution

1. The ____________________ is a general formula used to solve any quadratic equation.

2. The ____________________ is the quantity under the radical, $b^2 - 4ac$, in the quadratic formula.

3. If the discriminant is positive, but not the square of an integer, then there is (are) ____________________.

4. If the discriminant is zero, then there is (are) ____________________.

Objective 2 Solve quadratic equations by using the quadratic formula.

Solve using the quadratic formula. (All solutions for these equations are real numbers.)

1. $x^2 + 12x + 35 = 0$ **1.**________________

2. $x^2 + 5x - 14 = 0$ **2.**________________

3. $5t^2 - 13t + 6 = 0$ **3.**________________

Name: Date:
Instructor: Section:

4. $6m^2 - 17m + 12 = 0$ **4.**______________

5. $16x^2 - 9 = 0$ **5.**______________

6. $(z+2)^2 = 2(5z-2)$ **6.**______________

7. $5m^2 + 5m - 1 = 0$ **7.**______________

8. $7 - 6t - 5t^2 = 0$ **8.**______________

Solve using the quadratic formula. (All solutions for these equations are imaginary numbers.)

9. $x^2 - 6x + 10 = 0$ **9.**______________

10. $4x^2 + 4x + 5 = 0$ **10.**______________

11. $2x^2 - 5x + 4 = 0$ **11.**______________

12. $6x^2 + 7x = -3$ **12.**______________

13. $8x^2 + 5x + 10 = 3x^2 - 4x + 4$ **13.**______________

Name: Date:
Instructor: Section:

14. $3x^2 - 3x + 1 = 0$ **14.**________________

Objective 3 Use the discriminant to determine the number and type of solutions.

Use the discriminant to determine whether the equation has solutions that are

(a) two distinct rational numbers, *(b) exactly one rational number,*

(c) two distinct irrational numbers, *(d) two distinct nonreal complex numbers.*

Do not solve.

15. $4t^2 + 12t + 9 - 0$ **15.**________________

16. $5a^2 - 4a + 1 = 0$ **16.**________________

17. $p^2 - 2p + 4 = 0$ **17.**________________

18. $t^2 + 5t + 4 = 0$ **18.**________________

19. $2y^2 + 4y + 8 = 0$ **19.**________________

20. $3r^2 + 5r + 1 = 0$ **20.**________________

21. $m^2 - 4m - 4 = 0$ **21.**________________

22. $5y^2 - 5y + 2 = 0$ **22.**________________

Name: Date:
Instructor: Section:

Use the discriminant to decide whether the polynomial can be factored. If the polynomial can be factored, do so.

23. $4p^2 - 3p + 12$ **23.**_______________

24. $10p^2 + 21p + 8$ **24.**_______________

25. $12k^2 + 23k - 9$ **25.**_______________

26. $16r^2 + 30r + 5$ **26.**_______________

27. $14m^2 - 26m + 7$ **27.**_______________

28. $24x^2 + 38x + 15$ **28.**_______________

29. $30y^2 - 7y - 15$ **29.**_______________

30. $18n^2 - 4n - 13$ **30.**_______________

Name: Date:
Instructor: Section:

Chapter 11 QUADRATIC EQUATIONS, INEQUALITIES, AND FUNCTIONS

11.4 Equations in Quadratic Form

Learning Objectives
1 Solve an equation with fractions by writing it in quadratic form.
2 Use quadratic equations to solve applied problems.
3 Solve an equation with radicals by writing it in quadratic form.
4 Solve an equation that is quadratic in form by substitution.

Key Terms
Use the vocabulary terms listed below to complete each statement in exercises 1-3.

quadratic formula **quadratic in form** **assign a variable**

1. A nonquadratic equation that is written in the form $au^2 + bu + c = 0$, for $a \neq 0$ and an algebraic expression *u*, is called ____________________.

2. Any quadratic equation can be solved using the ____________________.

3. When solving an applied problem, read the problem and then ____________________ for the unknown.

Objective 1 Solve an equation with fractions by writing it in quadratic form.

Solve by first clearing the equation of fractions.

1. $1+\frac{1}{x}=\frac{6}{x^2}$

1.__________________

2. $\frac{2x}{x+1}=\frac{3}{x-1}$

2.__________________

3. $\frac{7}{x^2}+6=-\frac{23}{x}$

3.__________________

Name: Date:
Instructor: Section:

4. $4-\frac{8}{x-1}=-\frac{35}{x}$ **4.**________________

5. $2=\frac{1}{x}+\frac{28}{x^2}$ **5.**________________

6. $9-\frac{12}{x}=-\frac{4}{x^2}$ **6.**________________

7. $\frac{5x}{x+1}+\frac{6}{x+2}=\frac{3}{(x+1)(x+2)}$ **7.**________________

Objective 2 Use quadratic equations to solve applied problems.

Solve each problem by writing an equation with fractions and solving it.

8. Mike can row 3 miles per hour in still water. It takes him 3 hours and 36 minutes to go 3 miles upstream and return. Find the speed of the current. **8.**________________

9. Working together, Dale and Roger complete a job in 6 hours. It would take Dale 9 hours longer than Roger to do the job alone. How long would it take Roger alone? **9.**________________

Name: Date:
Instructor: Section:

10. The distance from Appletown to Medina is 45 miles, as is the distance from Medina to Westmont. Karl drove from Westmont to Medina, stopped at Medina for a hamburger, and then drove on to Appletown at 10 miles per hour faster. Driving time for the entire trip was 99 minutes. Find Karl's speed from Westmont to Medina.

10.________________

11. Two pipes together can fill a large tank in 10 hours. One of the pipes, used alone, takes 15 hours longer than the other to fill the tank. How long would each pipe used alone take to fill the tank?

11.________________

12. A jet plane traveling at a constant speed goes 1200 miles with the wind, then turns around and travels for 1000 miles against the wind. If the speed of the wind is 50 miles per hour and the total flight takes 4 hours, find the speed of the plane.

12.________________

Objective 3 Solve an equation with radicals by writing it in quadratic form.

Solve.

13. $x = \sqrt{x+6}$

13.________________

14. $\sqrt{7y-10} = y$

14.________________

15. $\sqrt{2}y = \sqrt{6-y}$

15.________________

Name: Date:
Instructor: Section:

16. $k-\sqrt{8k-15}=0$ **16.**________________

17. $2y=\sqrt{2(3y-1)}$ **17.**________________

18. $y=\sqrt{\frac{1-2y}{8}}$ **18.**________________

19. $\sqrt{3}y=\sqrt{28y-49}$ **19.**________________

20. $y=\sqrt{y+42}$ **20.**________________

Objective 4 Solve an equation that is quadratic in form by substitution.

Find all solutions for the equation.

21. $x^4-25x^2+144=0$ **21.**________________

22. $16m^4=25m^2-9$ **22.**________________

23. $(x+1)^2=10(x+1)+75$ **23.**________________

Name: Date:
Instructor: Section:

24. $c^4 - 13c^2 + 36 = 0$

24.________________

25. $(m+5)^2 + 6(m+5) + 8 = 0$

25.________________

26. $(3-r)^2 = -3(3-r) + 18$

26.________________

27. $x + 7\sqrt{x} = 8$

27.________________

28. $m^{-2} + m^{-1} - 20 = 0$

28.________________

29. $x^4 = -x^2 + 20$

29.________________

30. $\frac{1}{(x+6)^2} - \frac{7}{2(x+6)} = -\frac{3}{2}$

30.________________

Name: Date:
Instructor: Section:

Chapter 11 QUADRATIC EQUATIONS, INEQUALITIES, AND FUNCTIONS

11.5 Formulas and Further Applications

Learning Objectives
1. Solve formulas for variables involving squares and square roots.
2. Solve applied problems using the Pythagorean formula.
3. Solve applied problems using area formulas.
4. Solve applied problems using quadratic functions as models.

Key Terms
Use the vocabulary terms listed below to complete each statement in exercises 1-3.

square both sides **check** **quadratic function**

1. A function defined by an equation of the form $f(x) = ax^2 + bx + c$, for real numbers, a, b, and c, with $a \neq 0$, is a ____________________.

2. When solving an applied problem, always ____________________ the solution to see if it makes sense.

3. When solving equations involving square roots, ____________________ of the equation to eliminate the square roots.

Objective 1 Solve formulas for variables involving squares and square roots.

Solve for x.

1. $2p^2 = x^2 - k^2$ **1.**________________

2. $3m^2 = x^2 + r^2$ **2.**________________

3. $k^2 = 2x^2 - 9$ **3.**________________

Name: Date:
Instructor: Section:

4. $2m = m^2 + x^2 + x^2$ **4.**________________

5. $3y = y^2 - 2x^2$ **5.**________________

Solve the equation for the indicated variable.

6. $A = \dfrac{B}{C^2}$ for C **6.**________________

7. $A = \sqrt{bc}$ for c **7.**________________

8. $x = \dfrac{y}{\sqrt{z}}$ for z **8.**________________

9. $A = \dfrac{BC}{D^2}$ for D **9.**________________

Objective 2 Solve applied problems using the Pythagorean formula.

Solve the problem using a quadratic equation.

10. A 15-foot ladder is leaning against a building. The distance from the bottom of the ladder to the building is 3 feet less than the distance from the top of the ladder to the ground. How far is the bottom of the ladder from the building? **10.**________________

Name: Date:
Instructor: Section:

11. Two cars left an intersection at the same time, one heading south, the other heading east. Some time later the car traveling south had gone 18 miles farther than the car headed east. At that time they were 90 miles apart. How far had each car traveled?

11.________________

12. A child flying a kite has let out 45 feet of string to the kite. The distance from the kite to the ground is 9 feet more than the distance from the child to a point directly below the kite. How high up is the kite?

12.________________

13. The longest side of a right triangle is 4 centimeters longer than the next longest side. The third side is 16 centimeters in length. Find the length of the longest side.

13.________________

14. The longest side of a right triangle is 2 feet more than the middle side and the middle side is 1 foot less than twice the shortest side. Find the length of the shortest side.

14.________________

15. The width of a rectangle is 14 centimeters. The diagonal is 2 centimeters more than the length. Find the length of the rectangle.

15.________________

16. The length of a rectangle is 3 inches less than twice the width. The diagonal is 51 inches. Find the width of the rectangle.

16.________________

Name: Date:
Instructor: Section:

Objective 3 Solve applied problems using area formulas.

Solve the problem using a quadratic equation.

17. A rectangle has a length 1 meter less than twice its width. If 1 meter is cut from the length and added to the width, the figure becomes a square with an area of 16 square meters. Find the dimensions of the original rectangle.

17.________________

18. The area of a square is 81 square centimeters. If the same amount is added to one dimension and removed from the other, the resulting rectangle has an area 9 centimeters less than the area of the square. How much is added and subtracted?

18.________________

19. A rectangular piece of cardboard has a length that is 3 inches longer than the width. A square 1.5 inches on a side is cut from each corner. The sides are then turned up to form an open box with a volume of 162 cubic inches. Find the dimensions of the original piece of cardboard.

19.________________

20. A piece of plastic in the shape of a rectangle has a length 10 inches less than twice the width. A square 4 inches on a side is cut out of each corner, and the sides are turned up to form an open box with a volume of 160 cubic inches. Find the dimensions of the finished box.

20.________________

21. An open box is to be made from a rectangular piece of tin by cutting 2-inch squares out of the corners and folding up the sides. The length of the finished box is to be twice the width. The volume of the box will be 100 cubic inches. Find the dimensions of the rectangular piece of tin.

21.________________

Name: Date:
Instructor: Section:

22. A rectangular garden has dimensions of 12 feet by 5 feet. A gravel path of equal width is to be built around the garden. How wide can the path be if there is enough gravel for 138 square feet?

22.________________

23. A fish pond is 3 feet by 4 feet. How wide a strip of concrete can be poured around the pool if there is enough concrete for 44 square feet?

23.________________

Objective 4 Solve applied problems using quadratic functions as models.

The position of an object moving in a straight line is given by $s = 2t^2 - 3t$, *where s is in feet and t is the time in seconds the object has been in motion.*

24. How long will it take the object to move 2 feet?

24.________________

25. How long will it take the object to move 5 feet?

25.________________

26. How long (to the nearest tenth) will it take the object to move 6 feet?

26.________________

27. How long (to the nearest tenth) will it take the object to move 10 feet?

27.________________

The position of an object moving in a straight line is given by $s = t^2 - 8t$, *where s is in feet and t is in seconds.*

28. How long will it take the object to move 20 feet?

28.________________

Name: Date:
Instructor: Section:

29. How long (to the nearest tenth) will it take the object to move 10 feet?

29._______________

30. How long (to the nearest tenth) will it take the object to move 16 feet?

30._______________

Name: Date:
Instructor: Section:

Chapter 11 QUADRATIC EQUATIONS, INEQUALITIES, AND FUNCTIONS

11.6 Graphs of Quadratic Functions

Learning Objectives	
1	Graph a quadratic function.
2	Graph parabolas with horizontal and vertical shifts.
3	Use the coefficient of x^2 to predict the shape and direction in which a parabola opens.
4	Find a quadratic function to model data.

Key Terms

Use the vocabulary terms listed below to complete each statement in exercises 1-5.

parabola **vertex** **axis** **vertical shift** **horizontal shift**

1. The graph of $F(x) = x^2 + k$ has the same shape as the graph of $f(x) = x^2$ with a(n) ____________________ of k units up if $k > 0$, and $|k|$ units down if $k < 0$.

2. The graph of $F(x) = (x - h)^2$ has the same shape as the graph of $f(x) = x^2$ with a ____________________ of h units to the right if $h > 0$, and $|h|$ units to the left if $h < 0$.

3. The ____________________ of a parabola is the vertical or horizontal line through the vertex of the parabola.

4. The graph of a second-degree (quadratic) equation in two variables is a(n) ____________________.

5. The point on a parabola that has the smallest y-value (if the parabola opens up) or the largest y-value (if the parabola opens down) is called the ____________________ of the parabola.

Name: Date:

Instructor: Section:

Objectives 1 and 2 Graph a quadratic function; Graph parabolas with horizontal and vertical shifts.

Graph each parabola. Identify the vertex.

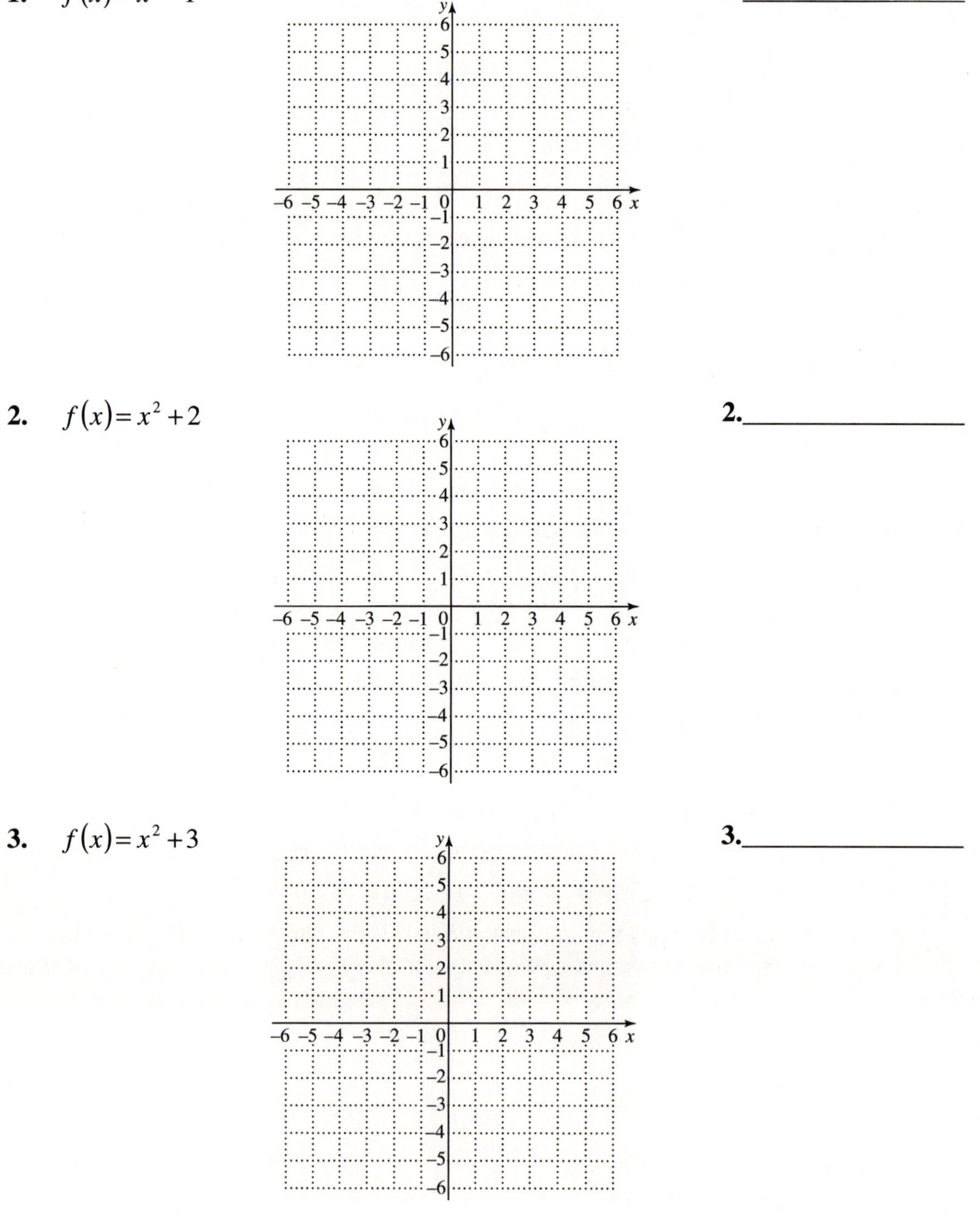

1. $f(x)=x^2-1$ **1.**________________

2. $f(x)=x^2+2$ **2.**________________

3. $f(x)=x^2+3$ **3.**________________

Name: Date:
Instructor: Section:

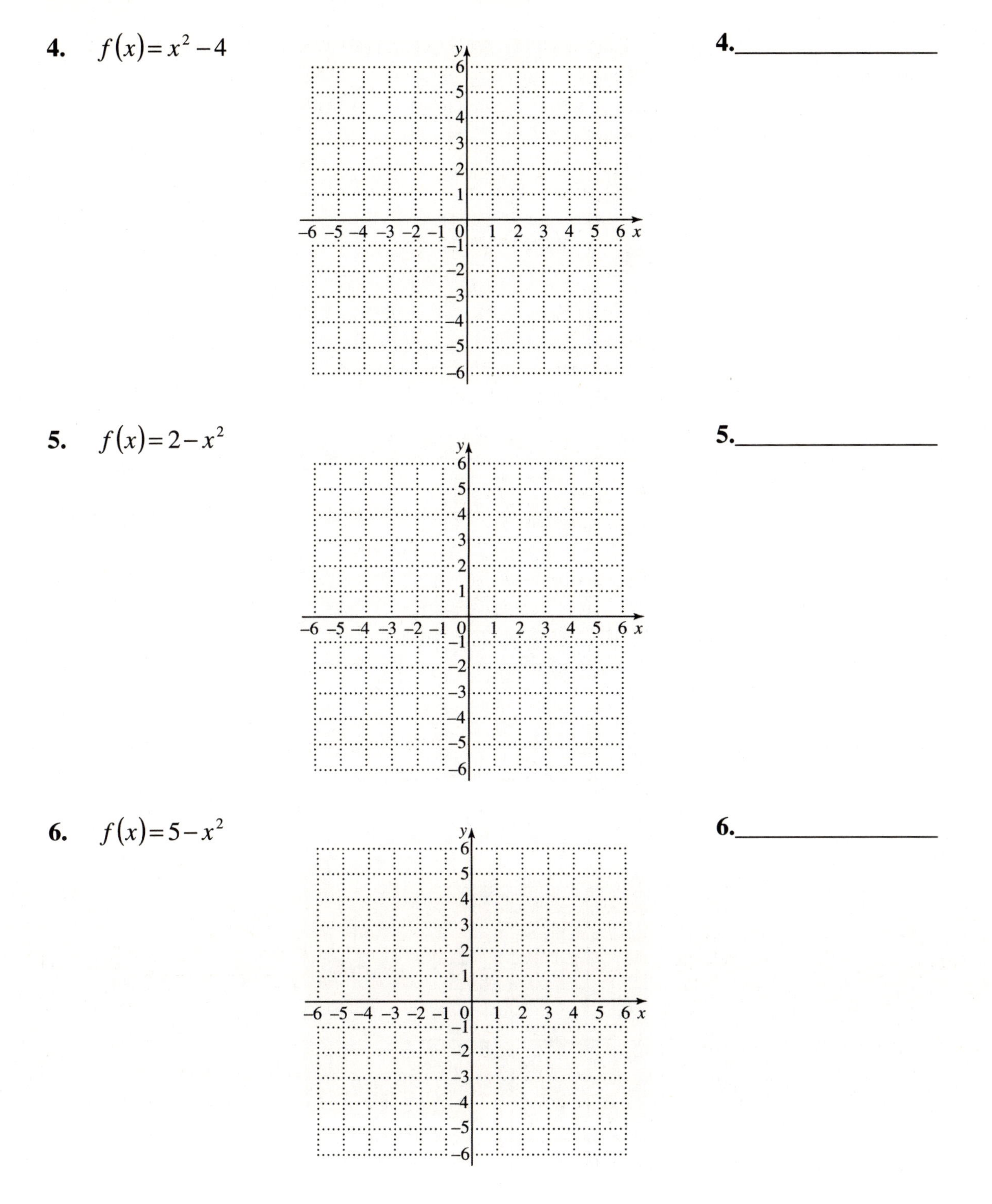

4. $f(x) = x^2 - 4$

4.________________

5. $f(x) = 2 - x^2$

5.________________

6. $f(x) = 5 - x^2$

6.________________

Name: Date:
Instructor: Section:

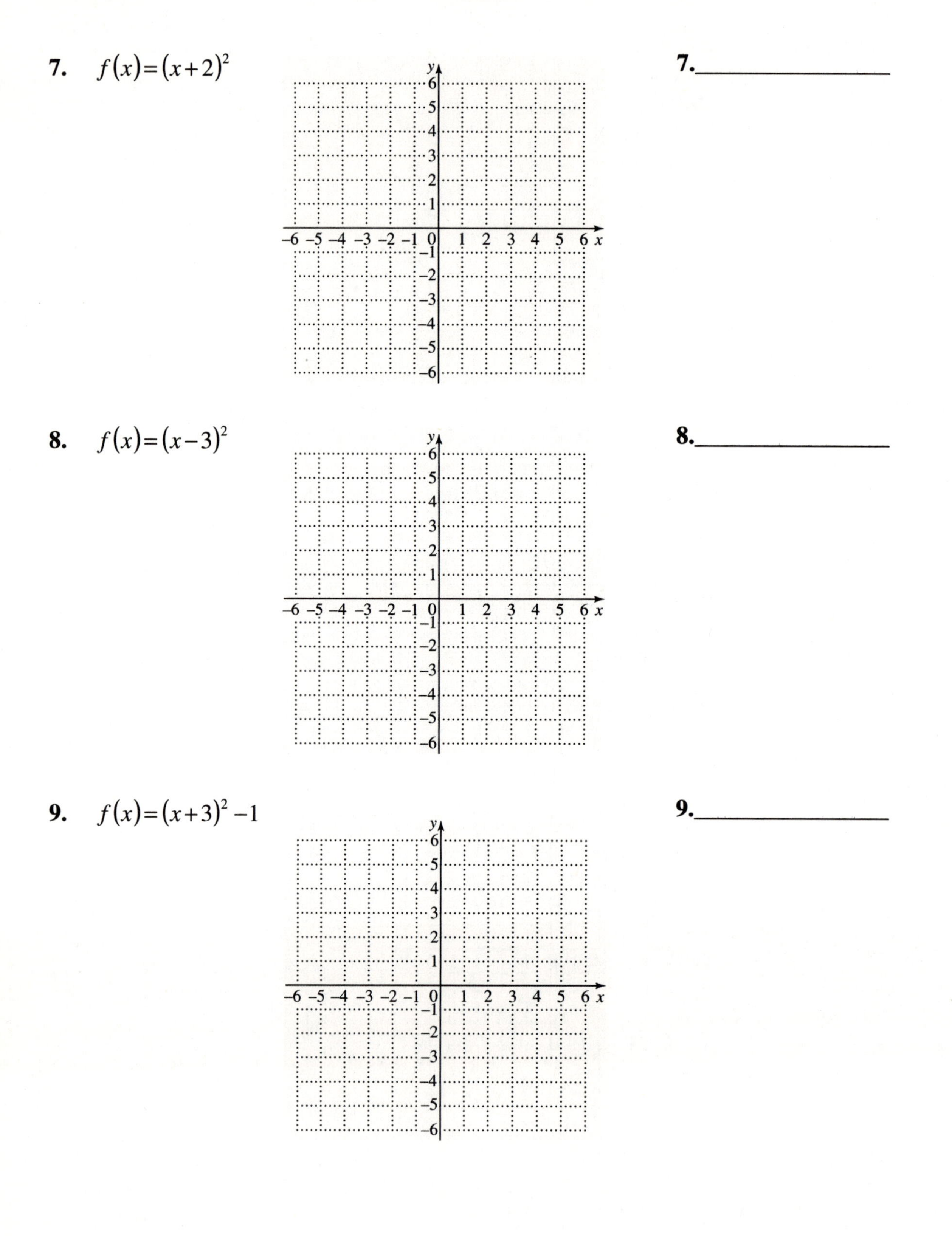

7. $f(x)=(x+2)^2$

7.________________

8. $f(x)=(x-3)^2$

8.________________

9. $f(x)=(x+3)^2-1$

9.________________

Name: Date:
Instructor: Section:

10. $f(x)=(x-1)^2+2$ **10.**________________

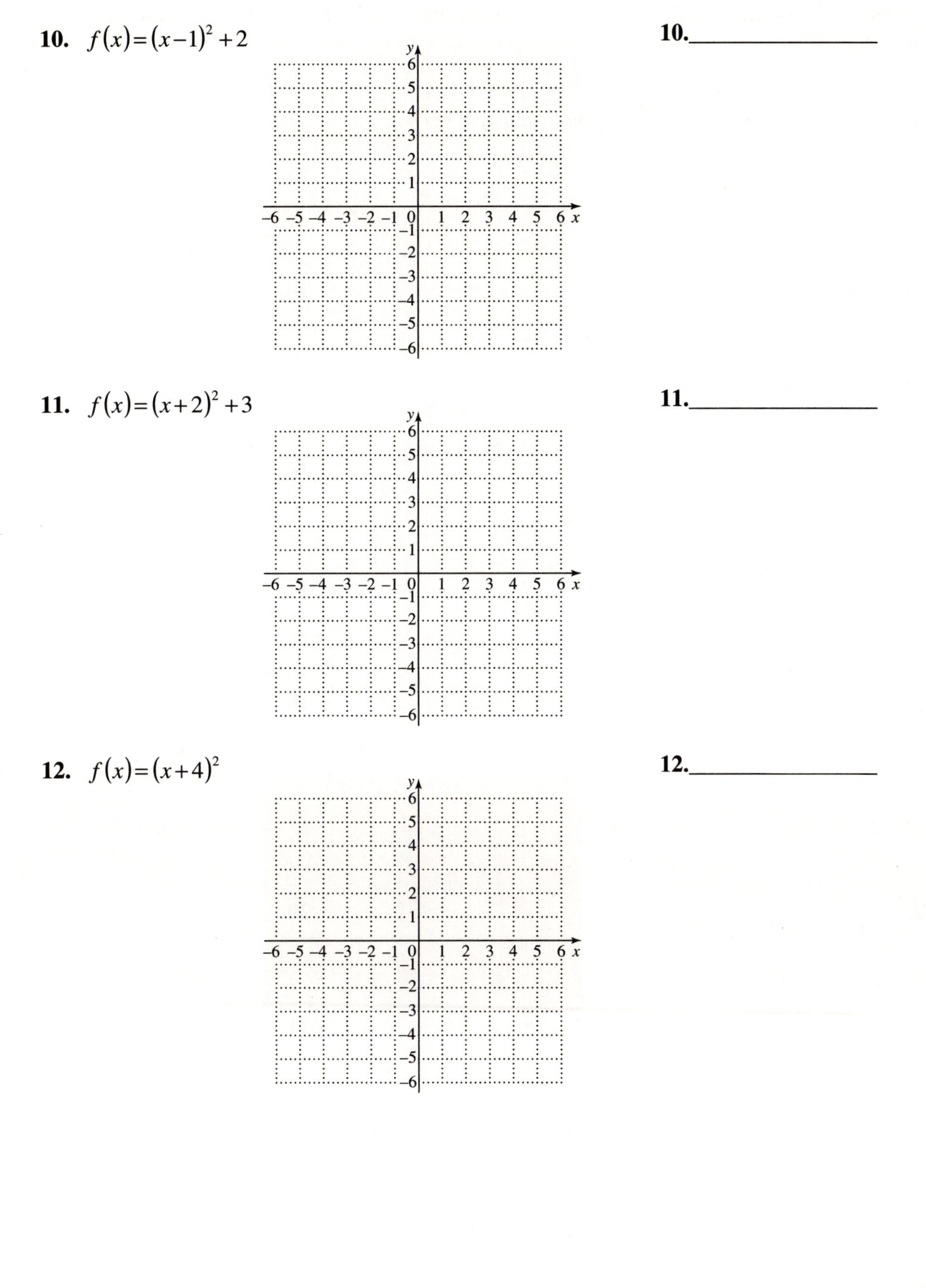

11. $f(x)=(x+2)^2+3$ **11.**________________

12. $f(x)=(x+4)^2$ **12.**________________

Name: Date:
Instructor: Section:

Objective 3 Use the coefficient of x^2 to predict the shape and direction in which a parabola opens.

For the parabola, tell whether the graph opens upward or downward and whether the graph is wider, narrower, or the same as the graph of $f(x)=x^2$.

13. $f(x)=\frac{1}{2}x^2$ **13.**________________

14. $f(x)=5x^2$ **14.**________________

15. $f(x)=-2x^2$ **15.**________________

16. $f(x)=-\frac{4}{3}x^2-1$ **16.**________________

17. $f(x)=\frac{3}{5}x^2+5$ **17.**________________

18. $f(x)=\frac{2}{3}(x-1)^2$ **18.**________________

19. $f(x)=-2(x+1)^2$ **19.**________________

20. $f(x)=-\frac{1}{3}(x+3)^2-4$ **20.**________________

21. $f(x)=\frac{5}{4}(x-1)^2+7$ **21.**________________

22. $f(x)=4-x^2$ **22.**________________

Name: Date:
Instructor: Section:

23. $f(x)=\frac{4}{3}(x-3)^2+4$ 23.________________

24. $f(x)=x^2-4$ 24.________________

25. $f(x)=-.5(x+3)^2$ 25.________________

26. $f(x)=4(x-3)^2+1$ 26.________________

Objective 4 Find a quadratic function to model data.

The number of ice cream cones sold by Reutter's 71 flavors ice cream parlor from 1993-1999 are shown in the following table. In the year column, 1 represents 1993, 2 represents 1994, and so on.

Year	*Number of Cones Sold*
1	*1775*
2	*4194*
3	*5063*
4	*5161*
5	*4663*
6	*4839*
7	*3710*

27. Plot the ordered pairs (year, number of cones sold). 27.________________

28. Use the graph to determine whether the coefficient *a* of x^2 in a quadratic model should be positive or negative. 28.________________

29. Determine a quadratic function that models these data by using a system of equations. Use the ordered pairs $(1, 1775)$, $(4, 5161)$, and $(7, 3710)$. 29.________________

30. Determine an appropriate domain for the function found in Question 29. 30.________________

Name: Date:
Instructor: Section:

Chapter 11 QUADRATIC EQUATIONS, INEQUALITIES, AND FUNCTIONS

11.7 More about Parabolas and Their Applications

Learning Objectives
1 Find the vertex of a vertical parabola.
2 Graph a quadratic function.
3 Use the discriminant to find the number of x-intercepts of a parabola with a vertical axis.
4 Use quadratic functions to solve problems involving maximum or minimum value.
5 Graph parabolas with horizontal axes.

Key Terms
Use the vocabulary terms listed below to complete each statement in exercises 1-3.

vertex **finding any intercepts** **horizontal axis**

1. The steps for graphing a quadratic function include, determining if the graph opens up or down, finding the vertex, and ___________________.

2. A parabola with a ___________________ has an equation of the form $x = ay^2 + by + c$.

3. The graph of a quadratic equation defined by $f(x) = ax^2 + bx + c \ (a \neq 0)$ has ___________________ $\left(\frac{-b}{2a}, f\left(\frac{-b}{2a}\right)\right)$.

Objectives 1 and 2 Find the vertex of a parabola; Graph a quadratic function.

For the function, find the vertex by completing the square. Then graph the function.

1. $f(x) = x^2 + 4x + 5$ **1.**_______________

Name: Date:
Instructor: Section:

2. $f(x) = x^2 - 6x + 4$ **2.**________________

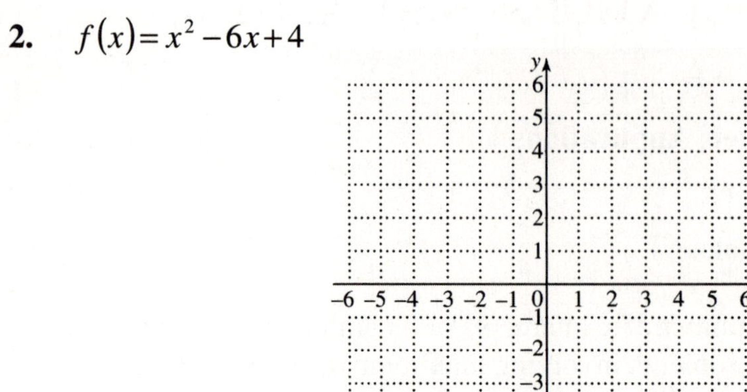

3. $f(x) = -x^2 + 8x - 10$ **3.**________________

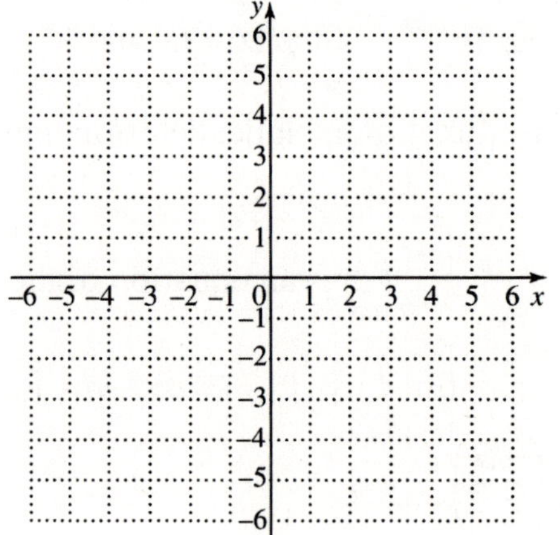

4. $f(x) = x^2 - 3x + 2$ **4.**________________

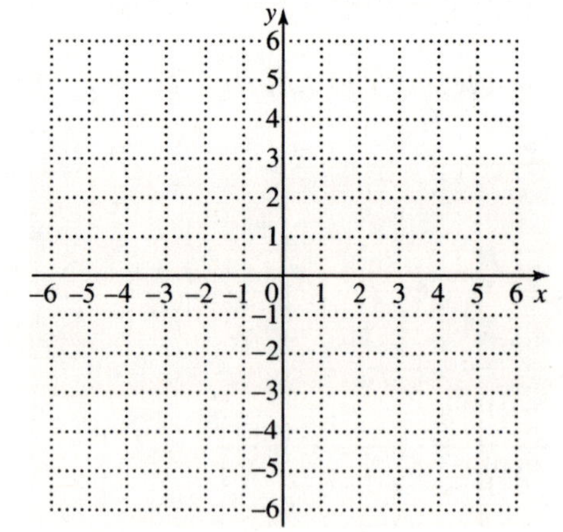

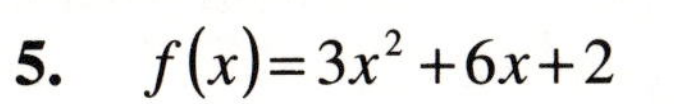

Name: Date:
Instructor: Section:

5. $f(x)=3x^2+6x+2$ **5.**________________

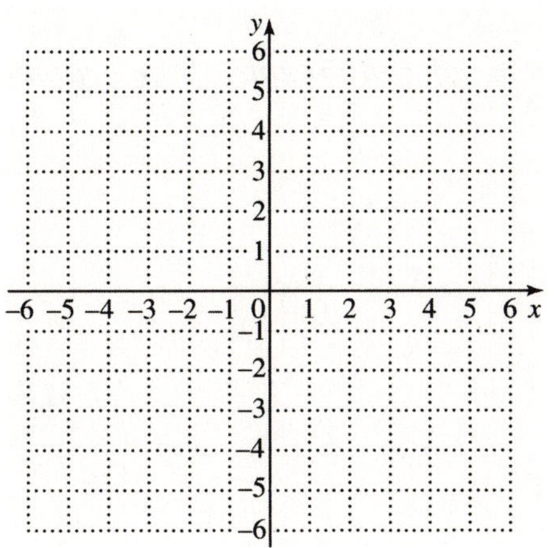

6. $f(x)=-2x^2-4x+1$ **6.**________________

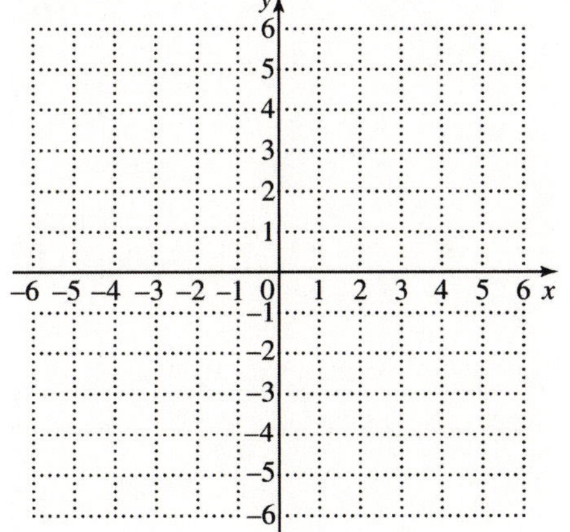

7. $f(x)=\frac{5}{4}x^2+5x+3$ **7.**________________

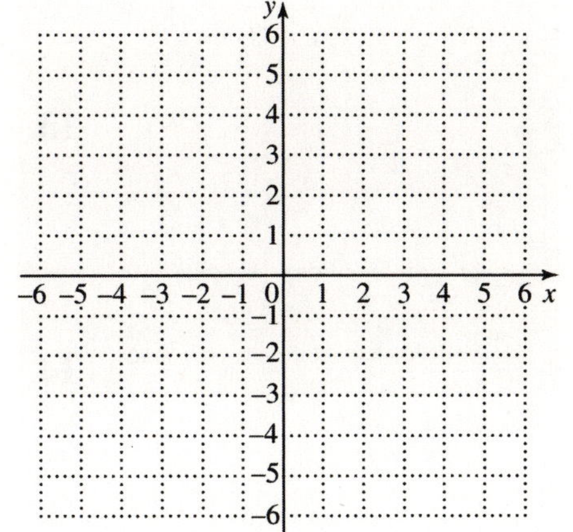

Name: Date:
Instructor: Section:

Objective 3 Use the discriminant to find the number of *x*-intercepts of a parabola with a vertical axis.

Use the discriminant to find the number of x-intercepts of the graph of the function.

8. $f(x)=5x^2-4x+1$ **8.**________________

9. $f(x)=x^2-2x+4$ **9.**________________

10. $f(x)=x^2+5x+4$ **10.**________________

11. $f(x)=2x^2+4x+8$ **11.**________________

12. $f(x)=x^2-4x-4$ **12.**________________

13. $f(x)=5x^2-5x+2$ **13.**________________

14. $f(x)=4x^2+12x+9$ **14.**________________

Name: Date:
Instructor: Section:

Objective 4 Use quadratic functions to solve problems involving maximum or minimum value.

Solve the problem.

15. A businessman has found that his daily profits are given by

$$P = -2x^2 + 120x + 4800,$$

where x is the number of units sold each day. Find the number he should sell daily to maximize his profit. What is the maximum profit?

15.________________

16. The same businessman has daily costs of

$$C = x^2 - 50x + 1625,$$

where x is the number of units sold each day. How many units must be sold to minimize his cost? What is the minimum cost?

16.________________

A projectile is fired upward so that its distance (in feet) above the ground t seconds after firing is as given. Find the maximum height it reaches and the number of seconds to reach that height.

17. $x = -16t^2 + 64t$

17.________________

18. $x = -16t^2 + 48t + 250$

18.________________

19. $s = -16t^2 + 80t + 156$

19.________________

Name: Date:
Instructor: Section:

20. The length and width of a rectangle have a sum of 48. What width will produce the maximum area? **20.**_______________

21. The perimeter of a rectangle is 36. What length will produce the maximum area? **21.**_______________

Objective 5 Graph parabolas with horizontal axes.

Graph the parabola. If necessary, first complete the square to find the vertex.

22. $x=3y^2$ **22.**_______________

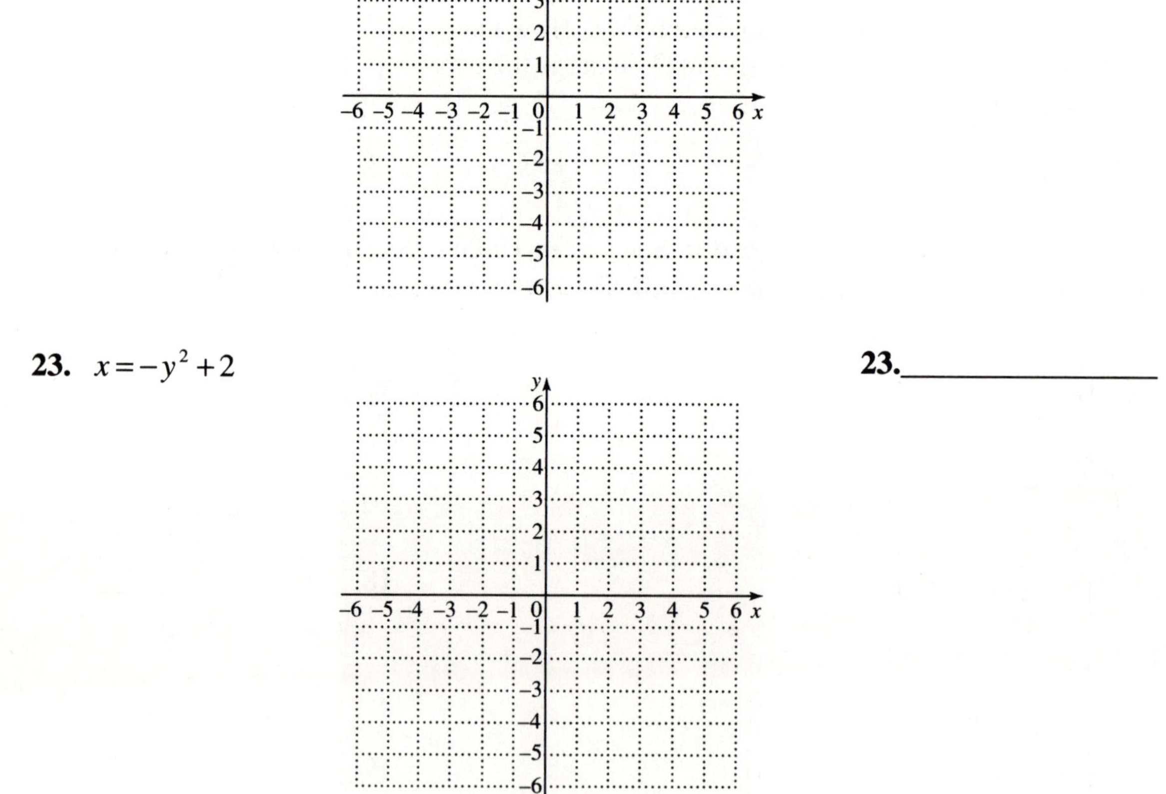

23. $x=-y^2+2$ **23.**_______________

Name: Date:
Instructor: Section:

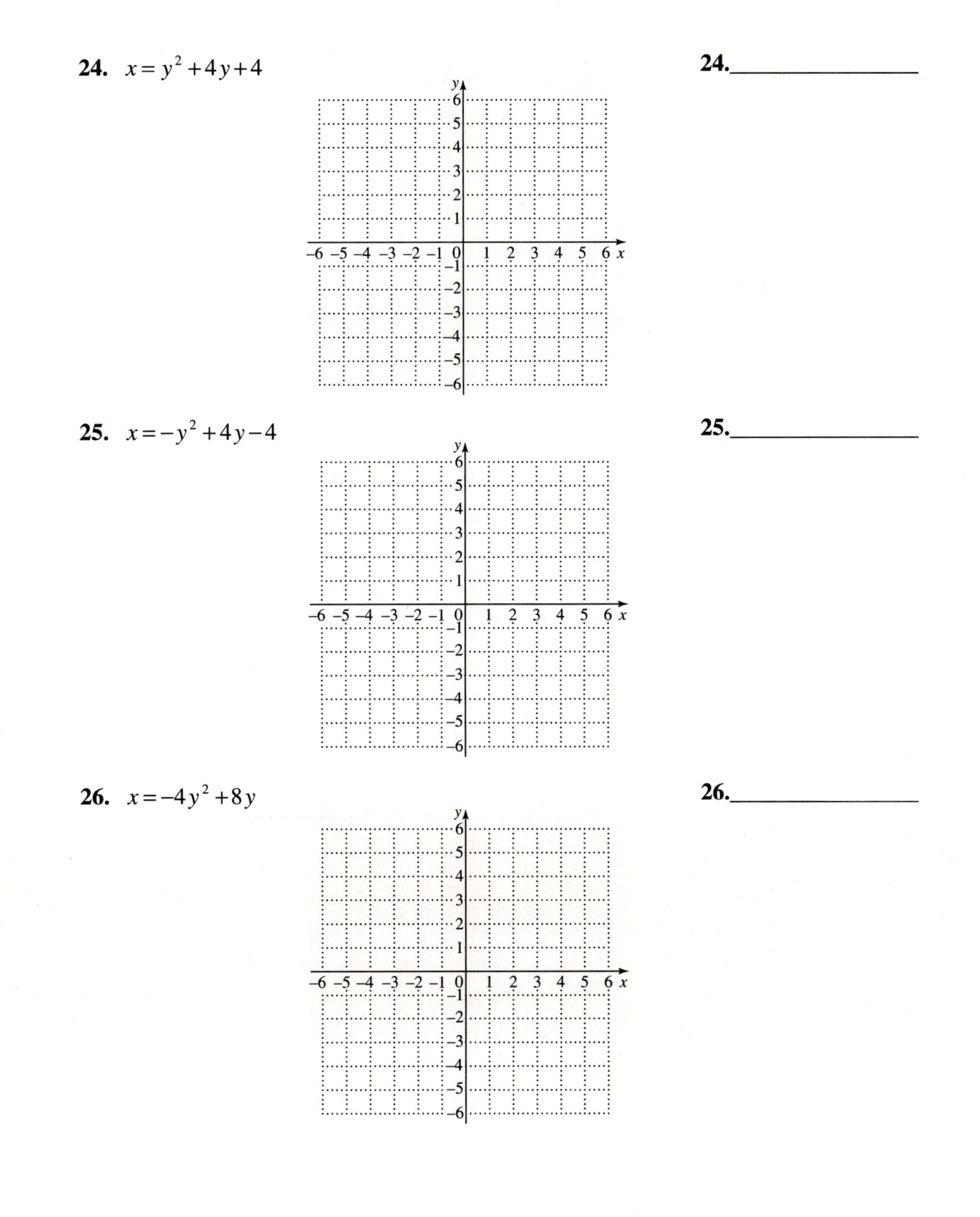

24. $x = y^2 + 4y + 4$

24.________________

25. $x = -y^2 + 4y - 4$

25.________________

26. $x = -4y^2 + 8y$

26.________________

Name: Date:
Instructor: Section:

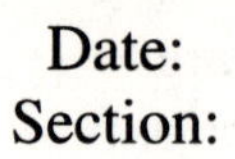

27. $x = 3y^2 + 6y - 2$

27.________________

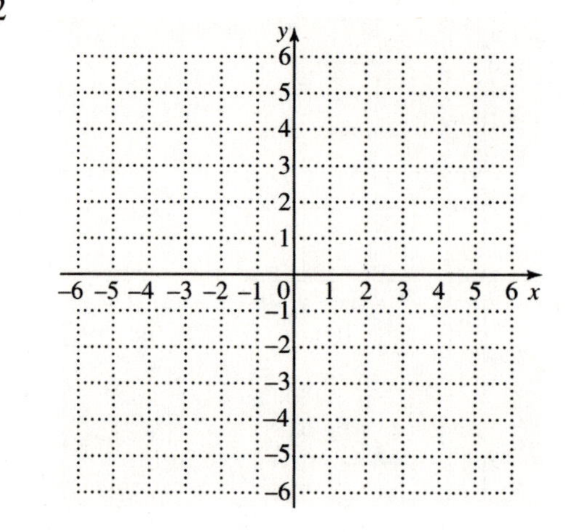

Find the vertex of the parabola.

28. $x = y^2 + 6y - 4$

28.________________

29. $x = 2y^2 + 4y + 3$

29.________________

30. $x = -4y^2 + 8y + 4$

30.________________

Name: Date:
Instructor: Section:

Chapter 11 QUADRATIC EQUATIONS, INEQUALITIES, AND FUNCTIONS

11.8 Quadratic and Rational Inequalities

Learning Objectives
1 Solve quadratic inequalities.
2 Solve polynomial inequalities of degree 3 or greater.
3 Solve rational inequalities.

Key Terms
Use the vocabulary terms listed below to complete each statement in exercises 1-3.

quadratic inequality **rational inequality** **test number**

1. A ____________________ can be written in the form $ax^2 + bx + c < 0$ (or with $\leq$ or $>$ or $\geq$), where a, b, and c are real numbers, with $a \neq 0$.

2. Solution intervals for a quadratic inequality can be found by substituting a ____________________ for x in the original inequality. If the result is *true* then all numbers in the given interval satisfy the inequality.

3. An inequality that involves rational expressions is called a ____________________.

Objective 1 Solve quadratic inequalities.

Solve each inequality.

1. $(x-2)(x+3) \geq 0$ **1.**__________________

2. $(m-5)(m+2) < 0$ **2.**__________________

3. $(r+3)(r-2) \geq 0$ **3.**__________________

4. $k^2 + 7k + 12 > 0$ **4.**__________________

Name: Date:
Instructor: Section:

5. $a^2 - 3a - 18 \leq 0$ **5.**________________

6. $2y^2 + 5y < 3$ **6.**________________

7. $6r^2 + 7r + 2 > 0$ **7.**________________

Solve each inequality.

8. $(x-1)^2 \geq -3$ **8.**________________

9. $(2k+5)^2 \leq -1$ **9.**________________

10. $(3z-2)^2 \leq 0$ **10.**________________

11. $(4m+1)^2 \leq 0$ **11.**________________

12. $(2a+9)^2 < 0$ **12.**________________

13. $(r-3)^2 > 0$ **13.**________________

14. $(y+5)^2 > 0$ **14.**________________

Name: Date:
Instructor: Section:

Objective 2 Solve polynomial inequalities of degree 3 or greater.

Solve each inequality.

15. $(x+1)(x-2)(x+4)\le 0$ **15.**________________

16. $(k+5)(k-1)(k+3)\le 0$ **16.**________________

17. $(y-4)(y+3)(y+1)\le 0$ **17.**________________

18. $(p-6)(p-4)(p-2)>0$ **18.**________________

19. $(2x-1)(2x+3)(3x+1)\le 0$ **19.**________________

20. $(4b+1)(6b-1)(3b-7)>0$ **20.**________________

21. $(4q-3)(2q-7)(3q-10)\ge 0$ **21.**________________

22. $(z+4)(3z+7)(2z+9)<0$ **22.**________________

Name: Date:
Instructor: Section:

Objective 3 Solve rational inequalities.

Solve each inequality.

23. $\frac{5}{y+2}<0$ 23.________________

24. $\frac{3}{a+4}>0$ 24.________________

25. $\frac{6}{3r-2}\geq 1$ 25.________________

26. $\frac{4}{3q+5}\geq -3$ 26.________________

27. $\frac{-5}{2x-3}\leq 2$ 27.________________

28. $\frac{-3}{4m-3}\geq 1$ 28.________________

29. $\frac{y}{y+1}\geq 3$ 29.________________

30. $\frac{r}{r-2}\geq 4$ 30.________________

Name: Date:
Instructor: Section:

Chapter 12 INVERSE, EXPONENTIAL, AND LOGARITHMIC FUNCTIONS

12.1 Inverse Functions

Learning Objectives

1 Decide whether a function is one-to-one and, if it is, find its inverse.
2 Use the horizontal line test to determine whether a function is one-to-one.
3 Find the equation of the inverse of a function.
4 Graph f^{-1} from the graph of f.
5 Use a graphing calculator to graph inverse functions.

Key Terms

Use the vocabulary terms listed below to complete each statement in exercises 1-4.

one-to-one function **inverse** **horizontal line test** **interchange *x* and *y***

1. The first step to find the inverse of a function is to ________________.

2. If f is a one-to-one function, then the ________________ of f is the set of all ordered pairs of the form (y, x) where (x, y) belongs to f.

3. A(n) ________________ is a function in which each x-value corresponds to only one y-value and each y-value corresponds to just one x-value.

4. The ________________ states that a function is one-to-one if every horizontal line intersects the graph of the function at most once.

Objective 1 Decide whether a function is one-to-one and, if it is, find its inverse.

Decide whether or not the function is one-to-one. If it is, find its inverse.

1. $\{(1, 0), (2, 0), (3, 5), (4, 1)\}$ **1.**________________

2. $\{(2, -1), (-2, 1), (1, 3), (-1, -3)\}$ **2.**________________

3. $\{(4, 0), (2, 3), (0, 0), (3, 5)\}$ **3.**________________

4. $\{(-1, 1), (-2, 2), (-3, 3)\}$ **4.**________________

5. $\{(-3, 1), (-2, 2), (-1, 3), (0, 4)\}$ **5.**__________________

6. $\{(-3, 2), (-2, 3), (4, 1), (-4, -1)\}$ **6.**__________________

Objective 2 Use the horizontal line test to determine whether a function is one-to-one.

Use the horizontal line test to determine whether the graph is the graph of a one-to-one function.

7. **7.**__________________

8. **8.**__________________

9. **9.**__________________

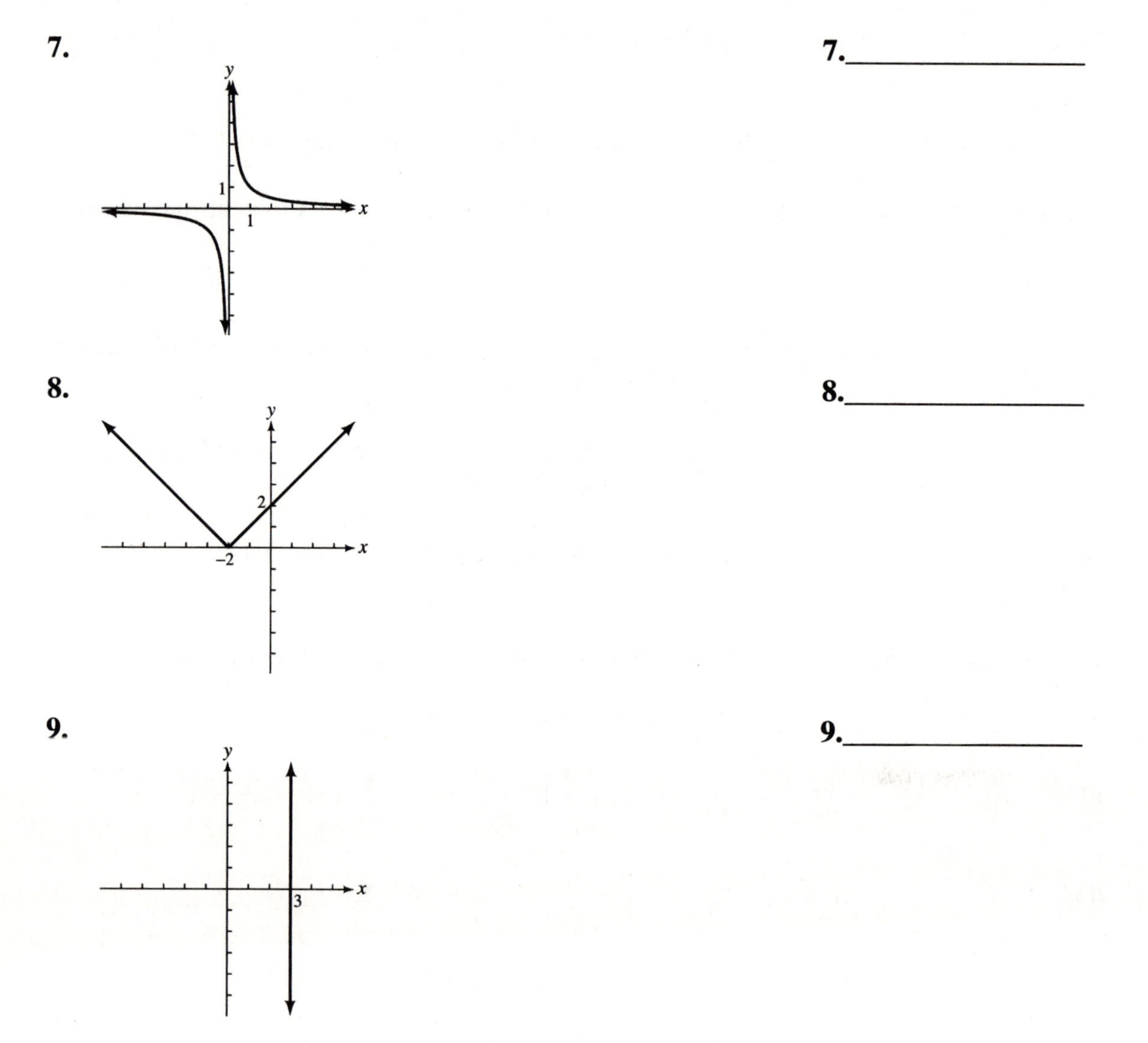

Name: Date:
Instructor: Section:

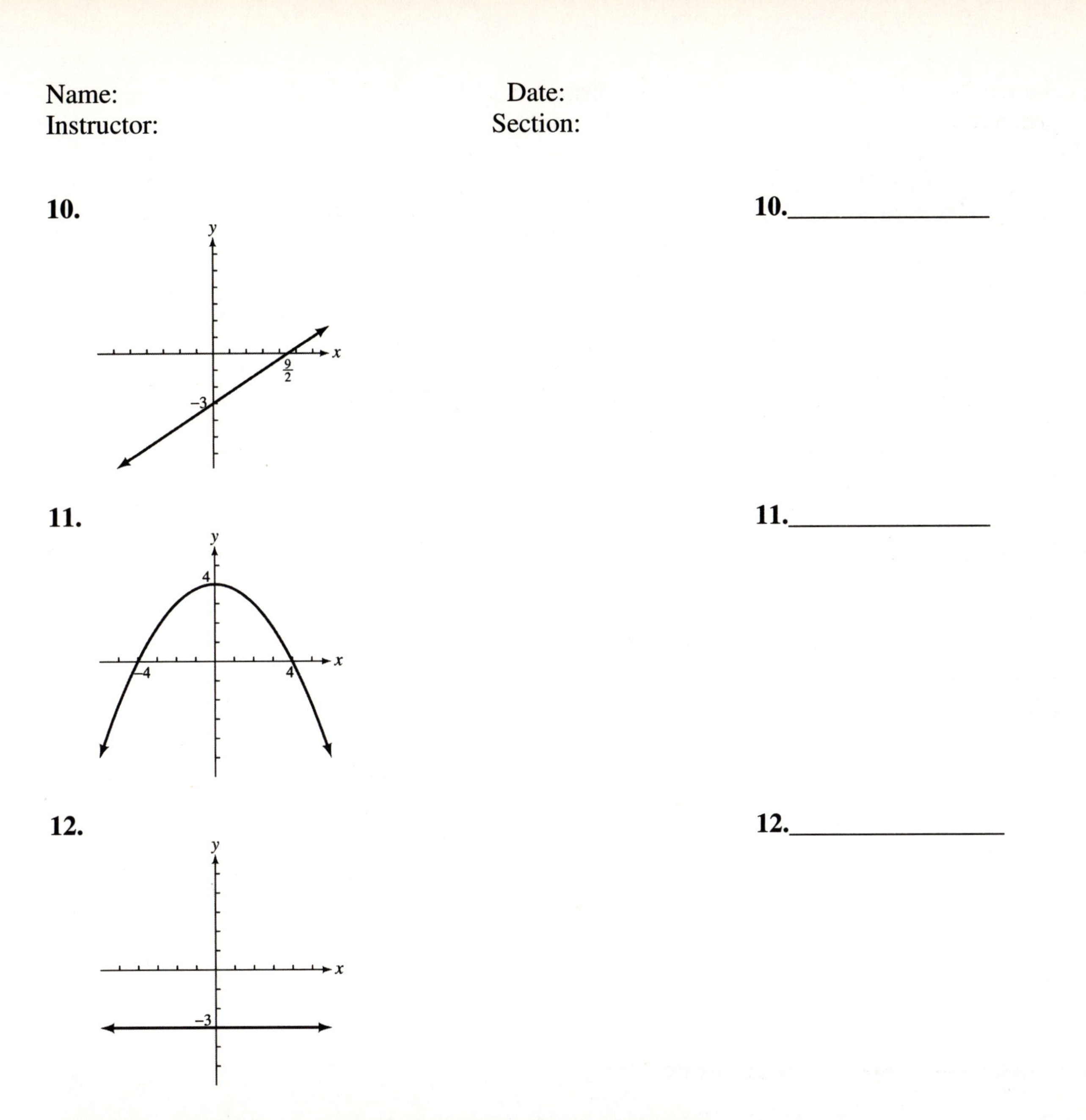

10. 10.________________

11. 11.________________

12. 12.________________

Objective 3 Find the equation of the inverse of a function.

Decide whether the equation defines a one-to-one function. If so, find the equation of the inverse.

13. $f(x)=2x-5$ 13.________________

14. $f(x)=x^2-1$ 14.________________

Name: Date:
Instructor: Section:

15. $f(x)=1-2x^2$ **15.**________________

16. $f(x)=\sqrt{x-1}, x\geq 1$ **16.**________________

17. $f(x)=x^3-1$ **17.**________________

18. $f(x)=4x^3+2$ **18.**________________

19. $f(x)=\dfrac{3}{x-1}$ **19.**________________

Objective 4 Graph f^{-1} from the graph of *f*.

The graphs of some functions are given below. For functions that are one-to-one, copy the graphs and graph the inverses with dashed curves on the same axes as the functions.

20. **20.**________________

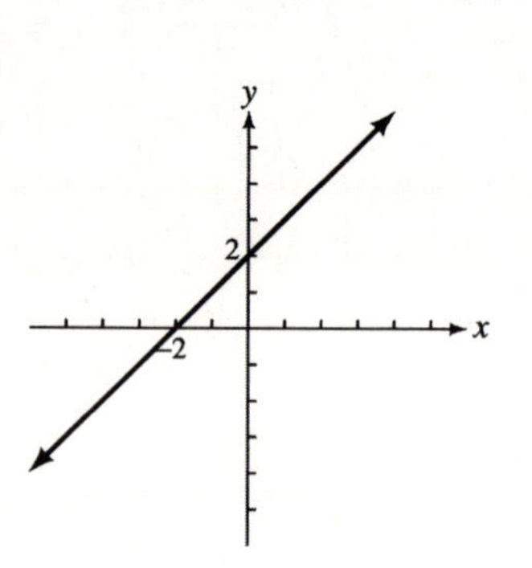

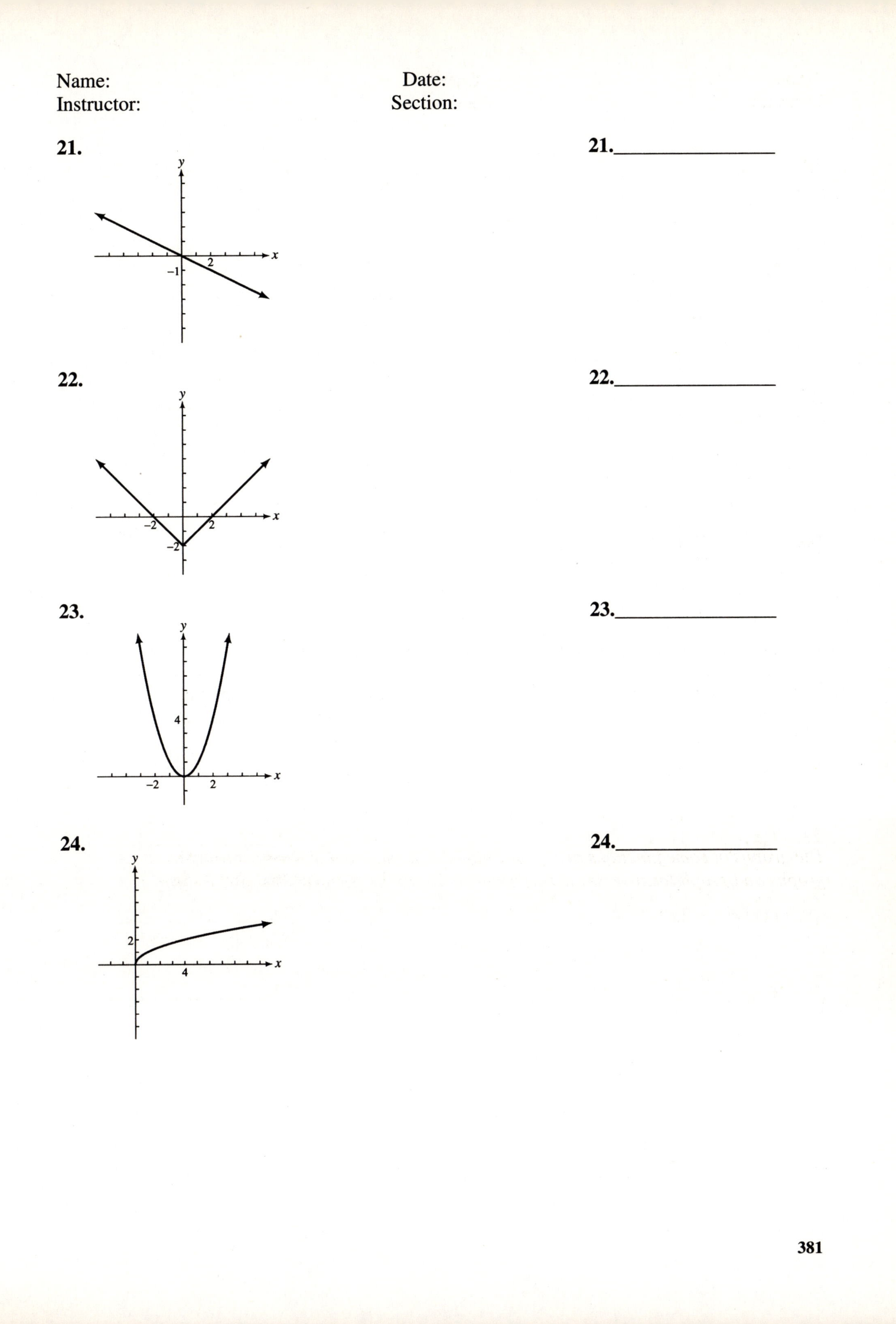

Name:
Date:
Instructor:
Section:
21.
21.
y
x
2
−1
22.
22.
y
x
−2
2
−2
23.
23.
y
x
4
−2
2
24.
24.
y
x
2
4
381

Name: Date:
Instructor: Section:

25. 25.________________

26. 26.________________

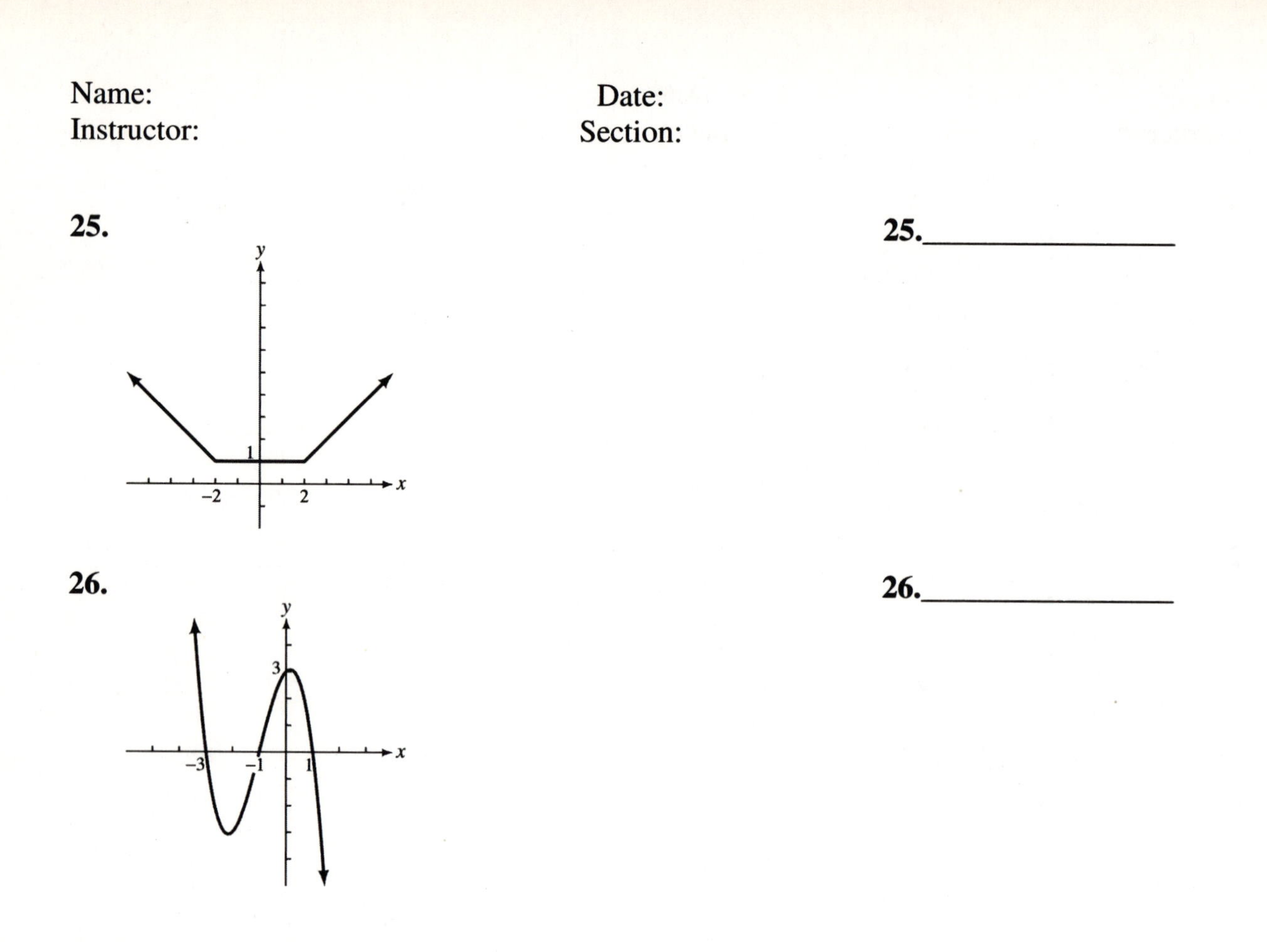

Objective 5 Use a graphing calculator to graph inverse functions.

Use a graphing calculator with the capability of drawing inverses to graph the function and its inverse in the same square window. (Note: The inverse may not be a function.)

27. $f(x)=2x-5$ **27.**________________

28. $f(x)=x^2+5x-2$ **28.**________________

29. $f(x)=x^3-2x^2+3$ **29.**________________

30. $f(x)=\sqrt{x+2}$ **30.**________________

Name: Date:
Instructor: Section:

Chapter 12 INVERSE, EXPONENTIAL, AND LOGARITHMIC FUNCTIONS

12.2 Exponential Functions

Learning Objectives	
1	Define an exponential function.
2	Graph an exponential function.
3	Solve exponential equations of the form $a^x = a^k$ for x.
4	Use exponential functions in applications involving growth or decay.

Key Terms

Use the vocabulary terms listed below to complete each statement in exercises 1-4.

exponential function **asymptote** **exponential equation**

same base

1. A(n) __________________ is a function defined by an expression of the form $f(x) = a^x$, where $a > 0$ and $a \neq 1$ for all real numbers x.

2. A(n) __________________ is an equation that has a variable as an exponent.

3. A line that a graph more and more close approaches as the graph gets farther away from the origin is called a(n) __________________ of the graph.

4. When solving an exponential equation, the first step is to check that each side has the __________________.

Objective 1 Define an exponential function.

Decide whether or not the function defines an exponential function.

1. $f(x) = 2^x$ **1.**______________

2. $f(x) = x^2$ **2.**______________

3. $f(x) = x + 2$ **3.**______________

4. $f(x) = (-2)^x$ **4.**______________

Name: Date:
Instructor: Section:

5. $f(x) = 2^{x+1}$ **5.**________________

6. $f(x) = 3^{2x}$ **6.**________________

7. $f(x) = 2x^3$ **7.**________________

Objective 2 Graph an exponential function.

Graph.

8. $f(x) = 3^x$ **8.**________________

9. $f(x) = 2^x$ **9.**________________

Name: Date:
Instructor: Section:

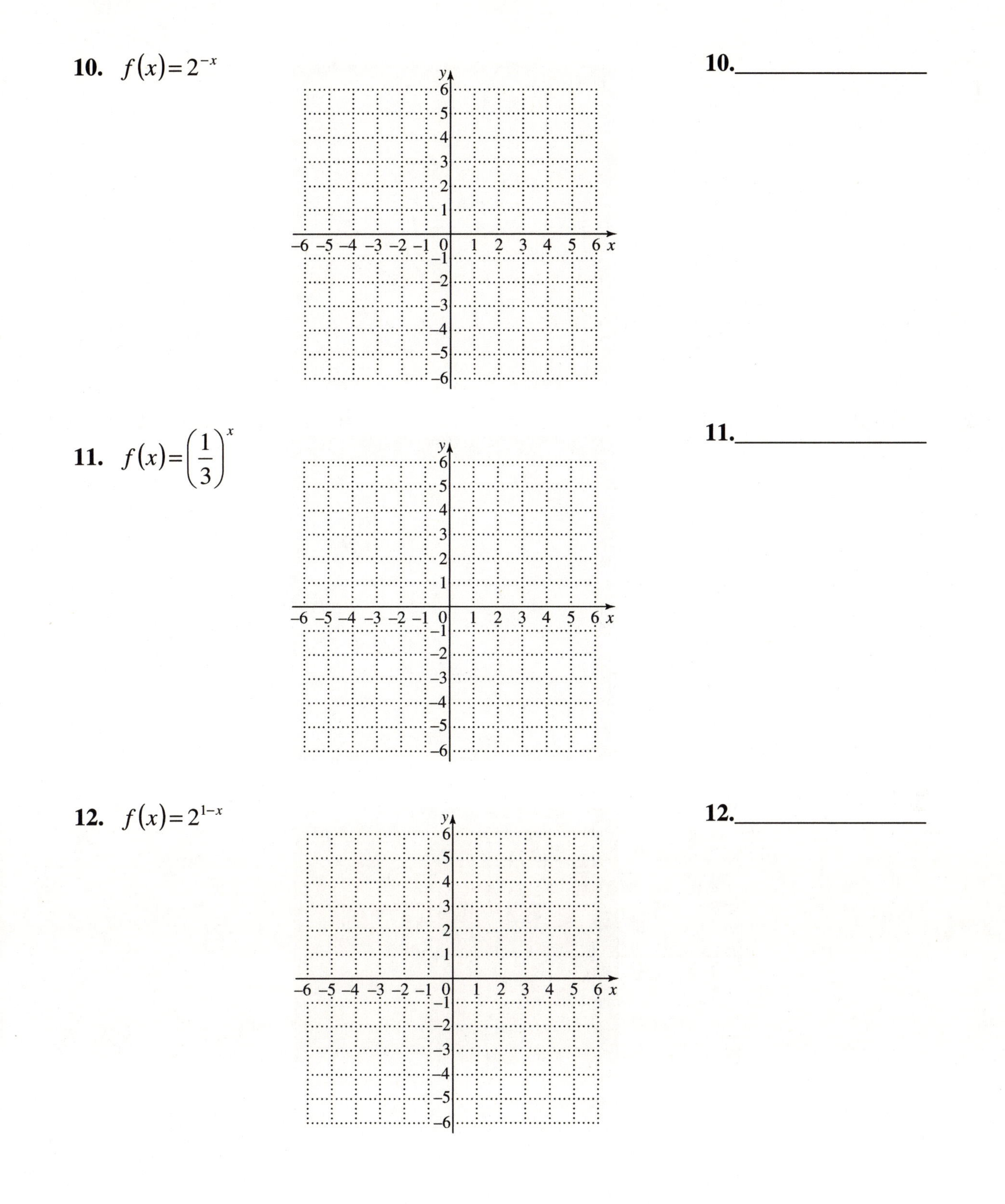

Name: Date:
Instructor: Section:

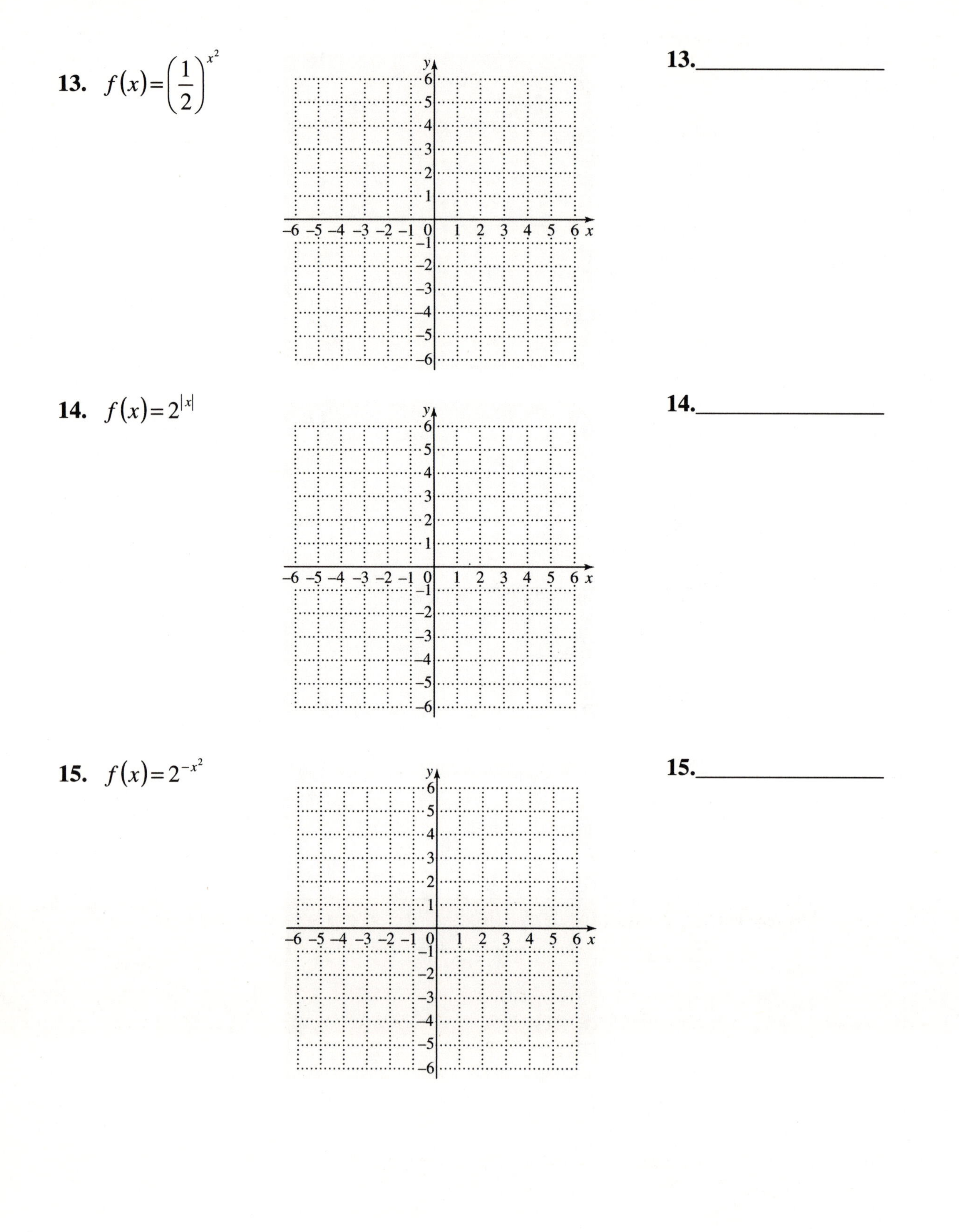

13. $f(x)=\left(\frac{1}{2}\right)^{x^2}$

13.________________

14. $f(x)=2^{|x|}$

14.________________

15. $f(x)=2^{-x^2}$

15.________________

Name: Date:
Instructor: Section:

Objective 3 Solve exponential equations of the form $a^x = a^k$ for x.

Solve the equation.

16. $16^x = 64$ **16.**________________

17. $27^k = 9$ **17.**________________

18. $25^p = 625$ **18.**________________

19. $10^{2x} = 100$ **19.**________________

20. $4^{k+2} = 32$ **20.**________________

21. $25^{1-t} = 5$ **21.**________________

22. $100^{2+t} = 1000$ **22.**________________

Objective 4 Use exponential function in applications involving growth or decay.

Solve the problem.

23. The population of Canadian geese that spend the summer at Gemini Lake each year has been growing according to **23.**________________

$$y = 56(2)^{.2x},$$

where x is the time in years from 1978. Find the number of geese in 1994.

Name: Date:
Instructor: Section:

24. The diameter in inches of a tree during a certain period grew according to

$$y = 2.5\left(9^{.05x}\right),$$

where x was the number of years after the start of this growth period. Find the diameter of the tree after 10 years.

24.________________

25. A culture of a certain kind of bacteria grows according to

$$y = 7750(2)^{.75x},$$

where x was the number of hours after 12 noon. Find the number of bacteria in the culture at 12 noon.

25.________________

26. An industrial city in Pennsylvania has found that its population is declining according to

$$y = 70{,}000(2)^{-.01x},$$

where x is the time in years from 1910. What is the city's anticipated population in the year 2010?

26.________________

27. A sample of a radioactive substance decays according to

$$y = 100\left(10^{-.2x}\right),$$

where y is the mass of the substance in grams and x is the time in hours after the original measurement. Find the mass of the substance after 10 hours.

27.________________

Name: Date:
Instructor: Section:

28. When a bactericide is placed in a certain culture of bacteria, the number of bacteria decreases according to

$$y = 3200(4)^{-.1x},$$

where x is the time in hours. Find the number of bacteria in the culture in 20 hours.

28.________________

29. Suppose the number of bacteria present in a certain culture after t minutes is given by the equation

$$Q(t) = Q_o\left(2^{.05t}\right),$$

where Q_0 represents the initial number of bacteria. If 5000 bacteria are present after 20 minutes, how many bacteria were present initially?

29.________________

30. The population of Evergreen Park is now 16,000. The population t years from now is given by the formula

$$P = 16{,}000\left(2^{\frac{t}{10}}\right).$$

What will be the population 40 years from now?

30.________________

Name: Date:
Instructor: Section:

Chapter 12 INVERSE, EXPONENTIAL, AND LOGARITHMIC FUNCTIONS

12.3 Logarithmic Functions

Learning Objectives	
1	Define a logarithm.
2	Convert between exponential and logarithmic forms.
3	Solve logarithmic equations of the form $\log_a b = k$ for a, b, or k.
4	Define and graph logarithmic functions.
5	Use logarithmic functions in applications involving growth or decay.

Key Terms
Use the vocabulary terms listed below to complete each statement in exercises 1-5.

logarithm **base** **logarithmic function** **rise** **fall**

1. The ____________________ is the number that is a repeated factor when written with an exponent.

2. A ____________________ is an exponent.

3. For the logarithmic function $f(x) = \log_a x$, when $a > 1$, the graph will ____________________ from left to right.

4. For the logarithmic function $f(x) = \log_a x$, when $a < 1$, the graph will ____________________ from left to right.

5. If a and x are positive numbers with $a \neq 1$, then $f(x) = \log_a x$ defines the ____________________ with base a.

Objective 1 Define a logarithm.

Simplify. (Example: $\log_3 9 = 2$*)*

1. $\log_2 8$ 1.________________

2. $\log_8 64$ 2.________________

Name: Date:
Instructor: Section:

3. $\log_4\left(\frac{1}{4}\right)$ 3.________________

4. $\log_3 \sqrt{3}$ 4.________________

5. $\log_{\frac{1}{2}} 4$ 5.________________

6. $\log_{81} 27$ 6.________________

Objective 2 Convert between exponential and logarithmic forms.

If the given equation is written in exponential form, rewrite it in logarithmic form. If the given equation is written in logarithmic form, rewrite it in exponential form.

7. $3^2 = 9$ 7.________________

8. $\log_3 27 = 3$ 8.________________

9. $\log_{16} 2 = \frac{1}{4}$ 9.________________

10. $10^{-2} = \frac{1}{100}$ 10.________________

11. $\log_{10} .001 = -3$ 11.________________

12. $2^{-7} = \frac{1}{128}$ 12.________________

Name: Date:
Instructor: Section:

Objective 3 Solve logarithmic equations of the form $\log_a b = k$ for a, b, or k.

Solve the equation.

13. $\log_2 64 = p$ 13.________________

14. $\log_3 x = -1$ 14.________________

15. $\log_m 25 = 2$ 15.________________

16. $\log_2 t = -5$ 16.________________

17. $\log_4 16 = y$ 17.________________

18. $\log_n .01 = -2$ 18.________________

Objective 4 Define and graph logarithmic functions.

Graph the function.

19. $y = \log_2 x$ 19.________________

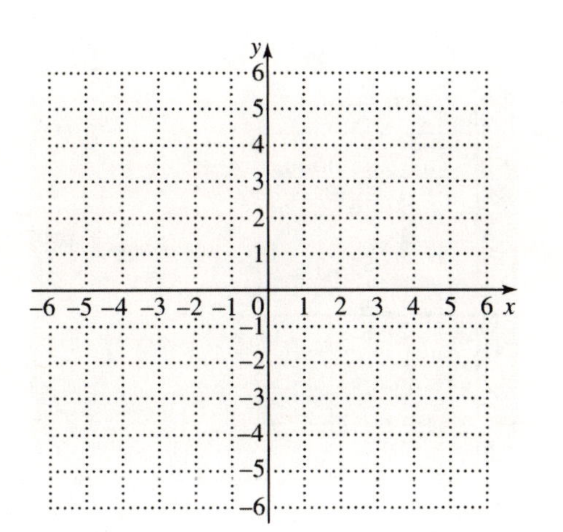

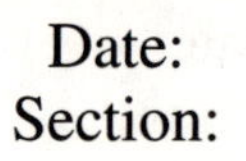

Name: Date:
Instructor: Section:

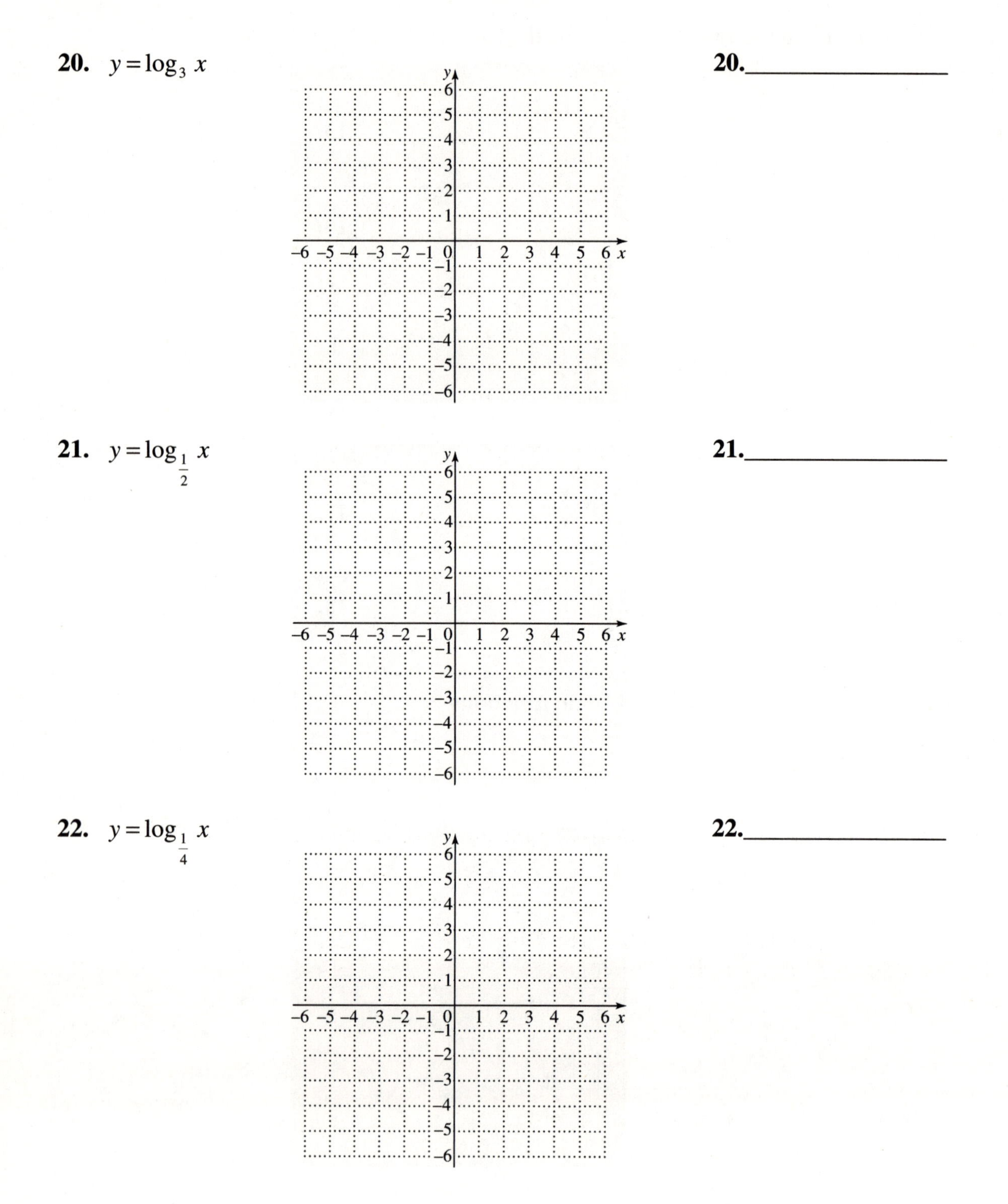

20. $y = \log_3 x$

20.________________

21. $y = \log_{\frac{1}{2}} x$

21.________________

22. $y = \log_{\frac{1}{4}} x$

22.________________

Name: Date:
Instructor: Section:

23. $y = \log_2 2x$

23.________________

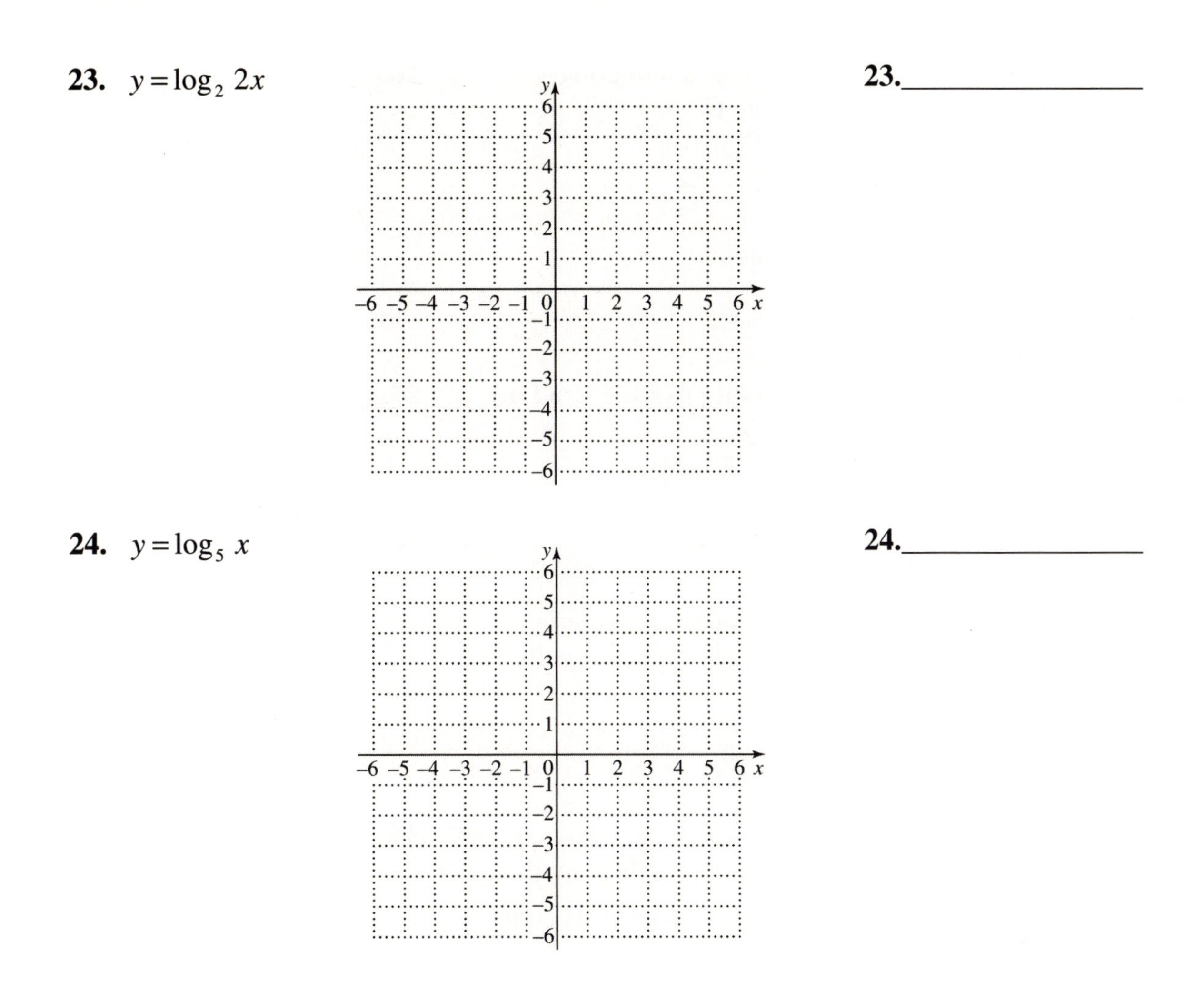

24. $y = \log_5 x$

24.________________

Objective 5 Use logarithmic functions in applications involving growth or decay.

Solve the problem.

25. After black squirrels were introduced to Williams Park, their population grew according to

25.________________

$$y = 14 \log_5 (x+5),$$

where x is the number of months after the squirrels were introduced. Find the number of squirrels after 20 months.

Name: Date:
Instructor: Section:

26. A manufacturer receives revenue y given in dollars for selling x units of an item according to

26.________________

$$y = 200 \log_3 (x+1).$$

Find the revenue for selling 26 units.

27. Under certain conditions, the velocity v of the wind in centimeters per second is given by

27.________________

$$v = 300 \log_2 \left(\frac{10x}{7}\right),$$

where x is the height in centimeters above the ground. Find the wind velocity at 11.2 centimeters above the ground.

28. A decibel is a measure of the loudness of a sound. A very faint sound is assigned an intensity of I_0; then another sound is given an intensity I found in terms of I_o, the faint sound. The decibel rating of the sound is given in decibels by

28.________________

$$d = 10 \log_{10} \frac{I}{I_0}.$$

Find the decibel rating of rock music that has intensity $I = 100{,}000{,}000{,}000 I_0$.

Name: Date:
Instructor: Section:

29. The number of students not completing intermediate algebra is given by

29.________________

$$f = 4 \log_7 (8s + 9),$$

where s is the number of sections of the class that is offered. If there are 5 sections of intermediate algebra this semester, how many students will not complete the course?

30. The number of fish in an aquarium is given by

30.________________

$$f = 5 \log_6 (3t + 6),$$

where t is time in months. Find the difference in the number of fish present between $t = 0$ and $t = 10$.

Name: Date:
Instructor: Section:

Chapter 12 INVERSE, EXPONENTIAL, AND LOGARITHMIC FUNCTIONS

12.4 Properties of Logarithms

Learning Objectives	
1	Use the product rule for logarithms.
2	Use the quotient rule for logarithms.
3	Use the power rule for logarithms.
4	Use properties to write alternative forms of logarithmic expressions.

Key Terms

Use the vocabulary terms listed below to complete each statement in exercises 1-4.

product rule for logarithms **quotient rule for logarithms**

power rule for logarithms **special properties**

1. The equations $b^{\log_b x} = x, x > 0$ and $\log_b b^x = x$ are referred to as ____________________ of logarithms.

2. The equation $\log_b \frac{x}{y} = \log_b x - \log_b y$ is referred to as the ____________________.

3. The equation $\log_b xy = \log_b x + \log_b y$ is referred to as the ____________________.

4. The equation $\log_b x^r = r \log_b x$ is referred to as the ____________________.

Objective 1 Use the product rule for logarithms.

Use the product rule for logarithms to express the logarithm as the sum of logarithms, or as a single number if possible.

1. $\log_3 (6)(5)$ **1.**________________

2. $\log_7 5m$ **2.**________________

3. $\log_6 6r$ **3.**________________

Name: Date:
Instructor: Section:

4. $\log_3 p^2$ **4.**________________

Use the product rule for logarithms to express the sum as a single logarithm.

5. $\log_4 7+\log_4 3$ **5.**________________

6. $\log_{10} 7+\log_{10} 9$ **6.**________________

7. $\log_7 11y+\log_7 2y+\log_7 3y$ **7.**________________

Objective 2 Use the quotient rule for logarithms.

Use the quotient rule for logarithms to express the logarithm as the difference of logarithms, or as a single number if possible.

8. $\log_2 \frac{7}{9}$ **8.**________________

9. $\log_3 \frac{m}{n}$ **9.**________________

10. $\log_{10} \frac{p}{r}$ **10.**________________

11. $\log_3 \frac{10}{x}$ **11.**________________

Use the quotient rule for logarithms to express the difference as a single logarithm.

12. $\log_2 7q^4-\log_2 5q^2$ **12.**________________

Name: Date:
Instructor: Section:

13. $\log_{10} 12m^3 - \log_{10} 7m^2$ 13.________________

14. $\log_7 60r^3 - \log_7 100r^7$ 14.________________

Objective 3 Use the power rule for logarithms.

Use the power rule for logarithms to rewrite the logarithm.

15. $\log_5 3^2$ 15.________________

16. $\log_7 6^4$ 16.________________

17. $\log_m 2^7$ 17.________________

18. $\log_b \sqrt{5}$ 18.________________

19. $\log_3 \sqrt[3]{7}$ 19.________________

20. $\log_2 \sqrt[3]{8}$ 20.________________

21. $\log_5 125^{\frac{1}{3}}$ 21.________________

Objective 4 Use properties to write alternative forms of logarithmic expressions.

Use the properties of logarithms to express the logarithm as a sum or difference of logarithms.

22. $\log_2 4p^3$ 22.________________

Name: Date:
Instructor: Section:

23. $\log_a \sqrt[3]{2k}$ **23.**________________

24. $\log_b \dfrac{2r}{r-1}$ **24.**________________

25. $\log_5 \dfrac{7m^3}{8y}$ **25.**________________

26. $\log_7 \dfrac{8r^7}{3a^3}$ **26.**________________

Use the properties of logarithms to express the sum or difference as a single logarithm.

27. $\log_2 5m + \log_2 3k$ **27.**________________

28. $\log_{10} 4k^2 j - \log_{10} 3kj^2$ **28.**________________

29. $\log_4 8y + \log_4 3y - \log_4 6y^3$ **29.**________________

30. $\log_2 (x-1) + \log_2 (x+1) - \log_2 (x^2 - 1)$ **30.**________________

Name: Date:
Instructor: Section:

Chapter 12 INVERSE, EXPONENTIAL, AND LOGARITHMIC FUNCTIONS

12.5 Common and Natural Logarithms

Learning Objectives
1 Evaluate common logarithms using a calculator.
2 Use common logarithms in applications.
3 Evaluate natural logarithms using a calculator.
4 Use natural logarithms in applications.
5 Use the change-of-base rule.

Key Terms
Use the vocabulary terms listed below to complete each statement in exercises 1-5.

common logarithm **always negative** **natural logarithm**

universal constant **change-of-base rule**

1. The common logarithm of a number between 0 and 1 is ___________________.

2. The number e is called a(n) ___________________ because of its importance in many areas of mathematics.

3. The equation $\log_a x = \dfrac{\log_b x}{\log_b a}$ is referred to as the ___________________.

4. A(n) ___________________ is a logarithm to base 10.

5. A(n) ___________________ is a logarithm to base e.

Objective 1 Evaluate common logarithms using a calculator.

Use a calculator to find the logarithm. Round the answer to the nearest ten thousandth.

1. $\log 57.23$ **1.**________________

2. $\log 7.355$ **2.**________________

3. $\log 843.71$ **3.**________________

Name: Date:
Instructor: Section:

4. log .091419 **4.**________________

5. log 280,037 **5.**________________

Objective 2 Use common logarithms in applications.

Find the pH of a solution with the given hydronium ion concentration. Round the answer to the nearest tenth.

6. 2.8×10^{-6} **6.**________________

7. 5.6×10^{-8} **7.**________________

8. 4.3×10^{-9} **8.**________________

9. 1.7×10^{-9} **9.**________________

Find the hydronium ion concentration of a solution with the given pH value.

10. 2.9 **10.**________________

11. 3.4 **11.**________________

12. 5.2 **12.**________________

13. 1.3 **13.**________________

Objective 3 Evaluate natural logarithms using a calculator.

Find the natural logarithm. Round the answer to the nearest ten thousandth.

14. ln .12 **14.**________________

15. ln 143 **15.**________________

Name: Date:
Instructor: Section:

16. ln 6 **16.**________________

17. ln 428 **17.**________________

18. ln .013 **18.**________________

Objective 4 Use natural logarithms in applications.

The population of a small town is

$$P(t) = 600e^{.01t},$$

where t represents time in years. Find the following to the nearest whole number.

19. $P(0)$ **19.**________________

20. $P(10)$ **20.**________________

A radioactive substance is decaying so that the amount present at time t in days is given by

$$Q(t) = 100e^{-.03t}.$$

Find the following to the nearest tenth.

21. $Q(10)$ **21.**________________

22. $Q(60)$ **22.**________________

Name: Date:
Instructor: Section:

Objective 5 Use the change-of-base rule.

Using the change-of-base rule, find the logarithm. Round the answer to the nearest thousandth.

23. $\log_{16} 27$ **23.**______________

24. $\log_{6} 3$ **24.**______________

25. $\log_{2} 5$ **25.**______________

26. $\log_{7} 28$ **26.**______________

27. $\log_{5} 180$ **27.**______________

28. $\log_{3} 142$ **28.**______________

29. $\log_{\frac{1}{4}} 11$ **29.**______________

30. $\log_{\frac{2}{3}} 5$ **30.**______________

Name: Date:
Instructor: Section:

Chapter 12 INVERSE, EXPONENTIAL, AND LOGARITHMIC FUNCTIONS

12.6 Exponential and Logarithmic Equations; Further Applications

Learning Objectives
1 Solve equations involving variables in the exponents.
2 Solve equations involving logarithms.
3 Solve applications of compound interest.
4 Solve applications involving base *e* exponential growth and decay.
5 Use a graphing calculator to solve exponential and logarithmic equations.

Key Terms
Use the vocabulary terms listed below to complete each statement in exercises 1-6.

$x = y$ **power rule** **single logarithm** **compound interest formula**

continuous compound interest formula **half-life**

1. The statements "If ____________________, then $b^x = b^y$" and its converse are useful in solving exponential equations.

2. When solving a logarithmic equation, transform the equation so that a ____________________ appears on one side.

3. The equation $A = Pe^{rt}$ is referred to as the ____________________.

4. The time it takes for a substance to decay to half of its original amount is called the ____________________.

5. When solving an exponential equation, take logarithms to the same base on both sides and then use the ____________________ of logarithms or the special property $\log_b b^x$.

6. The equation $A = P\left(1+\frac{r}{n}\right)^{nt}$ is referred to as the ____________________.

Objective 1 Solve equations involving variables in the exponents.

Solve the equation. Round the solution to the nearest hundredth.

1. $27^x = 5$ **1.**________________

Name: Date:
Instructor: Section:

2. $32^y = 6$ 2.________________

3. $7^m = 11$ 3.________________

4. $4^{m-3} = 6$ 4.________________

5. $8^{4-x} = 3$ 5.________________

6. $2^{3y-9} = 7$ 6.________________

Objective 2 Solve equations involving logarithms.

Solve the equation.

7. $\log(p-2) = \log 3$ 7.________________

8. $\log(2k+1) = \log 7$ 8.________________

9. $\log_2(x+1) - \log_2 x = \log_2 5$ 9.________________

10. $\log(-y) + \log 4 = \log(2y+5)$ 10.________________

11. $\log_p 10 = 4$ 11.________________

Name: Date:
Instructor: Section:

12. $\log_3 a = \log_3(a-1)+2$ **12.**________________

Objective 3 Solve applications of compound interest.

Solve the problem.

Find the final amount owed when the following amount is borrowed if interest is compounded annually. Use

$$A = P\left(1+\frac{r}{n}\right)^{nt},$$

where A is the amount owed, P is amount borrowed, r is the interest rate, $n=1$, *and t is the time in years.*

13. \$1000 for 3 years at 8% **13.**________________

14. \$25,000 for 5 years at 10% **14.**________________

15. \$5600 for 8 years at 11% **15.**________________

16. \$2700 for 10 years at 9% **16.**________________

17. \$8500 for 6 years at 4.75% **17.**________________

18. \$1300 for 10 years at 6% **18.**________________

Name: Date:
Instructor: Section:

Objective 4 Solve applications involving base *e* exponential growth and decay.

Solve the problem.

A radioactive substance decays according to the formula

$$y = 400e^{-.05t},$$

where y is the amount present in grams after t weeks.

19. How many grams were present initially? **19.**______________

20. How long would it take, to the nearest tenth of a week, for there to be 100 grams present? **20.**______________

21. What is the half-life of this substance, to the nearest tenth of a week? **21.**______________

A radioactive substance decays according to the formula

$$y = 600e^{-.08t},$$

where y is measured in pounds and t is measured in days?

22. How much will be present after 10 days? **22.**______________

23. How long will it take, to the nearest tenth of a day, for there to be 200 pounds present? **23.**______________

24. How long will it take, to the nearest tenth of a day, for there to be 150 pounds present? **24.**______________

Name: Date:
Instructor: Section:

Objective 5 Use a graphing calculator to solve exponential and logarithmic equations.

Use a graphing calculator to solve the equation, to the nearest hundredth.

25. $475e^{-.07x} = 200$ **25.**________________

26. $6000 = 2500(1.025)^x$ **26.**________________

27. $(1.018)^{4x} = 3.75$ **27.**________________

28. $250\log(x+1) = 700$ **28.**________________

29. $\ln(5x-1) = 39.7$ **29.**________________

30. $\ln(\sqrt{x}) = 4.3$ **30.**________________

Name: Date:
Instructor: Section:

Chapter 13 NONLINEAR FUNCTIONS, CONIC SECTIONS, AND NONLINEAR SYSTEMS

13.1 Additional Graphs of Functions

Learning Objectives

1 Recognize the graphs of elementary functions defined by $|x|$, $\frac{1}{x}$, and $\sqrt{x}$, and graph their translations.

2 Recognize and graph step functions.

Key Terms

Use the vocabulary terms listed below to complete each statement in exercises 1-5.

absolute value function **reciprocal function** **asymptotes**

square root function **greatest integer function**

1. The function defined by $f(x)=|x|$ with a graph that includes portions of two lines is called the ____________________.

2. When graphing the reciprocal function, the axes form the ____________________ of the function.

3. The function defined by $f(x)=[\![x]\!]$, where the symbol $[\![x]\!]$ is used to represent the greatest integer less than or equal to *x*, is called the ____________________.

4. The ____________________ is defined by $f(x)=\frac{1}{x}$.

5. The function defined by $f(x)=\sqrt{x}$, with $x\geq 0$, is called the ____________________.

Name: Date:
Instructor: Section:

Objective 1 Recognize the graphs of elementary functions defined by $|x|$, $\frac{1}{x}$, and $\sqrt{x}$, and graph their translations.

Graph each function.

1. $f(x)=|x-2|$

y
6
5
4
3
2
1
−6 −5 −4 −3 −2 −1 0 1 2 3 4 5 6 x
−1
−2
−3
−4
−5
−6

1.________________

2. $f(x)=|x+4|$

y
6
5
4
3
2
1
−6 −5 −4 −3 −2 −1 0 1 2 3 4 5 6 x
−1
−2
−3
−4
−5
−6

2.________________

3. $f(x)=|x-3|+3$

y
6
5
4
3
2
1
−6 −5 −4 −3 −2 −1 0 1 2 3 4 5 6 x
−1
−2
−3
−4
−5
−6

3.________________

Name: Date:
Instructor: Section:

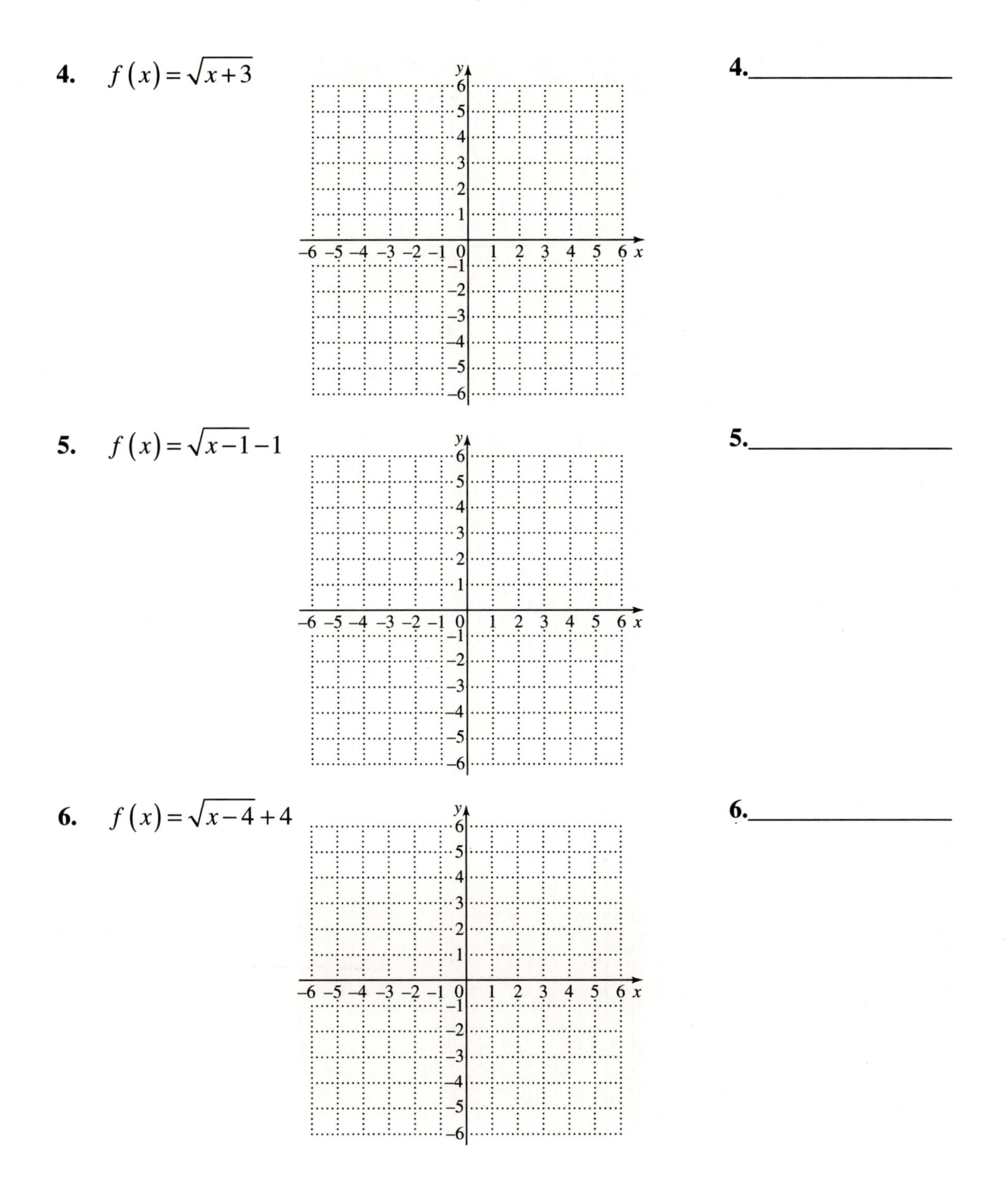

4. $f(x)=\sqrt{x+3}$

5. $f(x)=\sqrt{x-1}-1$

6. $f(x)=\sqrt{x-4}+4$

4.________________

5.________________

6.________________

Name: Date:
Instructor: Section:

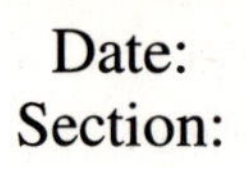

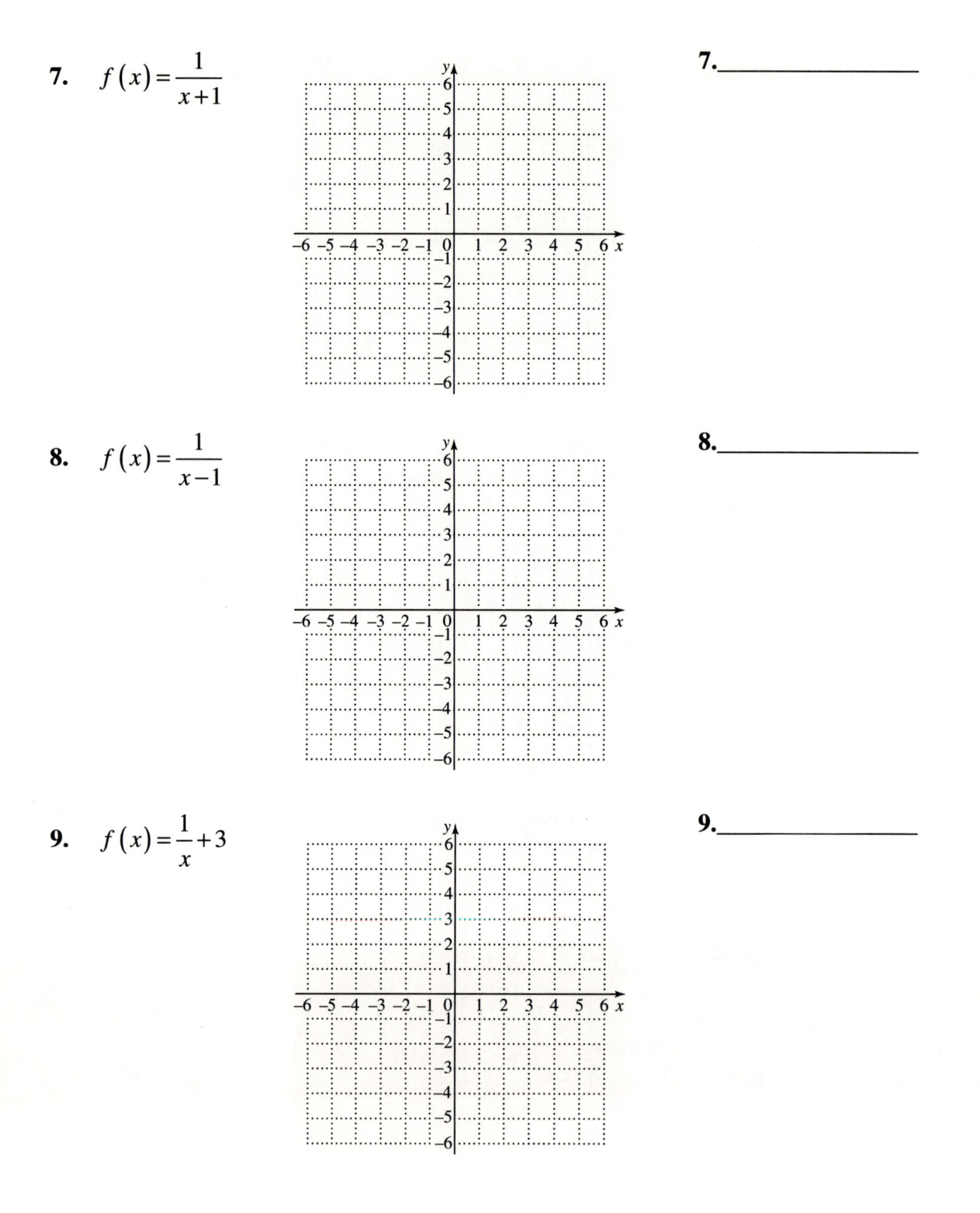

7. $f(x)=\dfrac{1}{x+1}$

7.________________

8. $f(x)=\dfrac{1}{x-1}$

8.________________

9. $f(x)=\dfrac{1}{x}+3$

9.________________

Name: Date:
Instructor: Section:

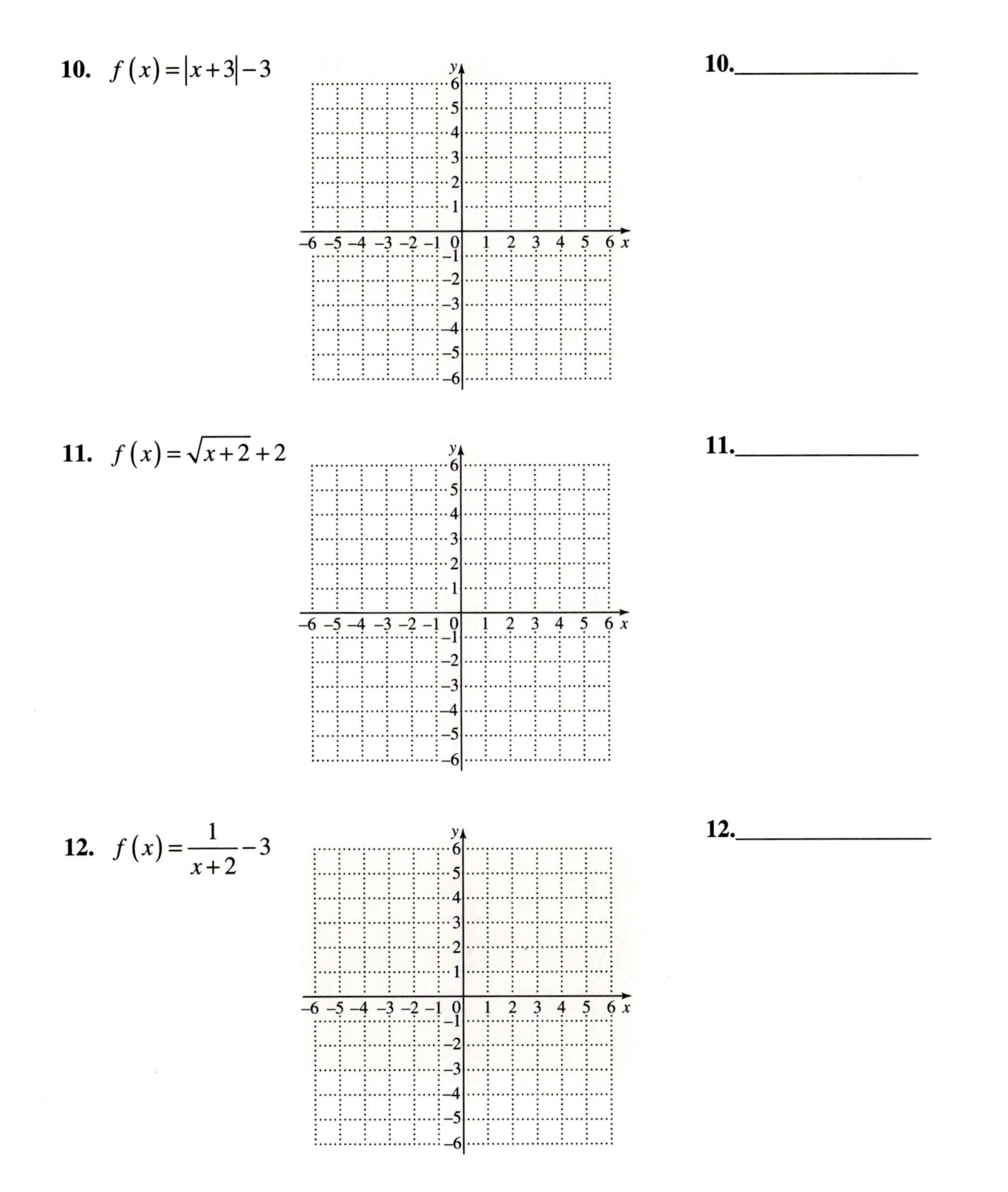

10. $f(x) = |x+3| - 3$

10. ________________

11. $f(x) = \sqrt{x+2} + 2$

11. ________________

12. $f(x) = \dfrac{1}{x+2} - 3$

12. ________________

Name: Date:
Instructor: Section:

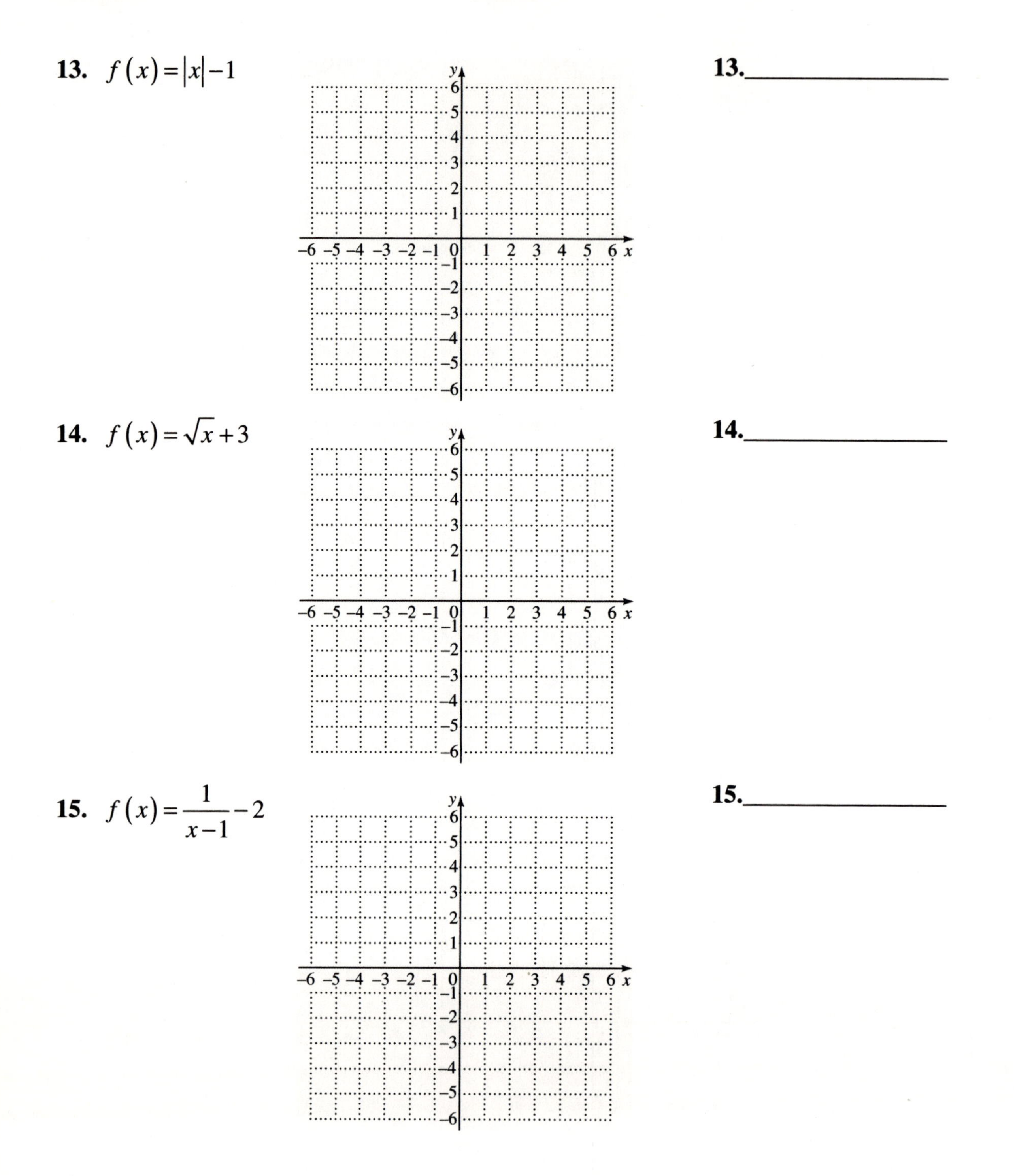

13. $f(x) = |x| - 1$

13.______________

14. $f(x) = \sqrt{x} + 3$

14.______________

15. $f(x) = \dfrac{1}{x-1} - 2$

15.______________

Name: Date:
Instructor: Section:

16. $f(x) = |x - 2| - 1$

16.________________

17. $f(x) = \sqrt{x+3} - 1$

17.________________

18. $f(x) = \dfrac{1}{x} - 1$

18.________________

Name: Date:
Instructor: Section:

19. $f(x) = |x| + 3$

19.________________

20. $f(x) = \sqrt{x-2} - 3$

20.________________

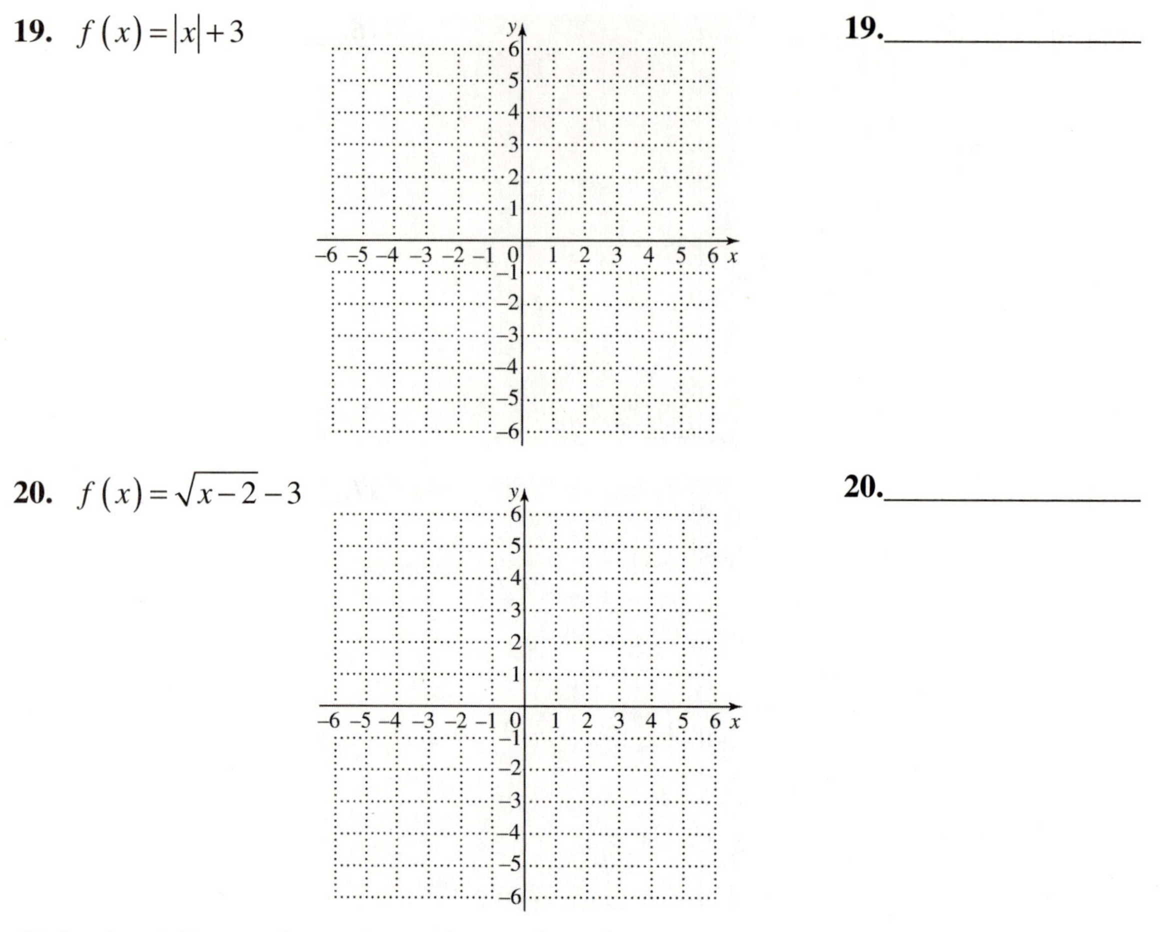

Objective 2 Recognize and graph step functions.

Graph each step function.

21. $f(x) = [\![x+3]\!]$

21.________________

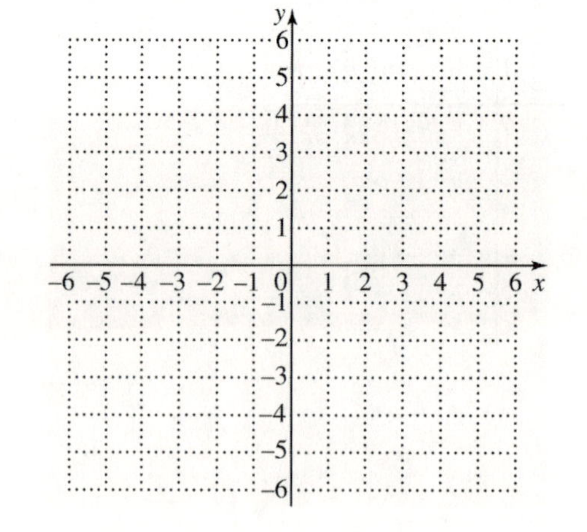

Name: Date:
Instructor: Section:

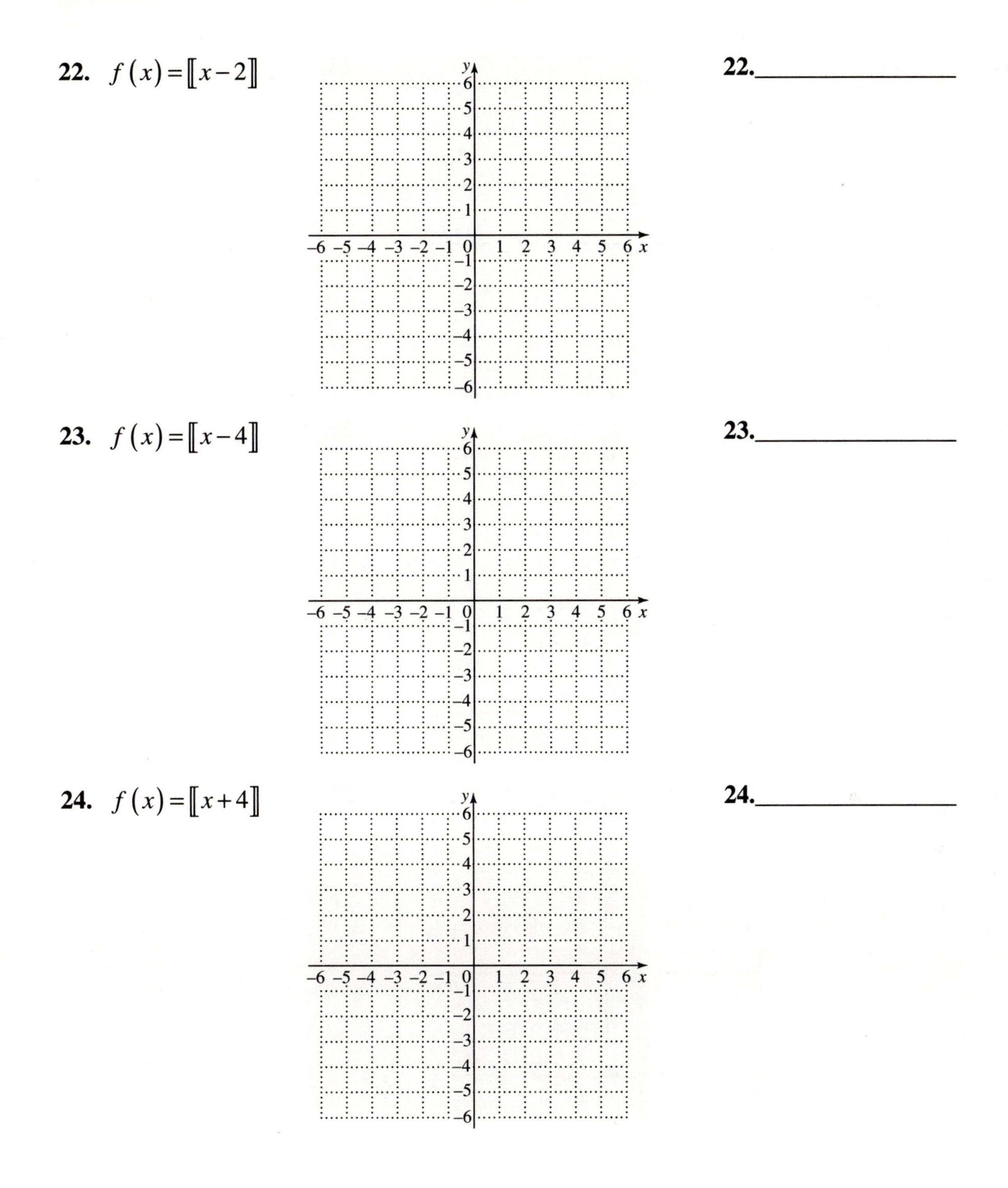

22. $f(x) = [\![x - 2]\!]$

22.________________

23. $f(x) = [\![x - 4]\!]$

23.________________

24. $f(x) = [\![x + 4]\!]$

24.________________

Name: Date:
Instructor: Section:

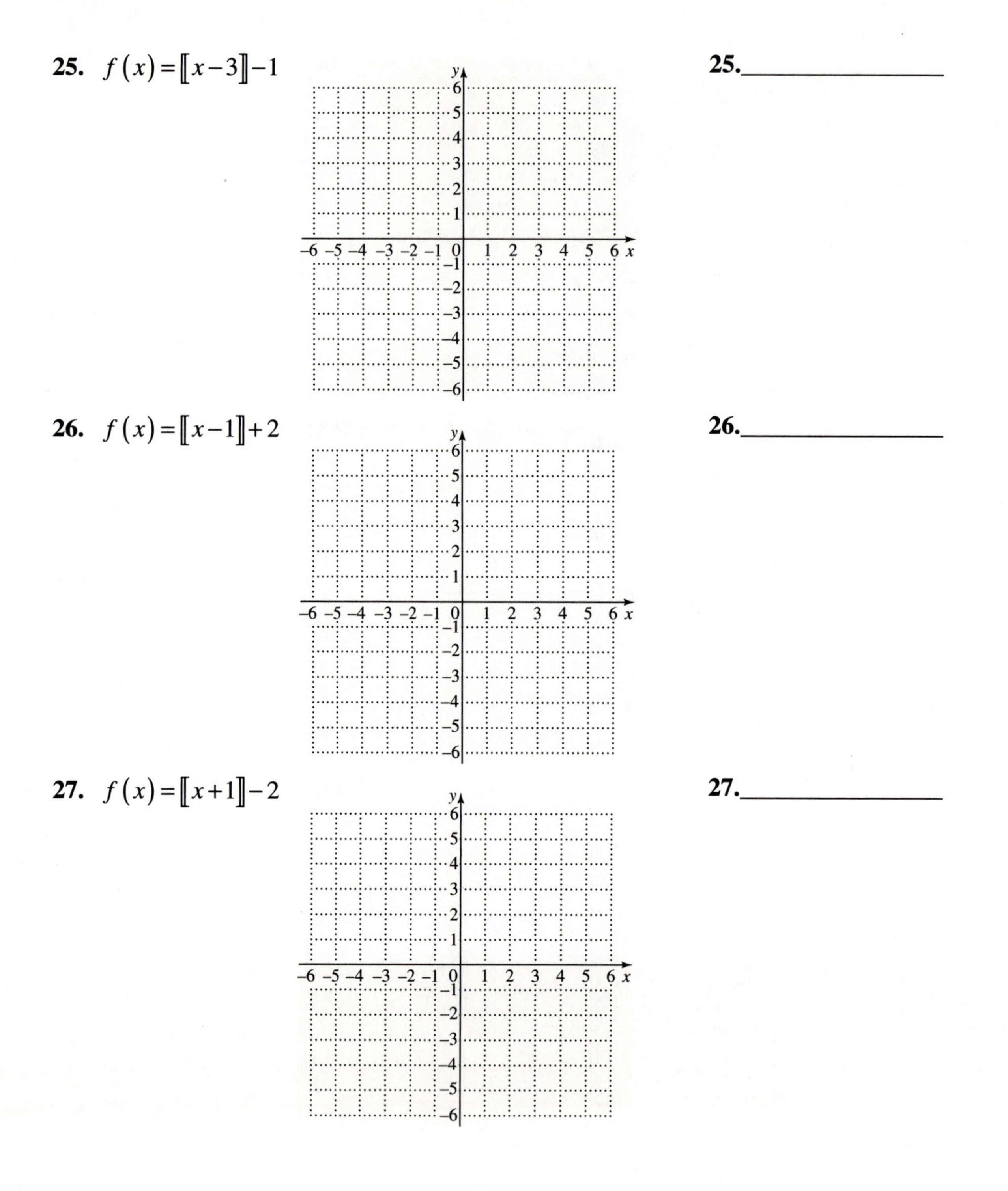

25. $f(x) = [\![x-3]\!] - 1$

25.________________

26. $f(x) = [\![x-1]\!] + 2$

26.________________

27. $f(x) = [\![x+1]\!] - 2$

27.________________

Name: Date:
Instructor: Section:

28. $f(x) = [\![x]\!] + 2$ **28.**________________

29. $f(x) = [\![x]\!] - 3$ **29.**________________

30. $f(x) = [\![x+2]\!] + 1$ **30.**________________

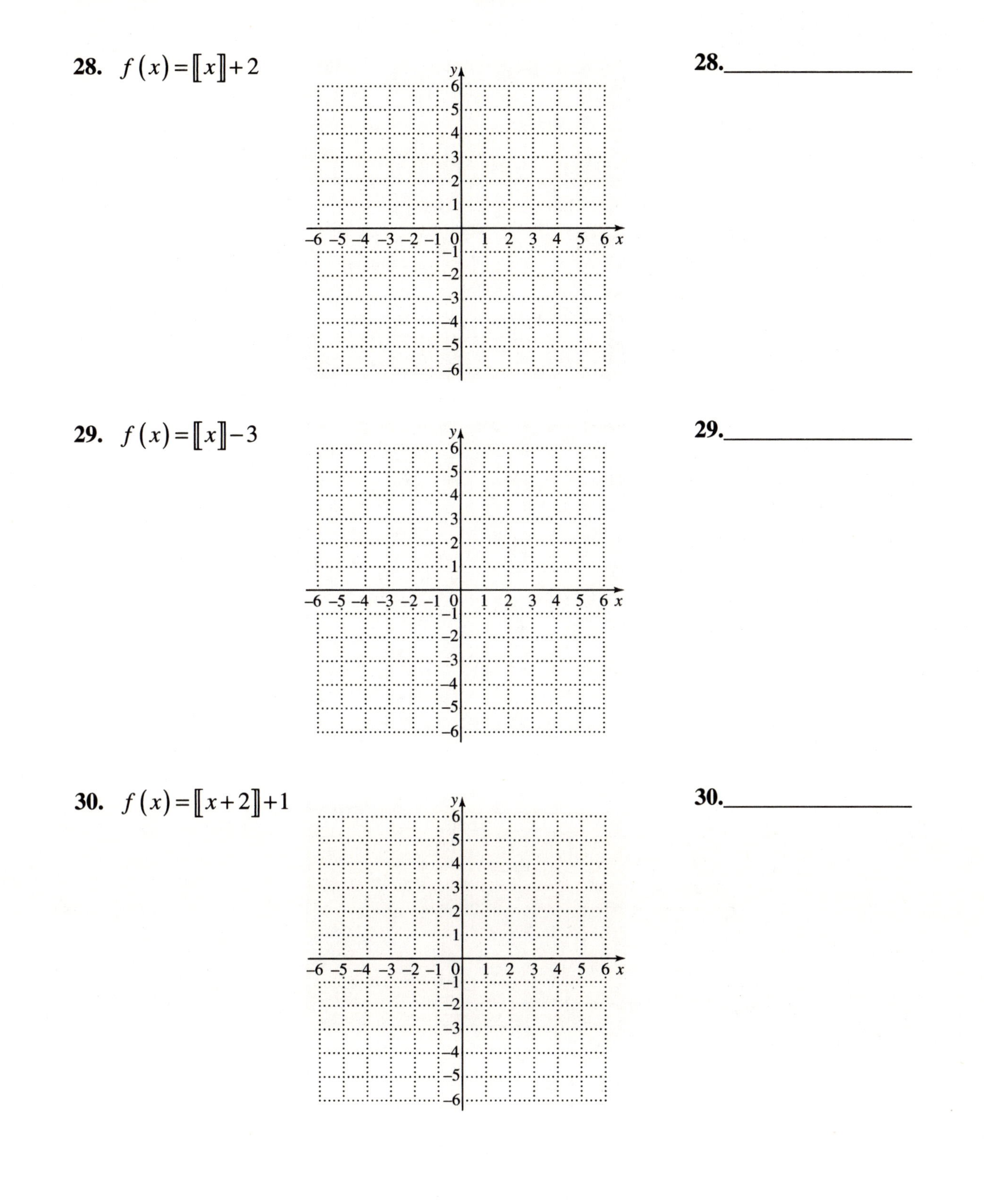

Name: Date:
Instructor: Section:

Chapter 13 NONLINEAR FUNCTIONS, CONIC SECTIONS, AND NONLINEAR SYSTEMS

13.2 The Circle and the Ellipse

Learning Objectives

1. Find an equation of a circle given the center and radius.
2. Determine the center and radius of a circle given its equation.
3. Recognize an equation of an ellipse.
4. Graph ellipses.
5. Graph circles and ellipses using a graphing calculator.

Key Terms

Use the vocabulary terms listed below to complete each statement in exercises 1-5.

conic sections **circle** **center-radius form** **ellipse** **foci**

1. The ____________________ of the equation of a circle with center (h, k) and radius r is $(x-h)^2+(y-k)^2=r^2$.

2. A(n) ____________________ is the set of all points in a plane that lie a fixed distance from a fixed point.

3. When a plane intersects an infinite cone at different angles, the figures formed by the intersections are called ____________________.

4. ____________________ are fixed points used to determine the points that form a parabola, and ellipse, or a hyperbola.

5. A(n) ____________________ is the set of all points in a plane the sum of whose distances from two fixed points is constant.

Objective 1 Find an equation of a circle given the center and radius.

Write an equation for the circle with the given center and radius.

1. Center $(2,-3)$; radius 5 **1.**________________

2. Center $(1, 4)$; radius 2 **2.**________________

Name: Date:
Instructor: Section:

3. Center $(0, 5)$; radius 3 **3.**________________

4. Center $(6, 2)$; radius 3 **4.**________________

5. Center $(-5, 4)$; radius 4 **5.**________________

6. Center $(7, 1)$; radius 2 **6.**________________

7. Center $(3, -4)$; radius 5 **7.**________________

Objective 2 Determine the center and radius of a circle given its equation.

Find the center and radius of the circle.

8. $x^2 + y^2 - 4x + 8y + 11 = 0$ **8.**________________

9. $x^2 + y^2 - 2x - 6y - 15 = 0$ **9.**________________

10. $x^2 + y^2 - 6x + 10y = 30$ **10.**________________

11. $x^2 + y^2 - 4x - 2y = 31$ **11.**________________

Name: Date:
Instructor: Section:

12. $x^2 + y^2 + 4x + 6y - 3 = 0$

12.______________

13. $x^2 + y^2 - 10x + 12y + 52 = 0$

13.______________

14. $x^2 + y^2 - 8x - 2y + 15 = 0$

14.______________

15. $x^2 + y^2 + 3x + 2y - 2 = 0$

15.______________

Objectives 3 & 4 Recognize an equation of an ellipse. Graph ellipses.

Graph the ellipse.

16. $\frac{x^2}{9} + \frac{y^2}{25} = 1$

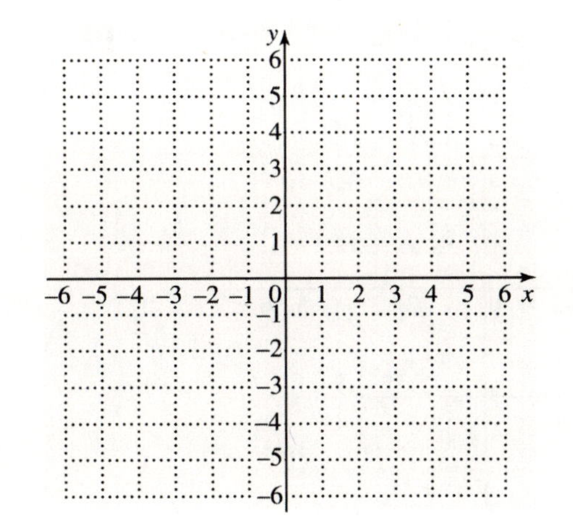

16.______________

Name: Date:
Instructor: Section:

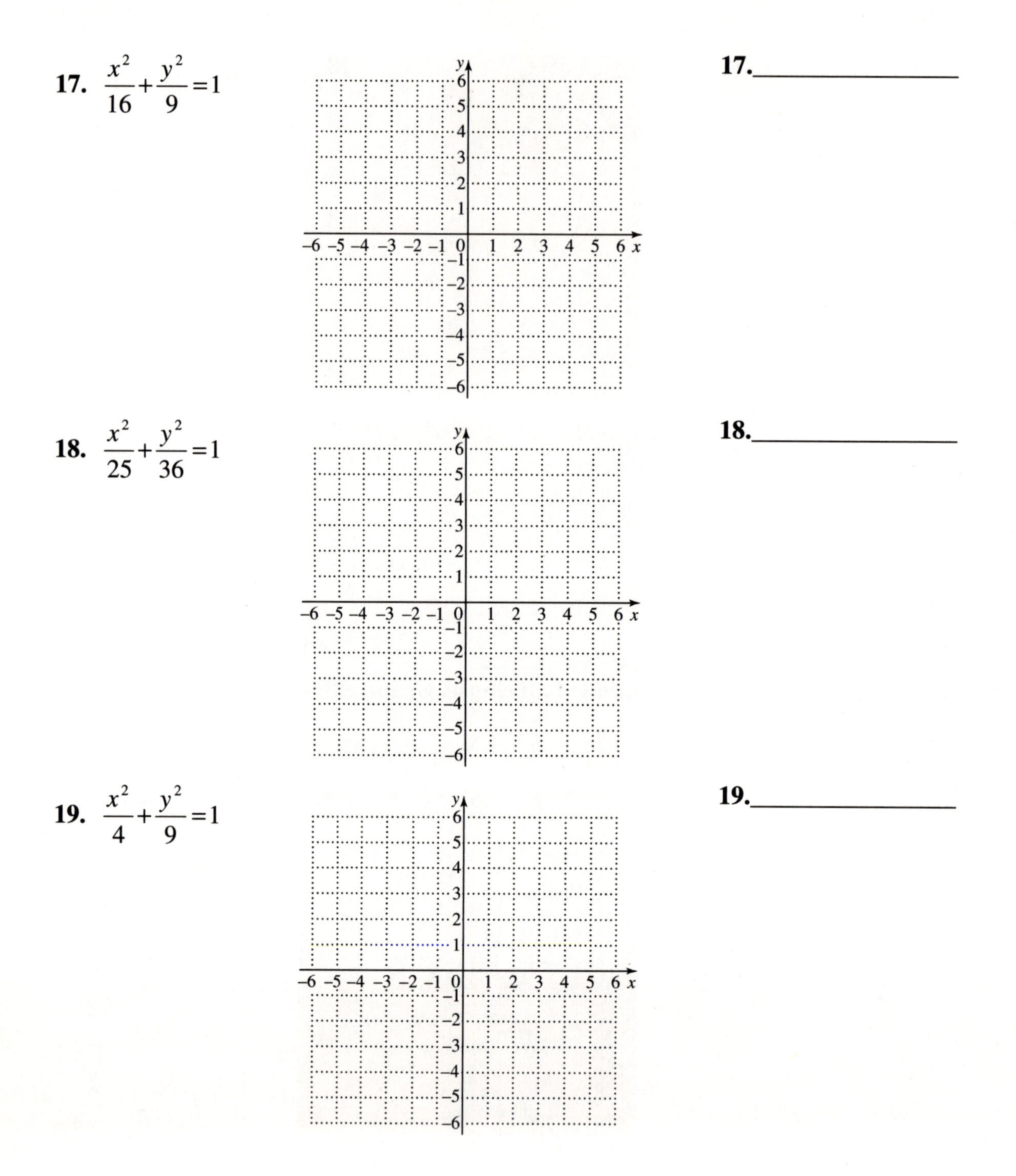

17. $\dfrac{x^2}{16}+\dfrac{y^2}{9}=1$

17.________________

18. $\dfrac{x^2}{25}+\dfrac{y^2}{36}=1$

18.________________

19. $\dfrac{x^2}{4}+\dfrac{y^2}{9}=1$

19.________________

Name: Date:
Instructor: Section:

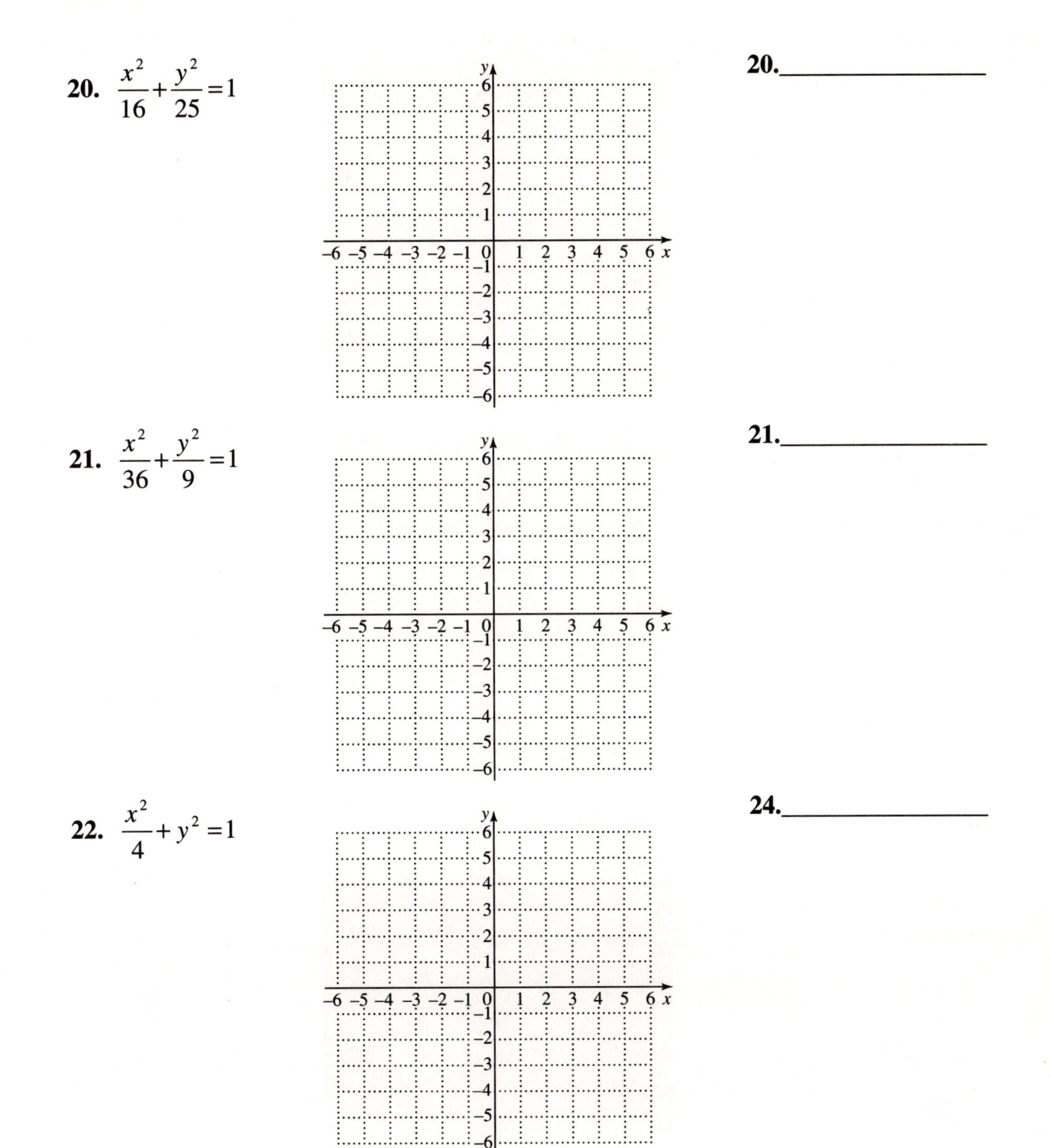

20. $\dfrac{x^2}{16}+\dfrac{y^2}{25}=1$

20.________________

21. $\dfrac{x^2}{36}+\dfrac{y^2}{9}=1$

21.________________

22. $\dfrac{x^2}{4}+y^2=1$

24.________________

Name: Date:
Instructor: Section:

Objective 5 Graph circles and ellipses using a graphing calculator.

For the equation, find the two functions that could be used to graph the conic section on a graphing calculator.

23. $x^2 + y^2 = 17$ **23.**________________

24. $(x+2.9)^2 + y^2 = 52$ **24.**________________

25. $\frac{x^2}{79} + \frac{y^2}{92} = 1$ **25.**________________

26. $\frac{(x+1.5)^2}{5} + \frac{y^2}{20} = 1$ **26.**________________

Use a graphing calculator to graph the indicated conic section.

27. Exercise 23 **27.**________________

28. Exercise 24 **28.**________________

29. Exercise 25 **29.**________________

30. Exercise 26 **30.**________________

Name: Date:
Instructor: Section:

Chapter 13 NONLINEAR FUNCTIONS, CONIC SECTIONS, AND NONLINEAR SYSTEMS

13.3 The Hyperbola and Functions Defined by Radicals

Learning Objectives
1 Recognize the equation of a hyperbola.
2 Graph hyperbolas by using asymptotes.
3 Identify conic sections by their equations.
4 Graph certain square root functions.

Key Terms

Use the vocabulary terms listed below to complete each statement in exercises 1-3.

hyperbola **asymptotes (of a hyperbola)** **fundamental rectangle**

1. The asymptotes of a hyperbola are the extended diagonals of its ____________________, with corners at the points (a, b), $(-a,b)$, $(-a,-b)$, and $(a,-b)$.

2. A(n) ____________________ is the set of all points in a plane such that the absolute value of the difference of the distances from two fixed points is constant.

3. The extended diagonals of the rectangle with vertices at the points (a, b), $(-a,b)$, $(-a,-b)$, and $(a,-b)$ are the ____________________.

Objectives 1 & 2 Recognize the equation of a hyperbola. Graph hyperbolas by using asymptotes.

Graph the hyperbola by using the asymptotes.

1. $\dfrac{x^2}{9} - \dfrac{y^2}{16} = 1$

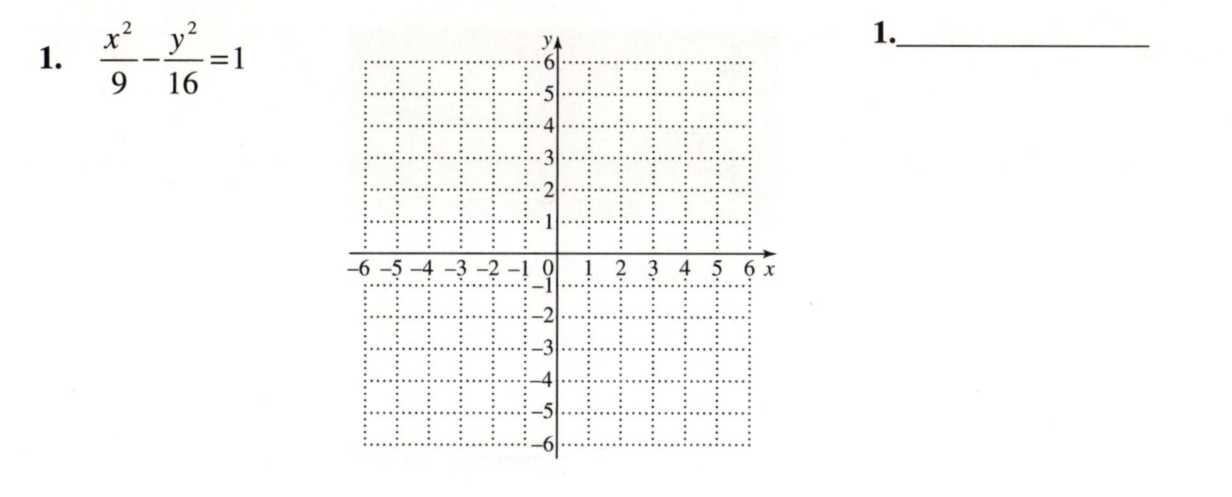

1.____________________

Name: Date:
Instructor: Section:

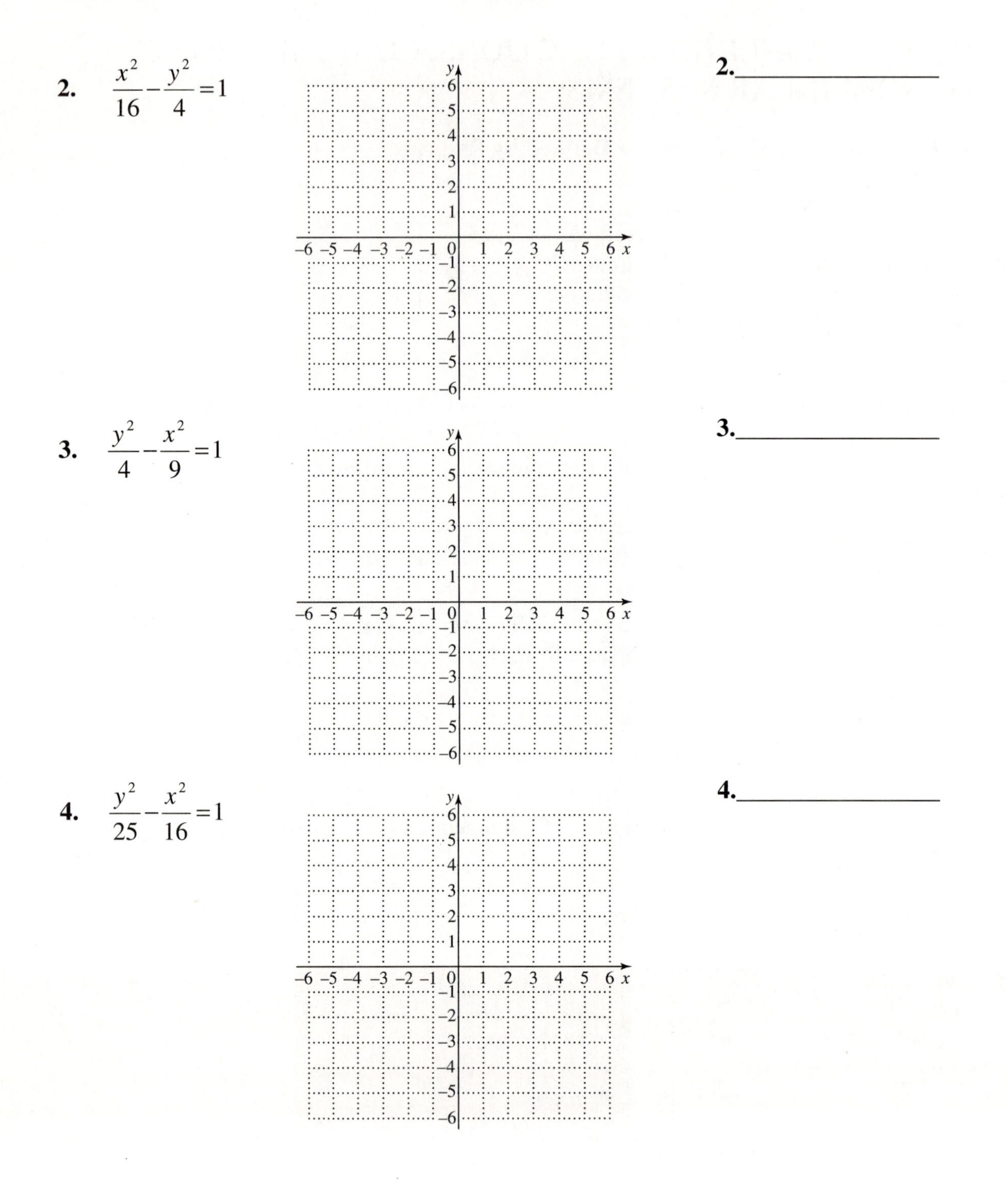

2. $\dfrac{x^2}{16}-\dfrac{y^2}{4}=1$

2.__________________

3. $\dfrac{y^2}{4}-\dfrac{x^2}{9}=1$

3.__________________

4. $\dfrac{y^2}{25}-\dfrac{x^2}{16}=1$

4.__________________

Name: Date:
Instructor: Section:

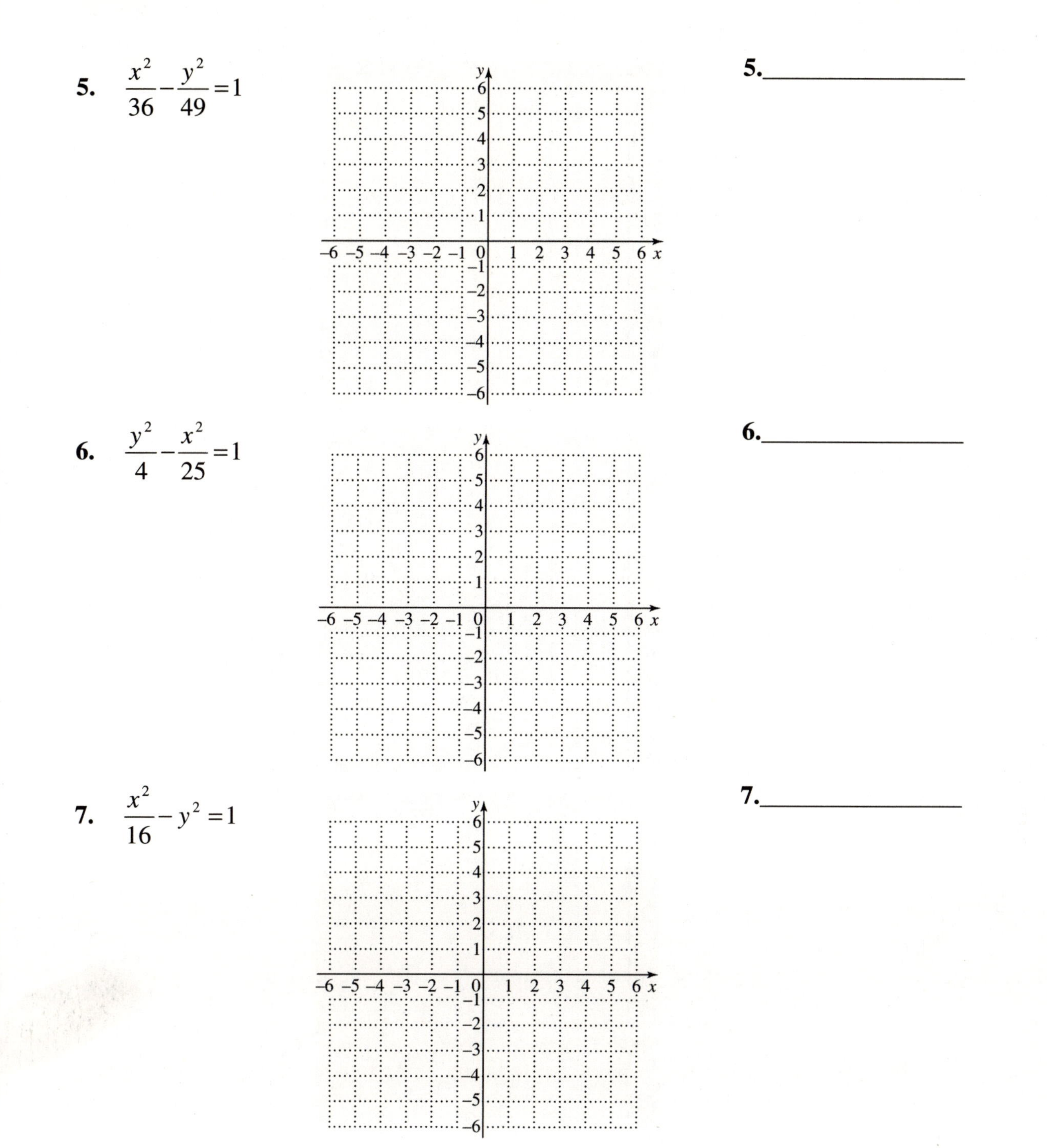

5. $\dfrac{x^2}{36}-\dfrac{y^2}{49}=1$

5.________________

6. $\dfrac{y^2}{4}-\dfrac{x^2}{25}=1$

6.________________

7. $\dfrac{x^2}{16}-y^2=1$

7.________________

Name: Date:
Instructor: Section:

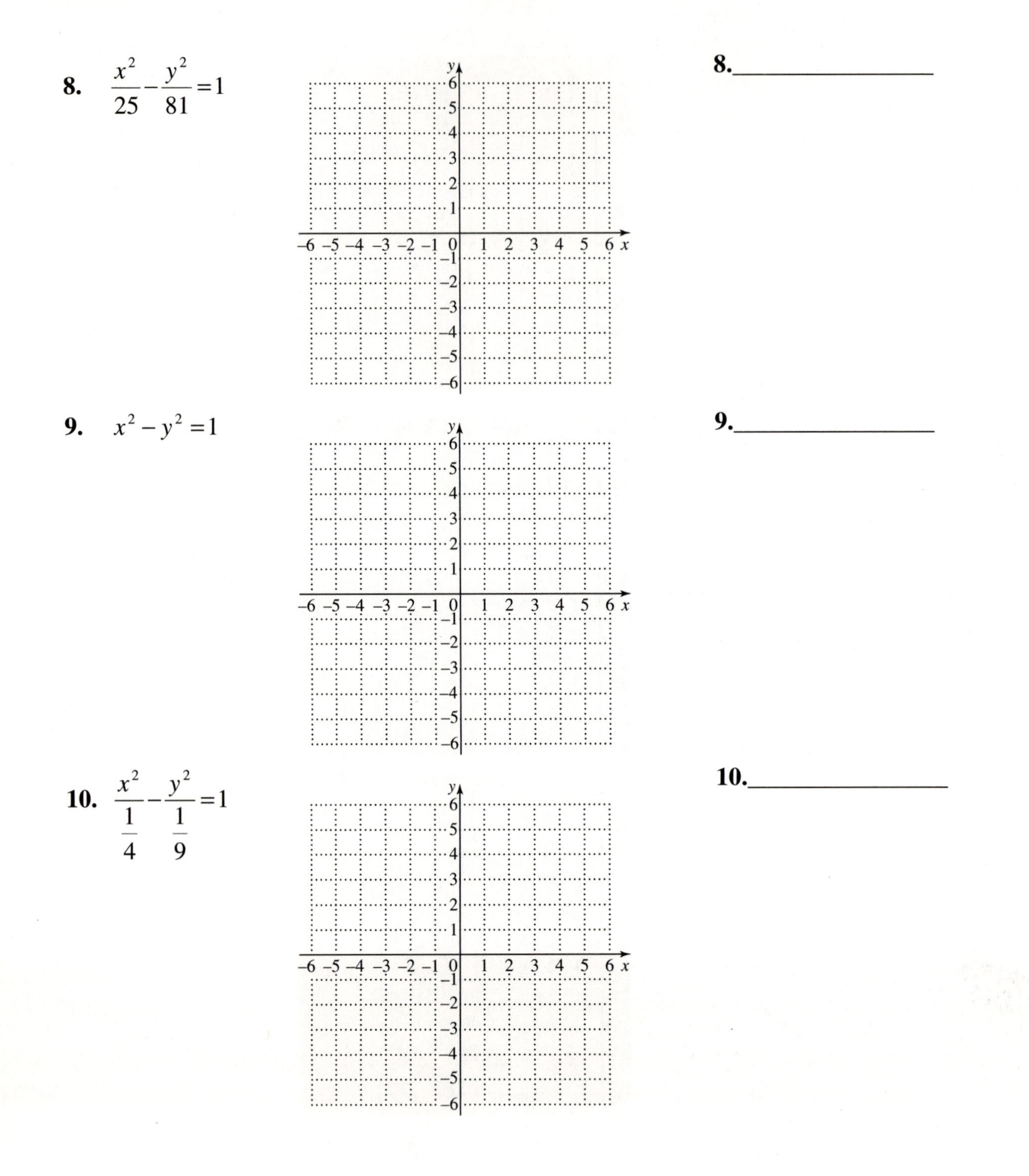

8. $\dfrac{x^2}{25} - \dfrac{y^2}{81} = 1$

8.________________

9. $x^2 - y^2 = 1$

9.________________

10. $\dfrac{x^2}{\frac{1}{4}} - \dfrac{y^2}{\frac{1}{9}} = 1$

10.________________

Name: Date:
Instructor: Section:

Objective 3 Identify conic sections by their equations.

Identify the graph of the equation as a parabola, a circle, an ellipse, or a hyperbola.

11. $x^2 = y^2 + 9$ **11.**________________

12. $2x^2 + y^2 = 16$ **12.**________________

13. $2x + y^2 = 16$ **13.**________________

14. $4x^2 - 9y^2 = 36$ **14.**________________

15. $25y^2 + 100 = 4x^2$ **15.**________________

16. $16x^2 + 9y = 144$ **16.**________________

17. $16x^2 + 16y^2 = 64$ **17.**________________

18. $2x^2 + 2y^2 = 8$ **18.**________________

19. $5x^2 = 25 - 5y^2$ **19.**________________

20. $x^2 = 25 - 5y^2$ **20.**________________

Name: Date:
Instructor: Section:

Objective 4 Graph certain square root functions.

Graph the square root function.

21. $f(x)=\sqrt{9-x^2}$

y
6
5
4
3
2
1
−6 −5 −4 −3 −2 −1 0 1 2 3 4 5 6 x
−1
−2
−3
−4
−5
−6

21.________________

22. $f(x)=\sqrt{25-x^2}$

y
6
5
4
3
2
1
−6 −5 −4 −3 −2 −1 0 1 2 3 4 5 6 x
−1
−2
−3
−4
−5
−6

22.________________

23. $f(x)=-\sqrt{16-x^2}$

y
6
5
4
3
2
1
−6 −5 −4 −3 −2 −1 0 1 2 3 4 5 6 x
−1
−2
−3
−4
−5
−6

23.________________

Name: Date:
Instructor: Section:

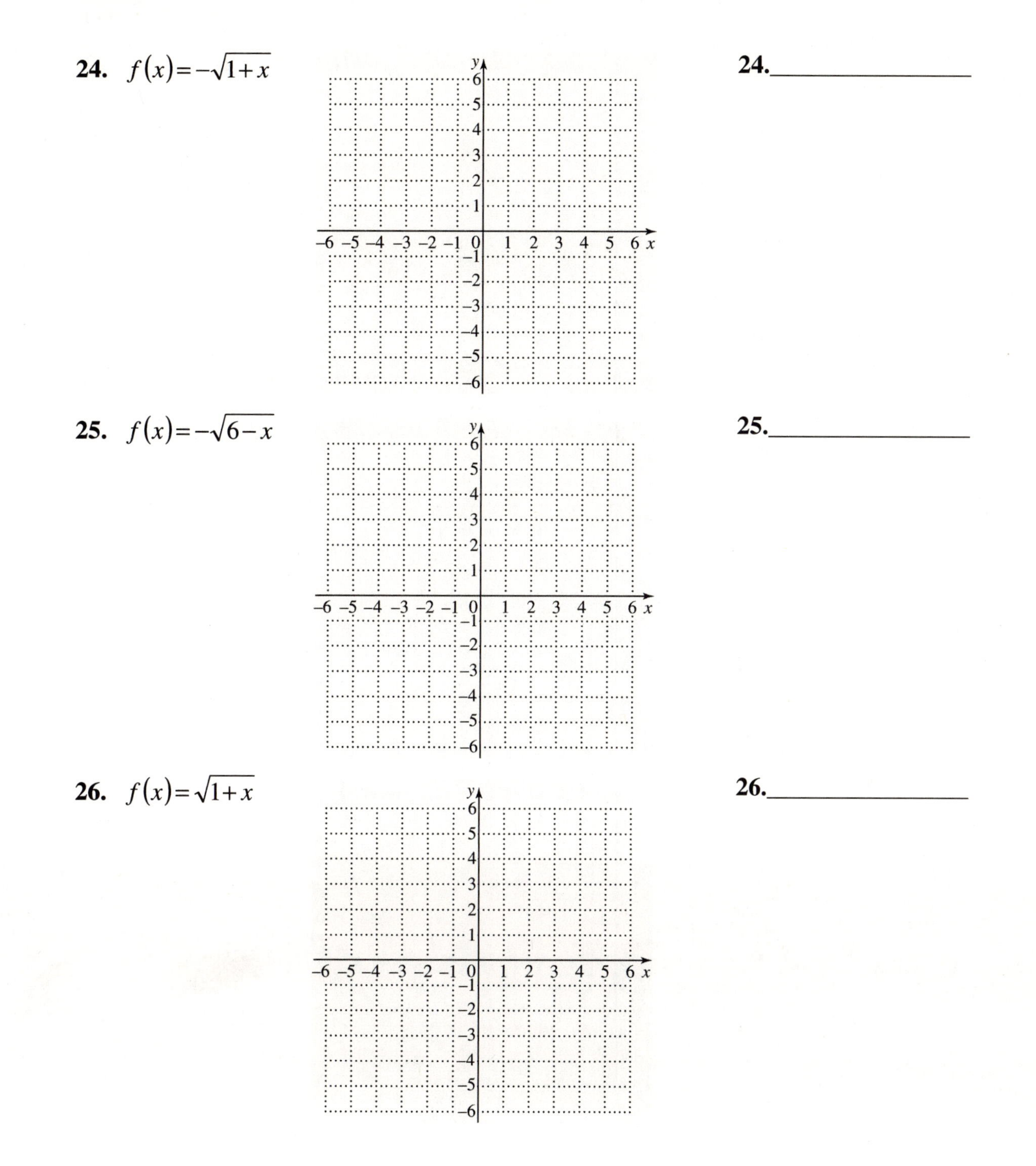

24. $f(x) = -\sqrt{1+x}$

24.________________

25. $f(x) = -\sqrt{6-x}$

25.________________

26. $f(x) = \sqrt{1+x}$

26.________________

Name: Date:
Instructor: Section:

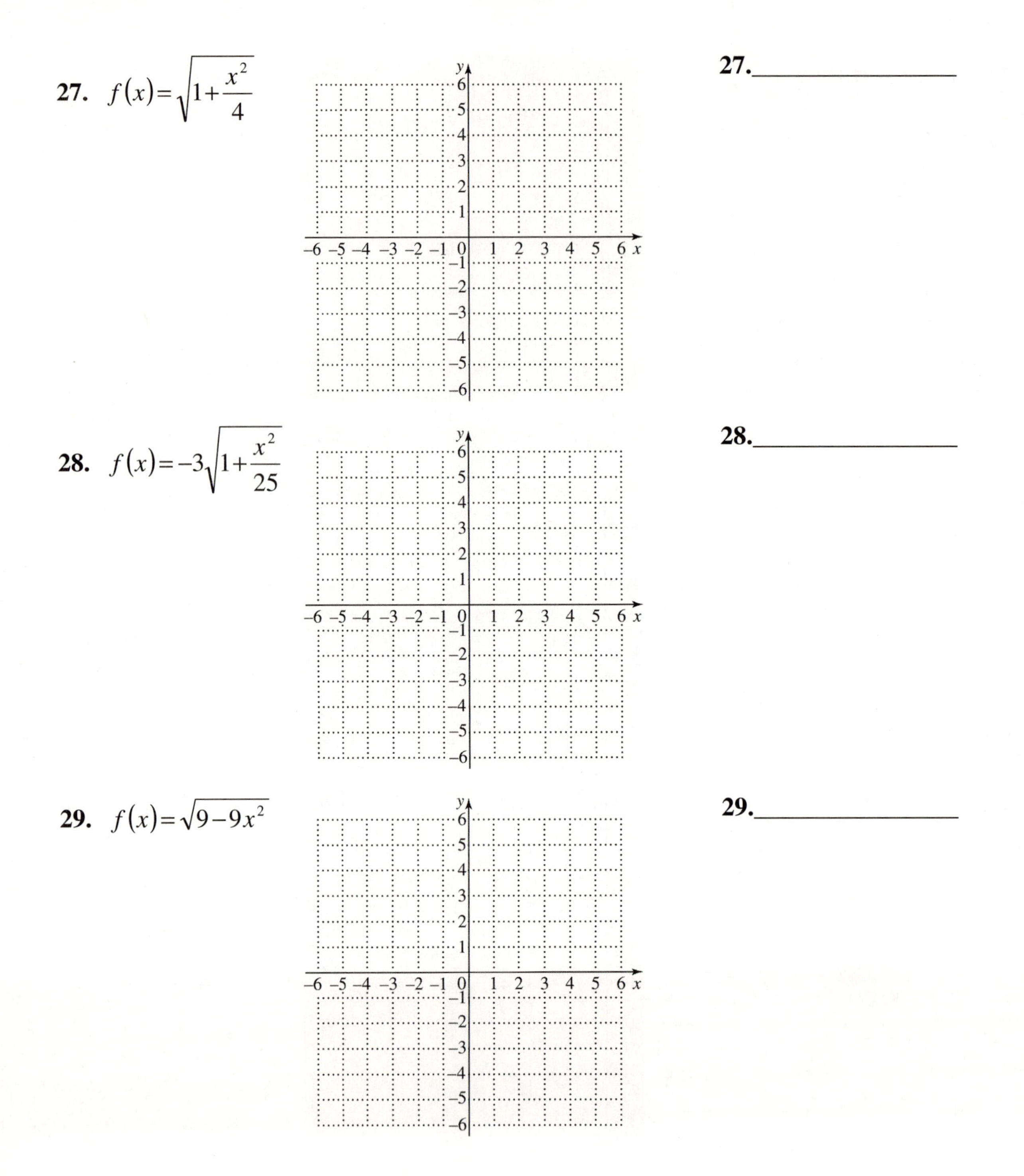

27.________________

28.________________

29.________________

Name: Date:
Instructor: Section:

30. $f(x) = -5\sqrt{1 - \dfrac{x^2}{9}}$

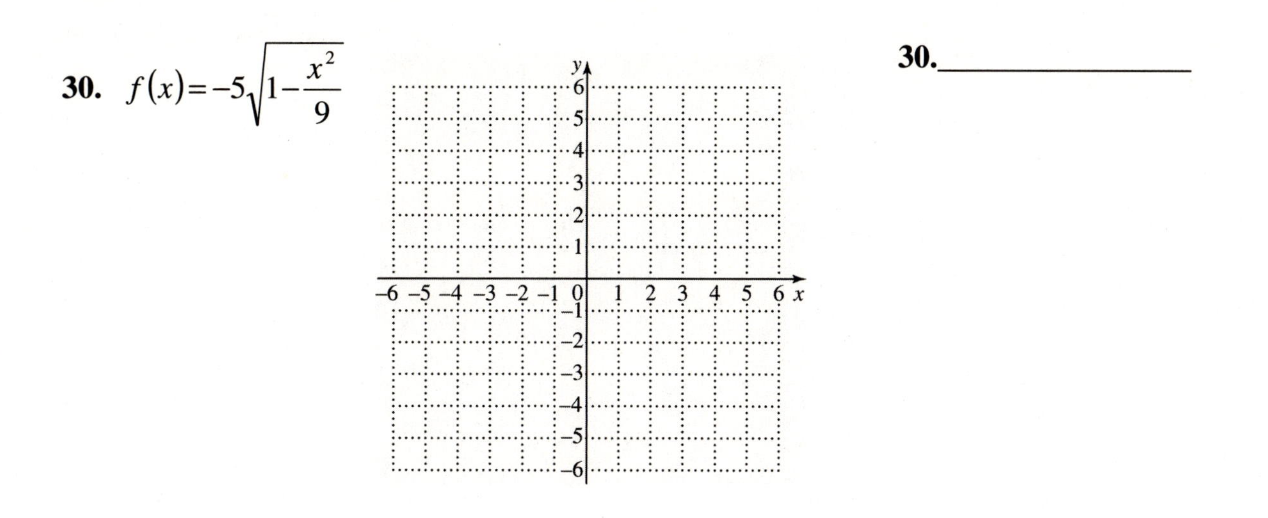

30.________________

Name: Date:
Instructor: Section:

Chapter 13 NONLINEAR FUNCTIONS, CONIC SECTIONS, AND NONLINEAR SYSTEMS

13.4 Nonlinear Systems of Equations

Learning Objectives

1. Solve a nonlinear system by substitution.
2. Use the elimination method to solve a system with two second-degree equations.
3. Solve a system that requires a combination of methods.
4. Use a graphing calculator to solve a nonlinear system.

Key Terms

Use the vocabulary terms listed below to complete each statement in exercises 1-4.

nonlinear system of equations **substitution method** **elimination method**

check all solutions

1. When solving nonlinear systems, it is possible to obtain extraneous solutions. Therefore, it is very important to ____________________.

2. When solving a nonlinear system of equations, the ____________________ is often used if one of the equations has a variable to the first power.

3. When solving a nonlinear system of equations, the ____________________ is often used if both equations are second degree.

4. A ____________________ is a system that includes at least one nonlinear equation.

Objective 1 Solve a nonlinear system by substitution.

Solve the system by the substitution method.

1. $x^2 + y^2 = 17$
$2x = y + 9$

1.____________________

2. $2x^2 - y^2 = -1$
$2x + y = 7$

2.____________________

Name: Date:
Instructor: Section:

3. $4x^2+3y^2 = 7$
$2x \;-5y = -7$

3.________________

4. $x^2 = 2y^2+2$
$y = 3x+7$

4.________________

5. $y = x^2-3x-8$
$x = y+3$

5.________________

6. $x = y^2+3y$
$5y = x$

6.________________

7. $xy = -6$
$x+y = 1$

7.________________

8. $xy = 24$
$y = 2x+2$

8.________________

Objective 2 Use the elimination method to solve a system with two second-degree equations.

Solve the system by the elimination method.

9. $x^2+y^2 = 10$
$2x^2-y^2 = -7$

9.________________

Name: Date:
Instructor: Section:

10. $2x^2 + y^2 = 54$
$x^2 - 3y^2 = 13$

10.________________

11. $2x^2 - 3y^2 = -19$
$4x^2 + y^2 = 25$

11.________________

12. $5x^2 - y^2 = 55$
$2x^2 + y^2 = 57$

12.________________

13. $x^2 + 2y^2 = 11$
$2x^2 - y^2 = 17$

13.________________

14. $3x^2 + 2y^2 = 30$
$2x^2 + y^2 = 17$

14.________________

15. $5x^2 + y^2 = 6$
$2x^2 - 3y^2 = -1$

15.________________

16. $4x^2 - 3y^2 = -8$
$2x^2 + y^2 = 5$

16.________________

Name: Date:
Instructor: Section:

Objective 3 Solve a system that requires a combination of methods.

Solve the system.

17. $x^2 + xy + y^2 = 43$
$x^2 + 2xy + y^2 = 49$

17.________________

18. $x^2 + xy - y^2 = 5$
$-x^2 + 3xy + y^2 = 3$

18.________________

19. $x^2 + 2xy + 3y^2 = 6$
$x^2 + 4xy + 3y^2 = 8$

19.________________

20. $4x^2 - 2xy + 4y^2 = 64$
$x^2 + y^2 = 13$

20.________________

21. $3x^2 - 4xy + 2y^2 = 59$
$-3x^2 + 5xy - 2y^2 = -65$

21.________________

22. $x^2 + 3xy + 2y^2 = 12$
$-x^2 + 8xy - 2y^2 = 10$

22.________________

23. $x^2 + 5xy - y^2 = 20$
$x^2 - 2xy - y^2 = -8$

23.________________

Name: Date:
Instructor: Section:

Objective 4 Use a graphing calculator to solve a nonlinear system.

(Round answers to the nearest hundredth.)

24. $4x^2 - 6y^2 = 24$
$2x^2 + 8y^2 = 32$

24.________________

25. $9x^2 + 16y^2 = 144$
$x^2 - y^2 = 4$

25.________________

26. $1.05x^2 - .93y^2 = 3.97$
$.99x^2 + 1.12y^2 = 4.03$

26.________________

27. $y = .97x^2 + 1.09$
$1.1x + .92y = 3.05$

27.________________

28. $1.99x^2 + 1.01y^2 = 6.02$
$2.1x - .97y = 0$

28.________________

29. $4.1x^2 + 2.9y^2 = 12.85$
$y = 1.73x$

29.________________

30. $x^2 + y^2 = 15.95$
$x^2 + y^2 = 6.09$

30.________________

Name: Date:
Instructor: Section:

Chapter 13 NONLINEAR FUNCTIONS, CONIC SECTIONS, AND NONLINEAR SYSTEMS

13.5 Second-Degree Inequalities and Systems of Inequalities

Learning Objectives
1 Graph second-degree inequalities.
2 Graph the solution set of a system of inequalities.

Key Terms
Use the vocabulary terms listed below to complete each statement in exercises 1-2.

second-degree inequality **system of inequalities**

1. A ____________________ is an inequality with at least one variable of degree 2 and no variable with degree greater than 2.

2. A ____________________ consists of two or more inequalities to be solved at the same time.

Objective 1 Graph second-degree inequalities.

Graph the nonlinear inequality.

1. $x \geq y^2$

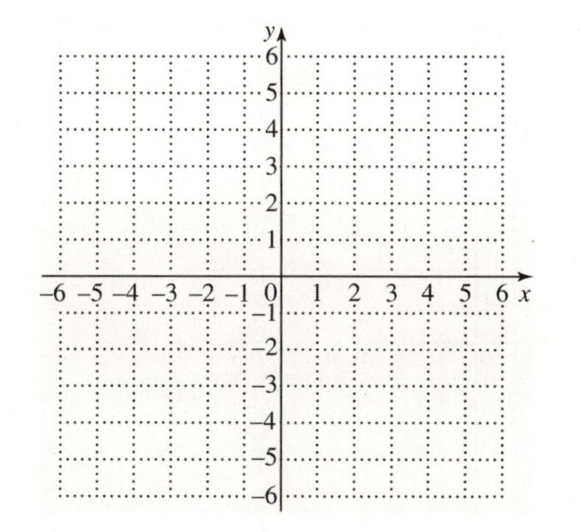

1.________________

Name: Date:
Instructor: Section:

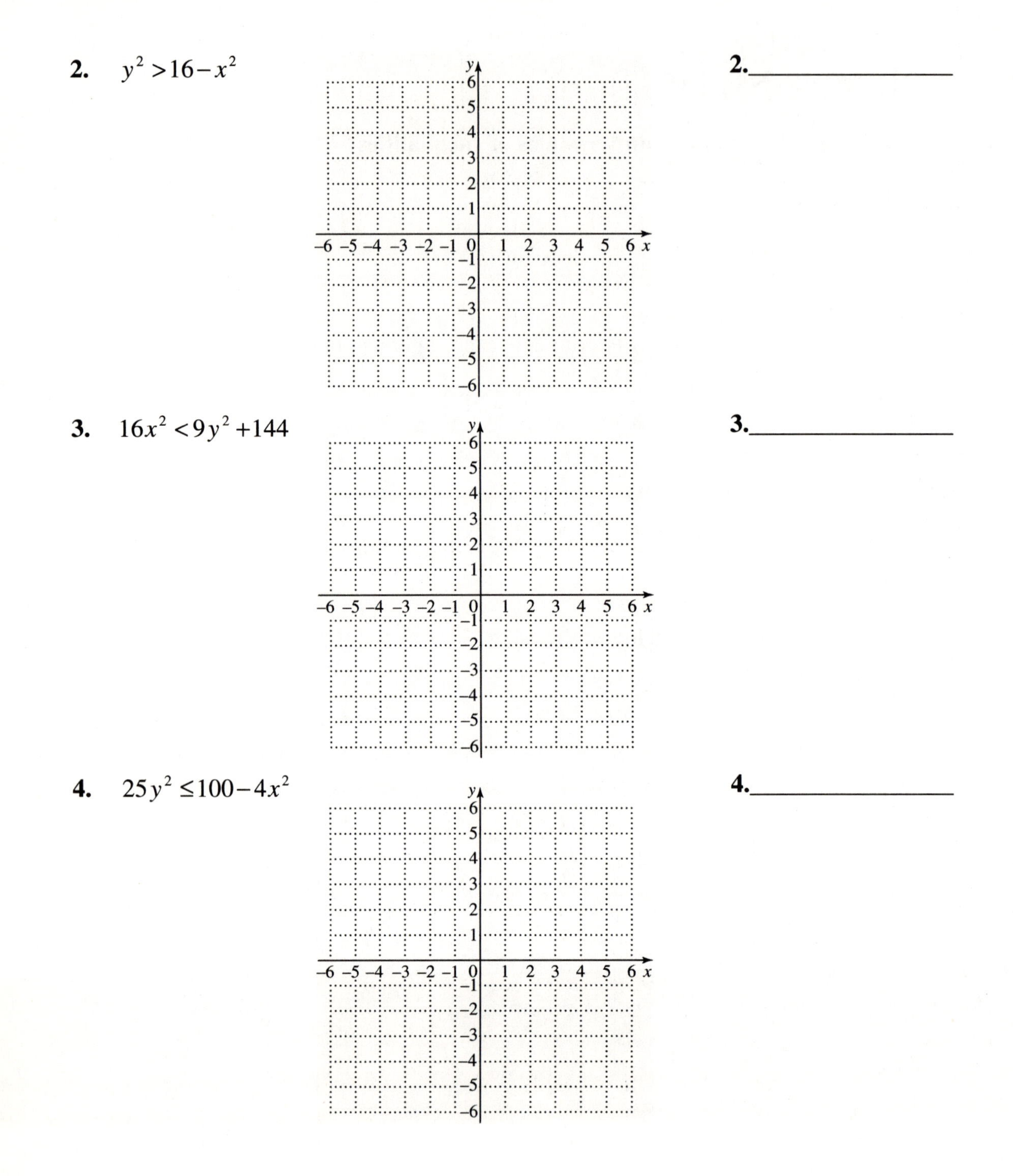

2. $y^2 > 16 - x^2$ **2.**________________

3. $16x^2 < 9y^2 + 144$ **3.**________________

4. $25y^2 \leq 100 - 4x^2$ **4.**________________

Name: Date:
Instructor: Section:

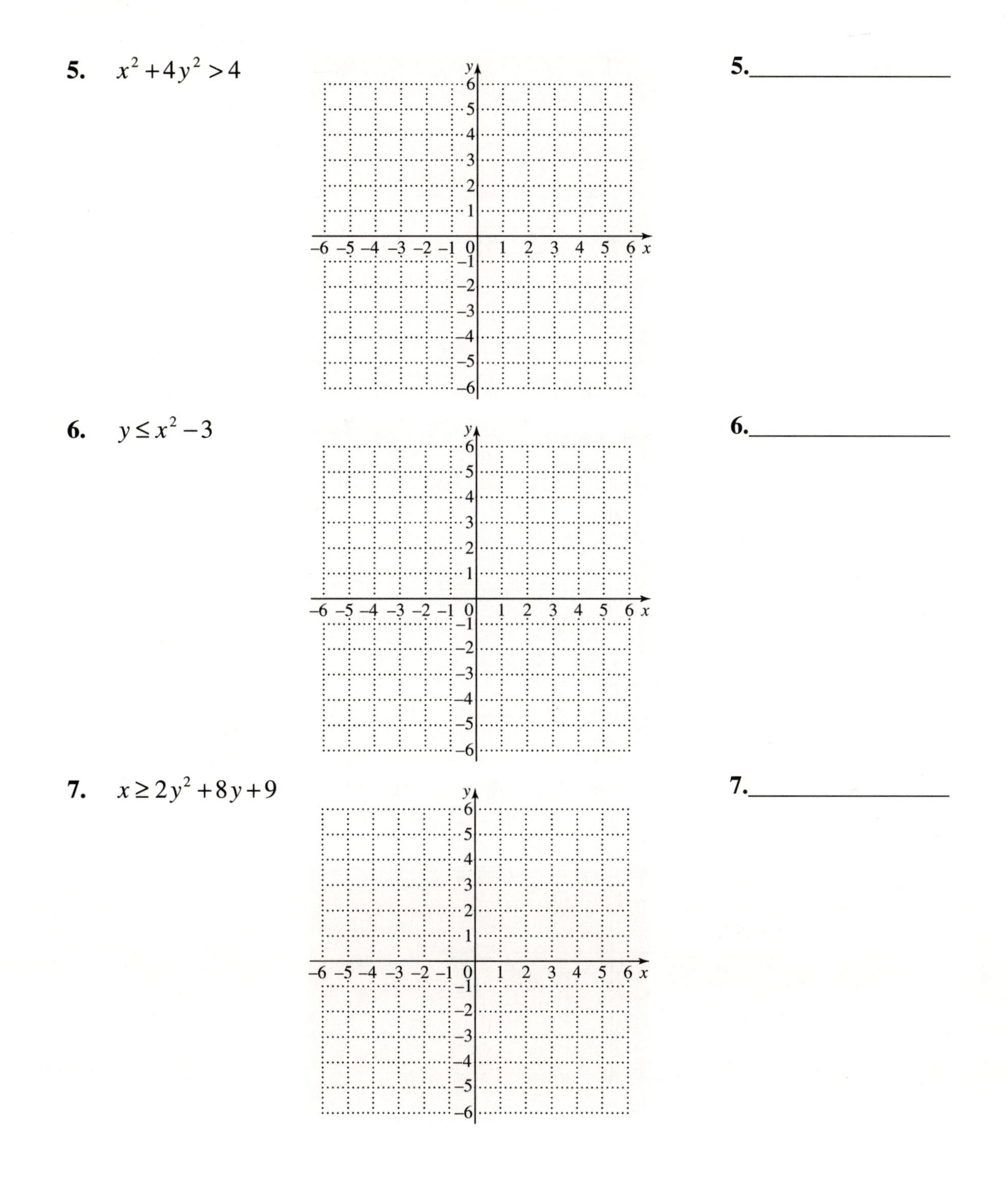

5. $x^2 + 4y^2 > 4$

5.________________

6. $y \le x^2 - 3$

6.________________

7. $x \ge 2y^2 + 8y + 9$

7.________________

Name: Date:
Instructor: Section:

8. $4y^2 \geq 196 + 49x^2$ **8.**________________

9. $x^2 + y^2 \leq 9$ **9.**________________

10. $x^2 + y^2 \geq 16$ **10.**________________

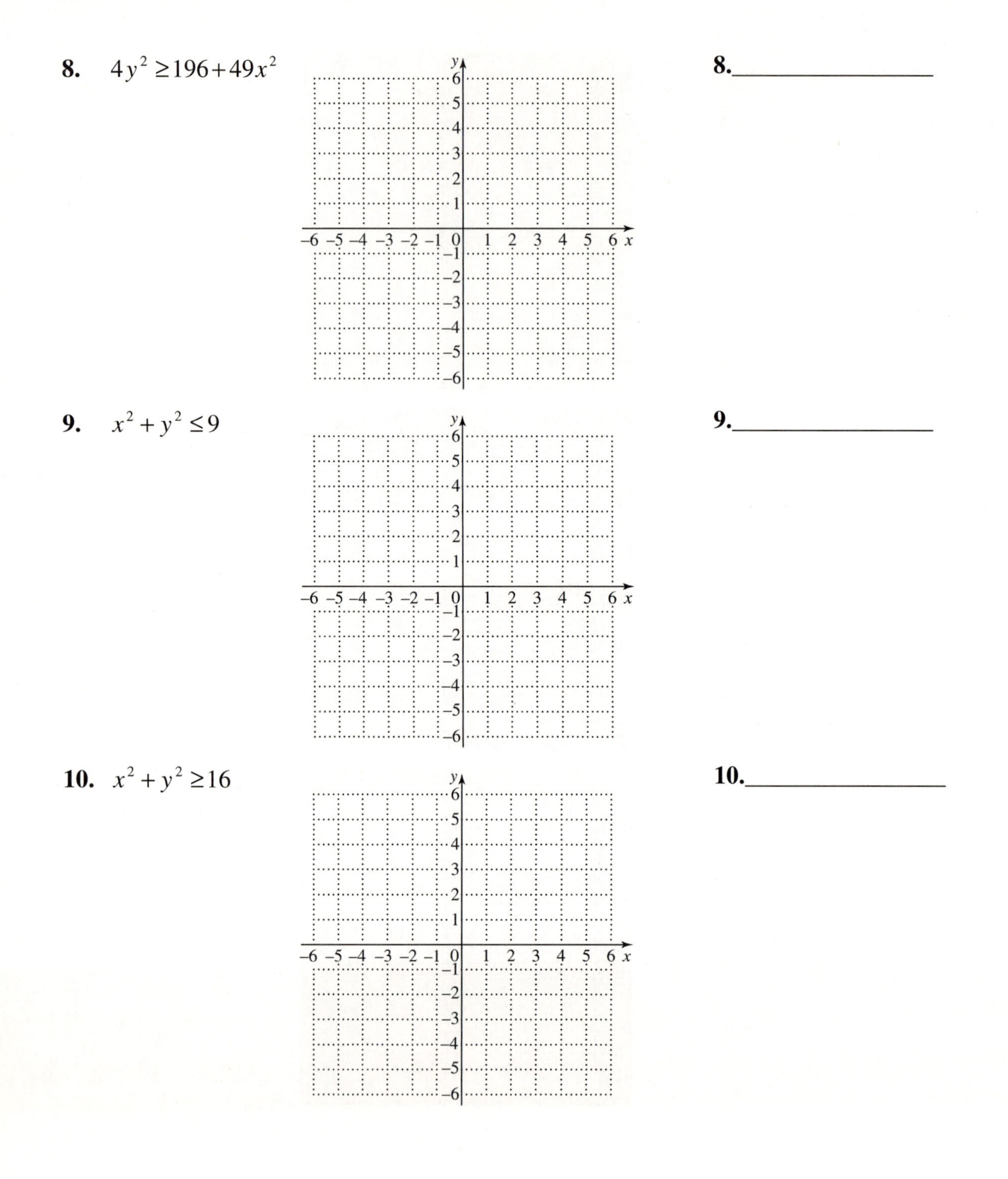

Name: Date:
Instructor: Section:

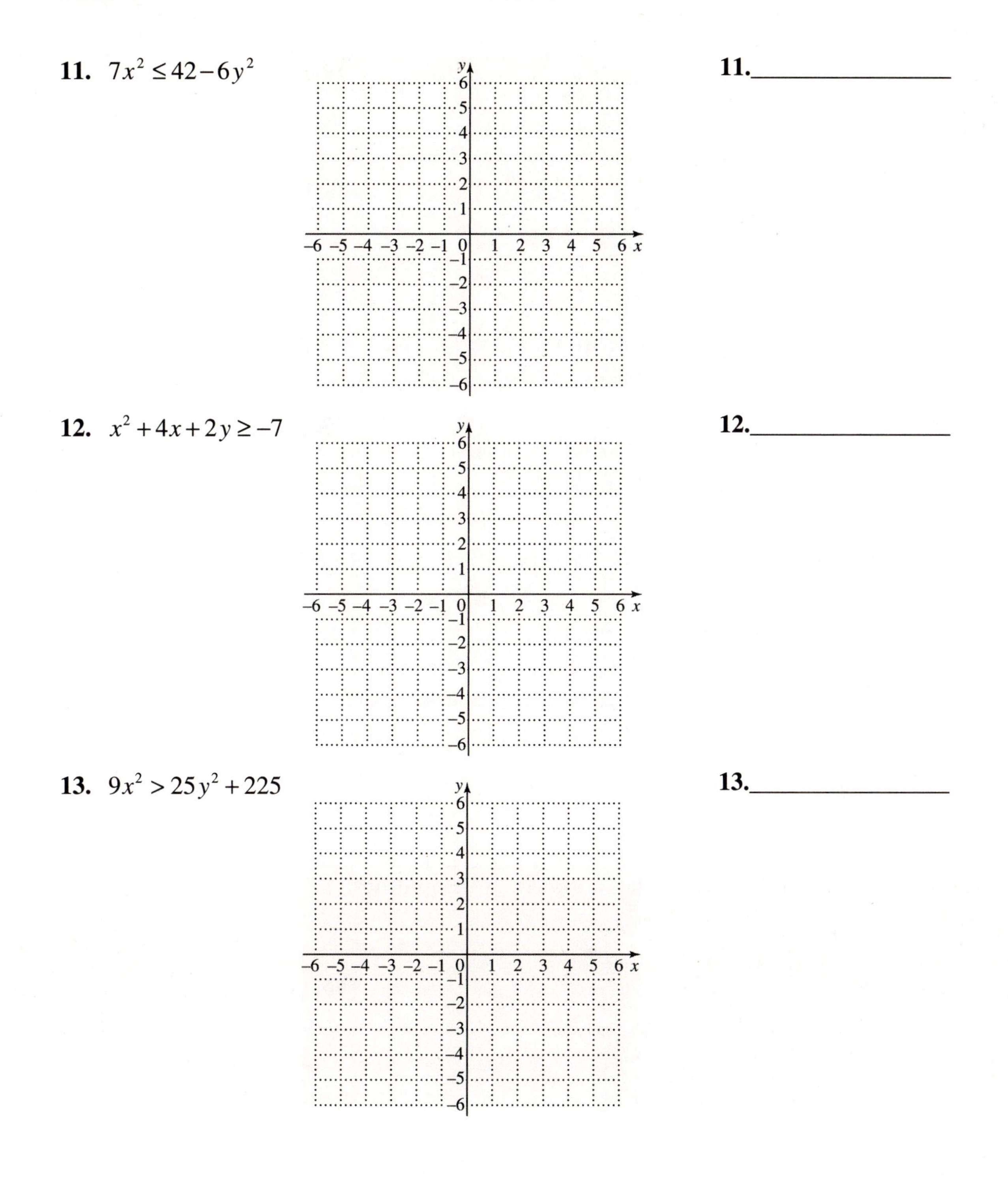

11. $7x^2 \le 42 - 6y^2$

11.________________

12. $x^2 + 4x + 2y \ge -7$

12.________________

13. $9x^2 > 25y^2 + 225$

13.________________

Name: Date:
Instructor: Section:

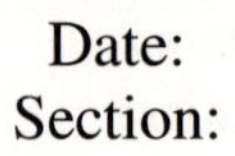

14. $y^2 > 25 - x^2$

14.________________

15. $5x^2 + 7y^2 < 35$

15.________________

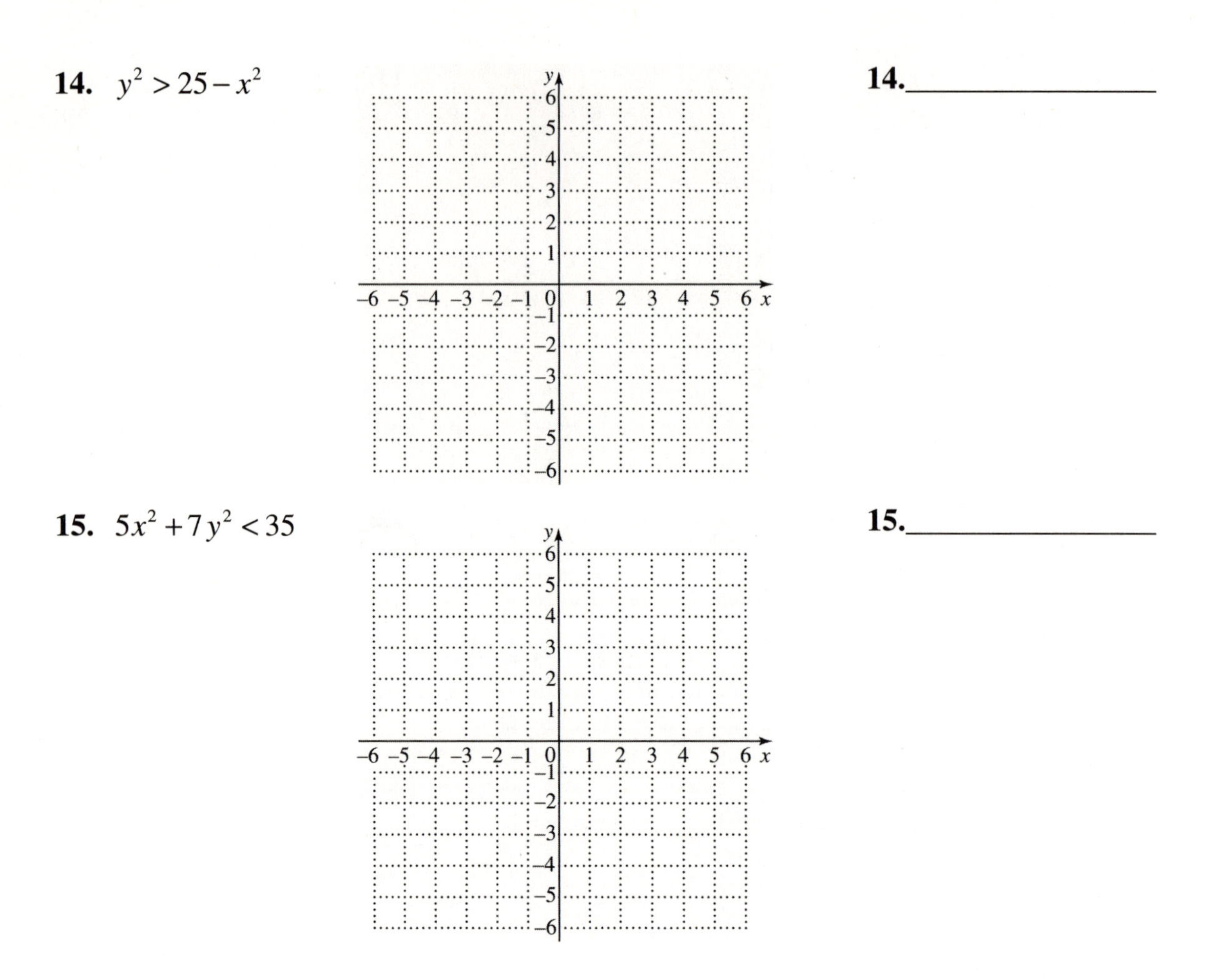

Objective 2 Graph the solution set of a system of inequalities.

Graph the system of inequalities.

16. $-x + y > 2$

$3x + y > 6$

16.________________

Name: Date:
Instructor: Section:

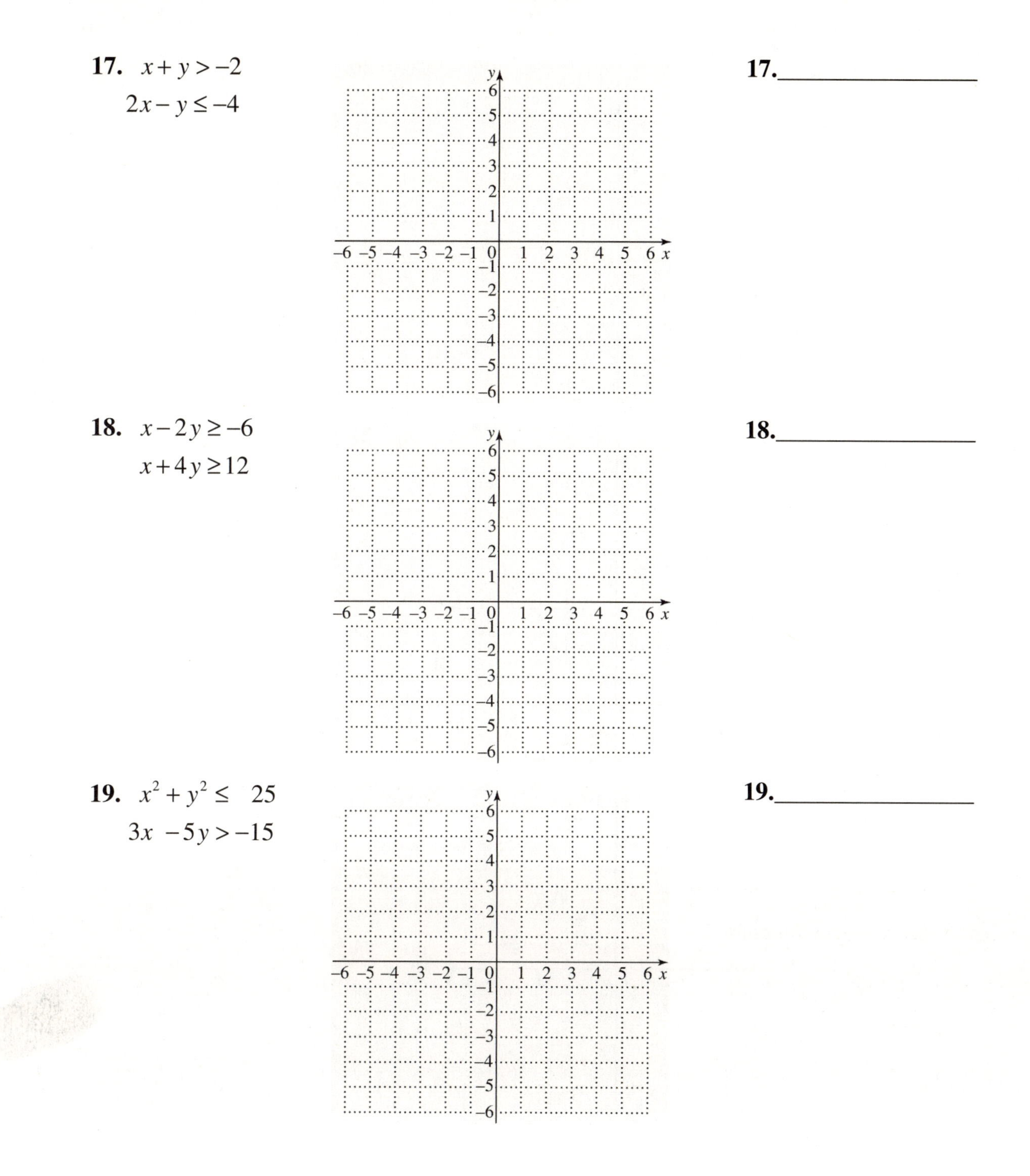

17. $x + y > -2$

$2x - y \le -4$

17. ________________

18. $x - 2y \ge -6$

$x + 4y \ge 12$

18. ________________

19. $x^2 + y^2 \le 25$

$3x - 5y > -15$

19. ________________

Name: Date:
Instructor: Section:

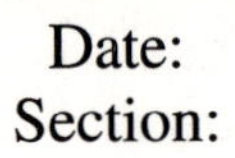

20. $9x^2 + 16y^2 < 144$
$y^2 - \quad x^2 > 4$

20.________________

21. $x^2 + y^2 < 9$
$y < x^2 - 3$

21.________________

22. $x^2 + y^2 \le 16$
$y \le x$

22.________________

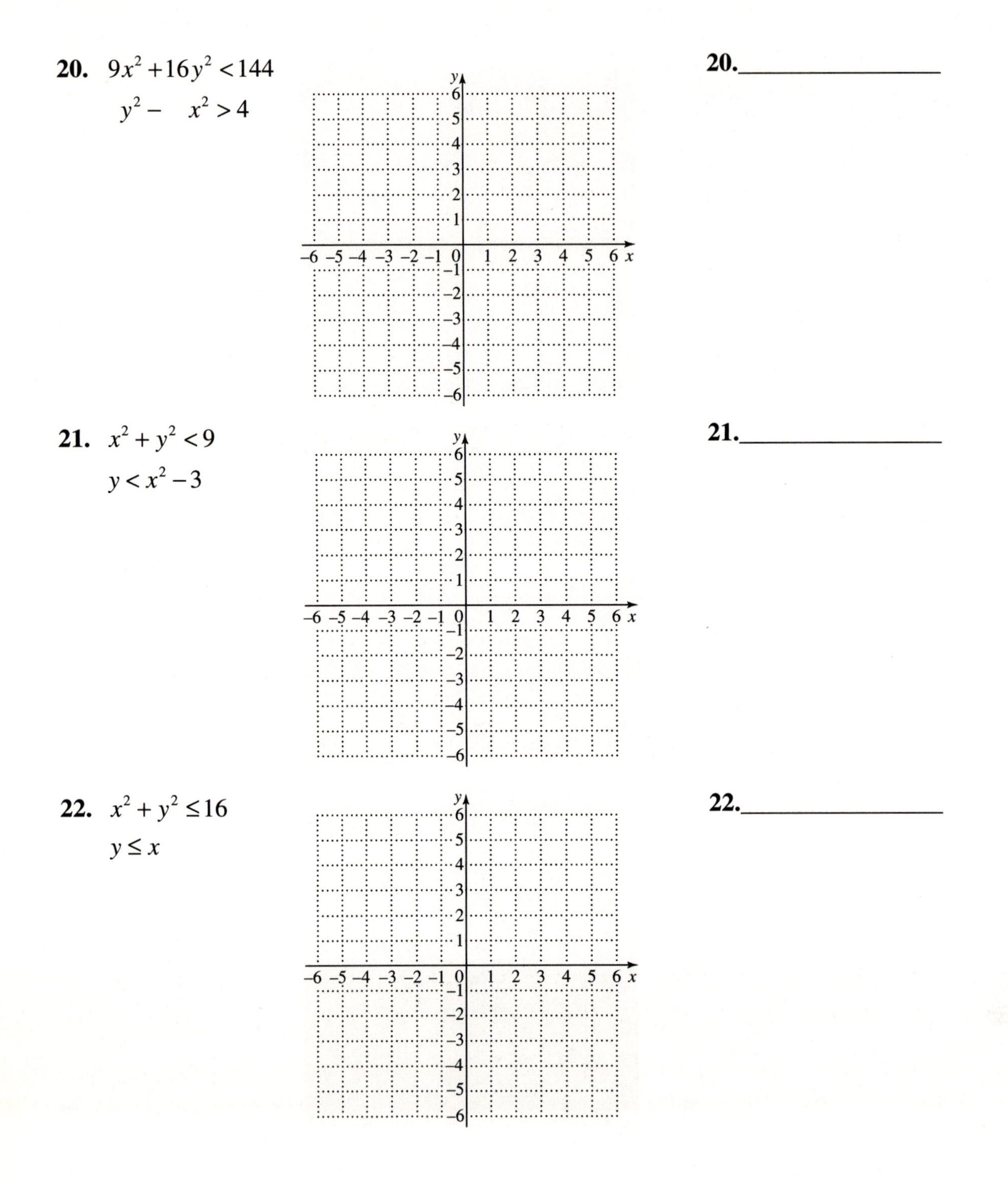

Name: Date:
Instructor: Section:

23. $x^2 + y^2 \le 25$
$x^2 + y^2 \ge 4$

23.________________

24. $4x^2 + 16y^2 \le 64$
$16x^2 + 4y^2 \ge 64$

24.________________

25. $y \le x^2 - 4x + 4$
$y \ge x^2 - 4x + 2$

25.________________

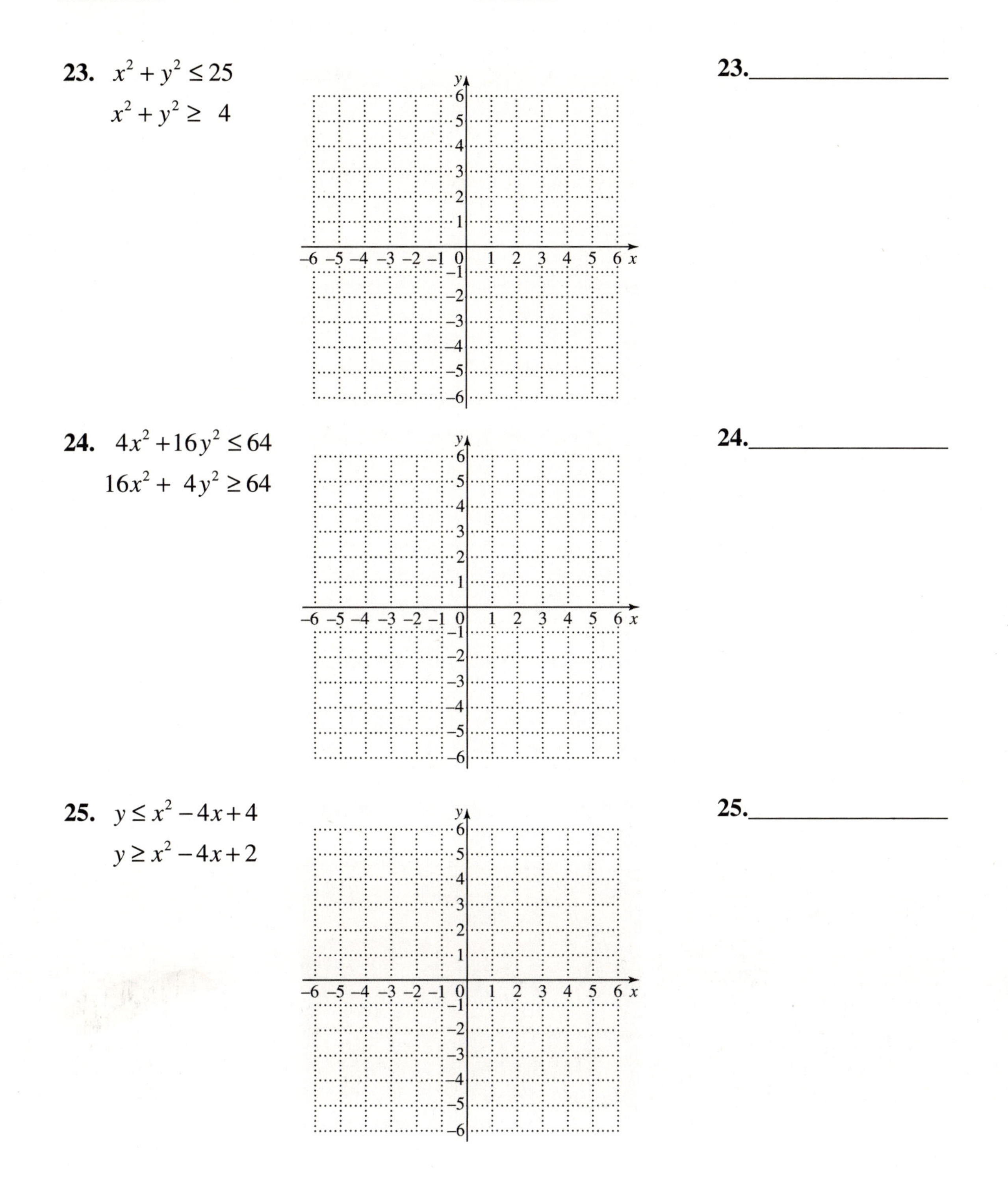

Name: Date:
Instructor: Section:

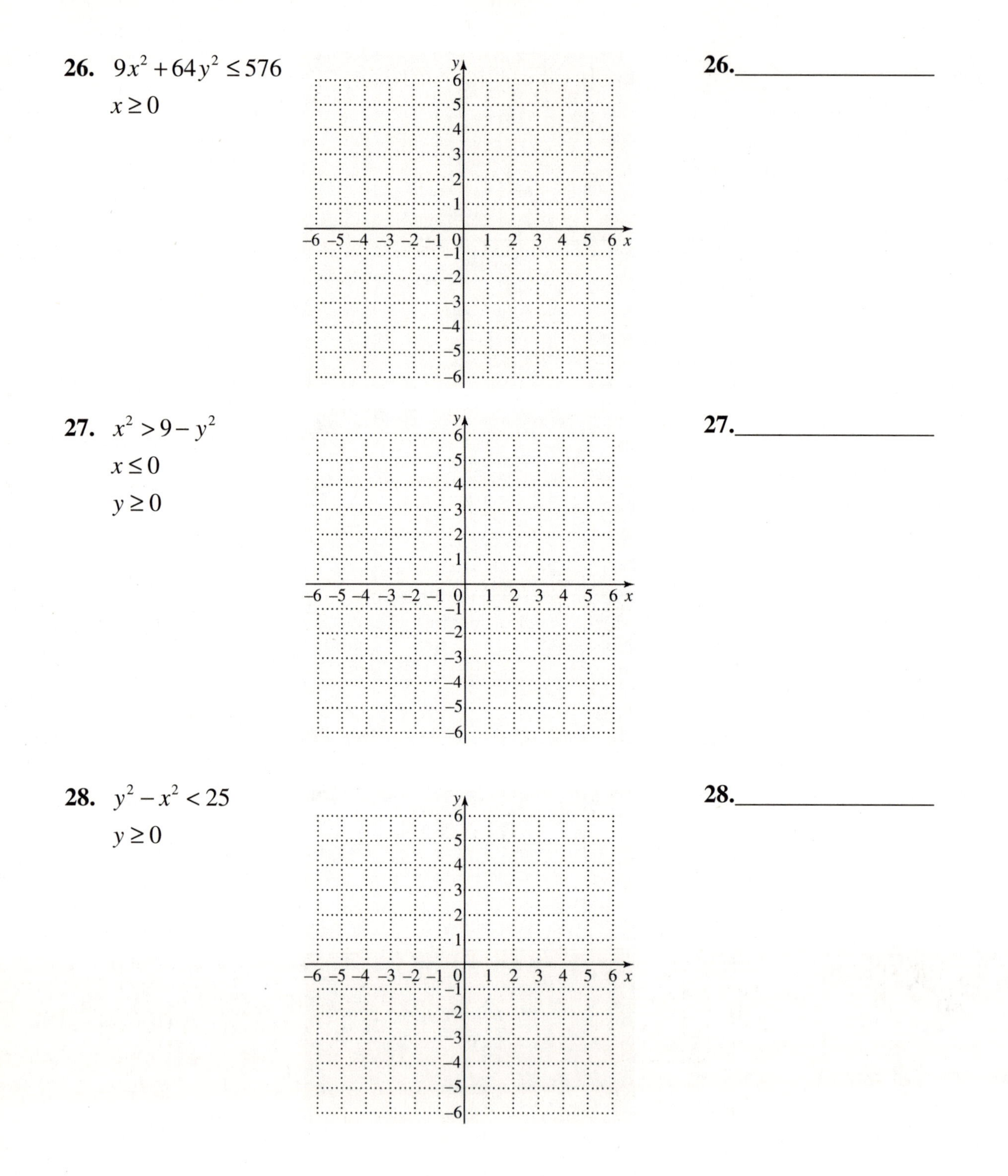

26. $9x^2 + 64y^2 \leq 576$
$x \geq 0$

26.________________

27. $x^2 > 9 - y^2$
$x \leq 0$
$y \geq 0$

27.________________

28. $y^2 - x^2 < 25$
$y \geq 0$

28.________________

Name: Date:
Instructor: Section:

29. $x^2 + 4y^2 \le 36$

$-5 < x < 2$

$y \ge 0$

29.________________

30. $y \ge x^2 + 2x + 1$

$4x - 5y \ge -16$

30.________________

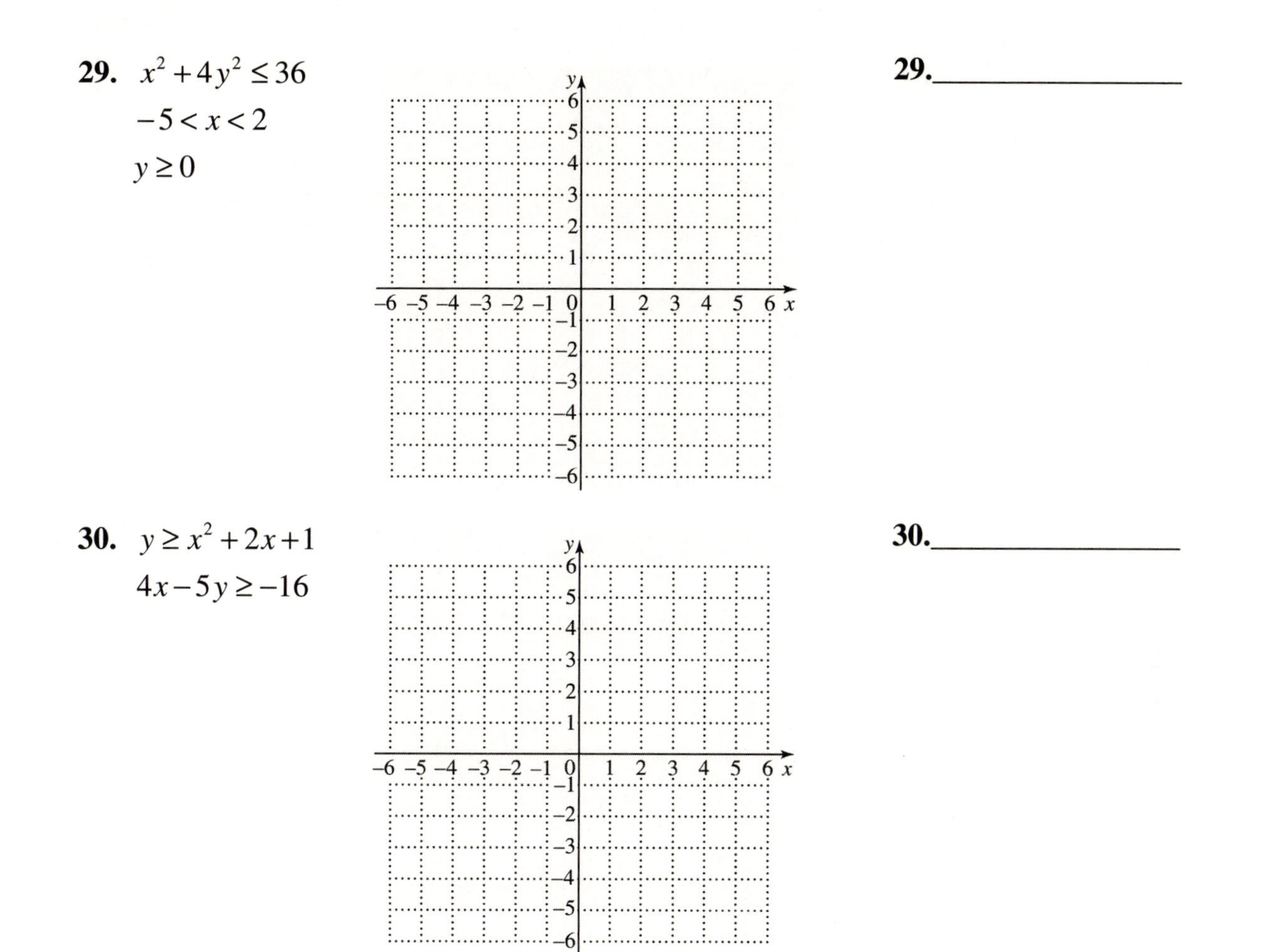

Name: Date:
Instructor: Section:

Chapter 14 SEQUENCES AND SERIES

14.1 Sequences and Series

Learning Objectives
1 Find the terms of a sequence, given the general term.
2 Find the general term of a sequence.
3 Use sequences to solve applied problems.
4 Use summation notation to evaluate a series.
5 Write a series with summation notation.
6 Find the arithmetic mean (average) of a group of numbers.

Key Terms
Use the vocabulary terms listed below to complete each statement in exercises 1-6.

general term **finite sequence** **series** **summation notation**

index of summation **arithmetic mean**

1. ____________________ is a compact way of writing a series using the general term of the corresponding sequence.

2. The expression a_n which defines a sequence, is called the ____________________ of the sequence.

3. The indicated sum of the terms of a sequence is called a ____________________.

4. A(n) ____________________ has a domain that includes only the first n positive integers.

5. The ____________________ of a group of numbers is the sum of all the numbers divided by the number of numbers. Using sigma notation, this value is given by $\frac{\sum_{i}^{n} x_i}{n}$, where x_i represents the individual numbers and n represents the number of numbers.

6. When using summation notation $\sum_{i}^{n} f(i)$, the letter i is called the ____________________.

Name: Date:
Instructor: Section:

Objective 1 Find the terms of a sequence, given the general term.

Write the first five terms of the sequence.

1. $a_n = 3n + 1$ **1.**________________

2. $a_n = \frac{1+n}{n}$ **2.**________________

3. $a_n = 2^n$ **3.**________________

4. $a_n = (-1)^n$ **4.**________________

Find the indicated term for the sequence.

5. $a_n = -2n;\ a_6$ **5.**________________

6. $a_n = \frac{3n-2}{5n+2};\ a_8$ **6.**________________

Objective 2 Find the general term of a sequence.

Find an expression for the general term, a_n, for the sequence.

7. 1, 2, 3, 4, 5, … **7.**________________

8. 4, 8, 12, 16, 20, … **8.**________________

Name: Date:
Instructor: Section:

9. 3, 5, 7, 9, 11, ... **9.**________________

10. $1, \frac{1}{2}, \frac{1}{3}, \frac{1}{4}, \frac{1}{5}, \ldots$ **10.**________________

11. $\sqrt{3}, 3, 3\sqrt{3}, 9, 9\sqrt{3}, \ldots$ **11.**________________

12. $-\frac{1}{5}, \frac{1}{10}, -\frac{1}{15}, \frac{1}{20}, -\frac{1}{25}, \ldots$ **12.**________________

Objective 3 Use sequences to solve applied problems.

Solve the problem.

13. A colony of bacteria doubles in weight every hour. If the colony weighs 2 grams at the beginning of an experiment, find the weight after 3 hours. **13.**________________

14. Ms. Burley is offered a new job with a salary of $20,000 per year and a $500 raise at the end of each year. Write a sequence showing her salary for each of the first five years. **14.**________________

15. A package of supplies is dropped to an isolated work site. The package falls 12 meters during the first second, 24 meters during the second, 36 meters during the third, and so on. What is the general term of this sequence? **15.**________________

Name: Date:
Instructor: Section:

Objective 4 Use summation notation to evaluate a series.

Write out the series and evaluate.

16. $\sum_{i=1}^{4}(2i+3)$

16.________________

17. $\sum_{i=1}^{7}(4i-3)$

17.________________

18. $\sum_{i=1}^{7}2i$

18.________________

19. $\sum_{i=1}^{4}\frac{-1}{2i}$

19.________________

20. $\sum_{i=2}^{6}(i+3)(i-4)$

20.________________

Objective 5 Write a series with summation notation.

Write in summation notation.

21. $2+3+4+5+6$

21.________________

22. $1+\frac{1}{2}+\frac{1}{3}+\frac{1}{4}+\frac{1}{5}+\frac{1}{6}$

22.________________

23. $1+4+9+16+25$

23.________________

Name: Date:
Instructor: Section:

24. $\frac{1}{5}+\frac{1}{8}+\frac{1}{11}+\frac{1}{14}+\frac{1}{17}$

24.________________

25. $1+2x+3x^2+4x^3+5x^4$

25.________________

Objective 6 Find the arithmetic mean (average) of a group of numbers.

Find the arithmetic mean for the collection of numbers. Round the answer to the nearest tenth.

26. 5, 7, 9, 11, 13

26.________________

27. 1, 2, 3, 4, 5, 6, 7, 8, 9, 10

27.________________

28. 2, 8, 18, 32, $-5, -7$

28.________________

29. $\frac{1}{3}, -\frac{1}{2}, \frac{1}{4}, \frac{7}{6}$

29.________________

30. 5, 7, 6, -6, 8, 4, $-9, -3$

30.________________

Name: Date:
Instructor: Section:

Chapter 14 SEQUENCES AND SERIES

14.2 Arithmetic Sequences

Learning Objectives

1. Find the common difference of an arithmetic sequence.
2. Find the general term of an arithmetic sequence.
3. Use an arithmetic sequence in an application.
4. Find any specified term or the number of terms of an arithmetic sequence.
5. Find the sum of a specified number of terms of an arithmetic sequence.

Key Terms

Use the vocabulary terms listed below to complete each statement in exercises 1-4.

arithmetic sequence (or arithmetic progression) **common difference**

general term **sum of the first *n* terms**

1. A(n) __________________ is a sequence in which each term after the first differs from the preceding term by a constant amount.

2. The __________________ d is the difference between any tow adjacent terms of an arithmetic sequence.

3. The __________________ of an arithmetic sequence is given by the formula $S_n = \frac{n}{2}(a_1 + a_2)$ or $S_n = \frac{n}{2}[2a_1 + (n-1)d]$.

4. The __________________ of an arithmetic sequence is given by the formula $a_n = a_1 + (n-1)d$.

Objective 1 Find the common difference of an arithmetic sequence.

Find the common difference for the arithmetic sequence.

1. 1, 5, 9, 13, ... **1.**________________

2. $-3, -1, 1, 3, 5, \ldots$ **2.**________________

3. $-9, -5, -1, 3, 7, \ldots$ **3.**________________

Name: Date:
Instructor: Section:

4. $\frac{1}{2}, \frac{3}{2}, \frac{5}{2}, \frac{7}{2}, \ldots$

4.________________

5. $\frac{1}{4}, -\frac{1}{4}, -\frac{3}{4}, -\frac{5}{4}, \ldots$

5.________________

Objective 2 Find the general term of an arithmetic sequence.

Find the general term of the arithmetic sequence.

6. $a_1 = -1,\ d = 2$

6.________________

7. $a_1 = 4,\ d = 3$

7.________________

8. $a_1 = -7,\ d = -5$

8.________________

9. $-8, -5, -2, 1, \ldots$

9.________________

10. $\frac{1}{2}, \frac{5}{6}, \frac{7}{6}, \frac{3}{2}, \frac{11}{6}, \ldots$

10.________________

11. $.8, .5, .2, \ldots, -29.8$

11.________________

Name: Date:
Instructor: Section:

Objective 3 Use an arithmetic sequence in an application.

Solve the problem.

12. Ben Ouellette's father has started a savings fund for Ben's college education. He makes an initial deposit of $1,000 and each month contributes an additional $50. How much money will be in the account after 60 months? (Disregard any interest.)

12.________________

13. Lisa Daer is starting a weight training program. She plans to lift 100 pounds per day the first week and add 25 pounds per day each week until she is lifting 275 pounds. In which week will this occur?

13.________________

Objective 4 Find any specified term or the number of terms of an arithmetic sequence.

Write the first five terms of the arithmetic sequence.

14. $a_1 = 2, \ d = 1$

14.________________

15. $a_1 = -5, \ d = -2$

15.________________

16. $a_1 = 10, \ d = 2$

16.________________

17. $a_1 = -1, \ d = 6$

17.________________

Name: Date:
Instructor: Section:

Find the number of terms in the arithmetic sequence.

18. 2, 5, 8, ..., 53 **18.**________________

19. $-2, -4, -6, \ldots, -124$ **19.**________________

Find the indicated term of the arithmetic sequence.

20. $a_1 = -5,\ d = 9;\ a_{12}$ **20.**________________

21. $a_1 = 4,\ d = -3;\ a_{10}$ **21.**________________

22. $a_1 = -\frac{1}{2},\ d = \frac{3}{2};\ a_{15}$ **22.**________________

Objective 5 Find the sum of a specified number of terms of an arithmetic sequence.

Find the sum of the first ten terms, S_{10}, of the arithmetic sequence.

23. $a_n = 3n - 5$ **23.**________________

24. $a_n = 2 - 3n$ **24.**________________

Name: Date:
Instructor: Section:

Find the sum of the arithmetic sequence.

25. $\sum_{i=1}^{12}(i+3)$ **25.**________________

26. $\sum_{i=1}^{20}(1-i)$ **26.**________________

27. $\sum_{i=1}^{999}\left(\frac{i+1}{2}\right)$ **27.**________________

28. $\sum_{i=1}^{10}(2i+3)$ **28.**________________

Find the sum of the first eight terms, S_8, *of the arithmetic sequence.*

29. $a_1 = 3,\ d = 2$ **29.**________________

30. $a_1 = 5,\ d = -7$ **30.**________________

Name: Date:
Instructor: Section:

Chapter 14 SEQUENCES AND SERIES

14.3 Geometric Sequences

Learning Objectives

1. Find the common ratio of a geometric sequence.
2. Find the general term of a geometric sequence.
3. Find any specified term of a geometric sequence.
4. Find the sum of a specified number of terms of a geometric sequence.
5. Apply the formula for the future value of an ordinary annuity.
6. Find the sum of an infinite number of terms of certain geometric sequences.

Key Terms

Use the vocabulary terms listed below to complete each statement in exercises 1-7.

geometric sequence (or geometric progression) **common ratio**

general term **sum of the first *n* terms** **annuity**

future value of an annuity **sum of an infinite geometric sequence**

1. The ____________________ of a geometric sequence is given by $S = \frac{a_1(r^n - 1)}{r - 1}$ where a_1 is the first term and r is the common ratio, with $r \neq 1$.

2. A(n) ____________________ r is the constant multiplier between adjacent terms in a geometric sequence.

3. The ____________________ of a geometric sequence is given by $a_n = a_1 r^{n-1}$ where a_1 is the first term and r is the common ratio.

4. A(n) ____________________ is a sequence in which each term after the first is a constant multiple of the preceding term.

5. A(n) ____________________ is a sequence of equal payments made at equal periods of time.

6. The ____________________ is given by $S = \frac{a_1}{1 - r}$ where a_1 is the first term and r is the common ratio, where $|r| < 1$.

7. The ____________________ is the sum of the compound amounts of all the payments, compounded to the end of the term.

Name: Date:
Instructor: Section:

Objective 1 Find the common ratio of a geometric sequence.

Find the common ratio of the geometric sequence.

1. $2, 4, 8, 16, 32, \ldots$ **1.**________________

2. $4, -8, 16, -32, 64, \ldots$ **2.**________________

3. $10, 10\sqrt{2}, 20, 20\sqrt{2}, 40, \ldots$ **3.**________________

4. $-\frac{1}{2}, \frac{1}{4}, -\frac{1}{8}, \frac{1}{16}, -\frac{1}{32}, \ldots$ **4.**________________

Objective 2 Find the general term of a geometric sequence.

Find the general term of the geometric sequence.

5. $6, 12, 24, 48, 96, \ldots$ **5.**________________

6. $-\frac{2}{3}, \frac{2}{9}, -\frac{2}{27}, \frac{2}{81}, \ldots$ **6.**________________

7. $\sqrt{3}, 3, 3\sqrt{3}, 9, 9\sqrt{3}, 27, \ldots$ **7.**________________

8. $-\frac{4}{5}, -\frac{4}{25}, -\frac{4}{125}, -\frac{4}{625}, \ldots$ **8.**________________

Name: Date:
Instructor: Section:

Objective 3 Find any specified term of a geometric sequence.

Find the indicated term of the geometric sequence.

9. $a_1 = 2,\ r = 3;\ a_6$ 9.________________

10. $a_1 = -4,\ r = 2;\ a_7$ 10.________________

11. $a_1 = \frac{2}{3},\ r = -3;\ a_6$ 11.________________

12. $a_1 = -1,\ r = 3;\ a_7$ 12.________________

Write the first five terms of the geometric sequence.

13. $a_1 = 3,\ r = 2$ 13.________________

14. $a_1 = -7,\ r = -3$ 14.________________

15. $a_1 = \frac{1}{2},\ r = 5$ 15.________________

Objective 4 Find the sum of a specified number of terms of a geometric sequence.

Find the sum of the first six terms of the geometric sequence.

16. $a_1 = 2,\ r = 3$ 16.________________

Name: Date:
Instructor: Section:

17. $a_1 = -5,\ r = \frac{1}{2}$

17.________________

18. $a_1 = \frac{2}{3},\ r = 3$

18.________________

Find the sum of the geometric sequence.

19. 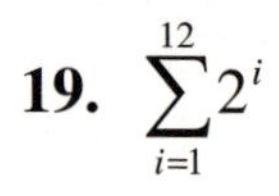 $\sum_{i=1}^{12} 2^i$

19.________________

20. $\sum_{i=1}^{7} 3(-2)^i$

20.________________

21. $\sum_{i=1}^{5} \frac{1}{2}\left(4^i\right)$

21.________________

22. 2, 4, 8, 16

22.________________

Objective 5 Apply the formula for the future value of an ordinary annuity.

Solve the problem.

23. Matthew is a professional wrestler who believes that his wrestling career will last ten years. To prepare for retirement, he deposits $14,000 at the end of each year for ten years into an account paying 7% compounded annually. How much will he have on deposit after ten years?

23.________________

Name: Date:
Instructor: Section:

24. Donald and Marla are millionaires. Donald decides that he wants to retire from the financial world in eight years, so he decides at the end of each year to put $250,000 in an off-shore account paying 16.9% compounded annually for 8 years. How much will Donald and Marla have to live on at the end of eight years?

24.________________

25. Gordon wants to buy a new car. If he puts $1,100 at the end of each year into an account paying 4.9% compounded annually for 3 years, how much will he have to spend on a car?

25.________________

Objective 6 Find the sum of an infinite number of terms of certain geometric sequences.

Find the sum, if possible, of the infinite geometric sequence.

26. $a_1 = 2,\ r = -\frac{1}{3}$

26.________________

27. $a_1 = 8,\ r = \frac{1}{4}$

27.________________

Find the sum, if possible, of the geometric sequence.

28. $\sum_{i=1}^{\infty} \left(\frac{1}{2}\right)^i$

28.________________

29. $\sum_{i=1}^{\infty} -5(.2)^i$

29.________________

30. $\sum_{i=1}^{\infty} -5\left(-\frac{5}{3}\right)^i$

30.________________

Name: Date:
Instructor: Section:

Chapter 14 SEQUENCES AND SERIES

14.4 The Binomial Theorem

Learning Objectives
1 Expand a binominal raised to a power.
2 Find any specified term of the expansion of a binomial.

Key Terms

Use the vocabulary terms listed below to complete each statement in exercises 1-3.

***n* factorial** **binomial theorem** ***r*th term of the binomial expansion**

1. The ____________________ is given by $\dfrac{n!}{\left[n-(r-1)\right]!(r-1)!}x^{n-(r-1)}y^{r-1}$.

2. The ____________________ is a formula used to expand a binomial raised to a power.

3. For any positive integer *n*, ____________________ has a value given by $n(n-1)(n-2)(n-3)\cdots(2)(1)=n!$.

Objective 1 Expand a binominal raised to a power.

Use the binomial theorem to expand the binomial.

1. $(x-3)^3$ **1.**________________

2. $(r+3)^3$ **2.**________________

3. $(r+2s)^3$ **3.**________________

4. $(2a-3b)^3$ **4.**________________

Name: Date:
Instructor: Section:

5. $(4m-5p)^3$ **5.**________________

6. $(8m^3-3n^2)^3$ **6.**________________

7. $\left(\frac{1}{2}z-\frac{1}{3}x\right)^3$ **7.**________________

8. $(p+3q)^4$ **8.**________________

9. $(3r-2s)^4$ **9.**________________

10. $(5a-2b^2)^4$ **10.**________________

11. $(3m^3-5n^2)^4$ **11.**________________

12. $(2y+3z)^5$ **12.**________________

Name: Date:
Instructor: Section:

13. $(3x+y)^5$

13.______________

14. $(r^4-3s^2)^5$

14.______________

15. $(y^2-x^2)^6$

15.______________

Objective 2 Find any specified term of the expansion of a binomial.

Find the first four terms in the expansion of the binomial.

16. $(y-3)^7$

16.______________

17. $(r+2)^7$

17.______________

18. $(2k+1)^8$

18.______________

19. $(3r+5)^8$

19.______________

20. $(3m+2p)^{10}$

20.______________

21. $(2z-y)^{10}$

21.______________

Name: Date:
Instructor: Section:

Find the indicated term of the expansion of the binomial.

22. Third term of $(m-3)^5$ **22.**________________

23. Fourth term of $(2k-1)^4$ **23.**________________

24. Fourth term of $(3y-2)^4$ **24.**________________

25. Fourth term of $(4x-3y)^5$ **25.**________________

26. Fourth term of $(2p+5q)^5$ **26.**________________

27. Fifth term of $(3r-1)^6$ **27.**________________

28. Fifth term of $(k-4)^6$ **28.**________________

29. Sixth term of $(m^3-3r)^7$ **29.**________________

30. Eighth term of $(3k-p^2)^9$ **30.**________________

Chapter 1 THE REAL NUMBER SYSTEM

1.1 Fractions

Key Terms

1. proper fraction
2. factor
3. least common denominator (LCD)
4. greatest common factor
5. improper fraction
6. basic principal of fractions
7. reciprocals
8. mixed number
9. composite
10. circle graph
11. difference
12. lowest terms
13. quotient
14. product
15. prime

Objective 1

1. composite 3. prime 5. $2 \cdot 2 \cdot 2 \cdot 13$

Objective 2

7. $\frac{5}{6}$ 9. $\frac{8}{25}$ 11. $\frac{4}{9}$

Objective 3

13. $\frac{8}{15}$ 15. 2 17. $\frac{11}{4}$ or $2\frac{3}{4}$

Objective 4

19. $\frac{25}{36}$ 21. $\frac{1}{2}$ 23. $\frac{17}{2}$ or $8\frac{1}{2}$

Objective 5

25. $3\frac{1}{4}$ bushels 27. 12 cakes

Objective 6

29. $\frac{43}{125}$

1.2 Exponents, Order of Operations, and Inequality

Key Terms

1. exponential expression 2. base 3. bar graph 4. order of operations

5. exponent 6. inequality 7. grouping symbols

Objective 1

1. 32 3. $\frac{8}{125}$ 5. .16

Objective 2

7. 145 9. $\frac{41}{36}$

Objective 3

11. 102 13. 68 15. 189

Objective 4

17. False 19. False

Objective 5

21. $7 = 13 - 5$ 23. $6 \leq 6$

Objective 6

25. $12 > 9$ 27. $\frac{3}{4} > \frac{2}{3}$

Objective 7

29. \$–126 million

1.3 Variables, Expressions, and Equations

Key Terms

1. variable 2. equation 3. set 4. solution

5. elements 6. algebraic expression

Objective 1

1. 5 3. 52 5. $-\frac{2}{3}$

Objective 2

7. $x+4$ 9. $2x-7$ 11. $\dfrac{x+4}{2x}$

Objective 3

13. Yes 15. No 17. No

Objective 4

19. $x+4=10$; 6 21. $5x=40$; 8 23. $\dfrac{10}{x}=3+x$; 2

Objective 5

25. Expression 27. Equation 29. Equation

1.4 Real Numbers and the Number Line

Key Terms

1. coordinate 2. absolute value 3. number line 4. additive inverses

5. set-builder notation 6. ordering numbers 7. signed numbers

Objective 1

1. natural numbers, whole numbers, integers, rational numbers, real numbers

3. rational numbers, real numbers

5. rational numbers, real numbers

7. 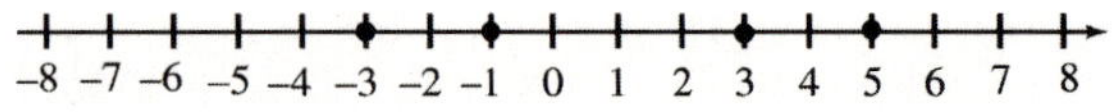 –8 –7 –6 –5 –4 –3 –2 –1 0 1 2 3 4 5 6 7 8

9.  $2\frac{1}{3}$ $6\frac{2}{3}$
–8 –7 –6 –5 –4 –3 –2 –1 0 1 2 3 4 5 6 7 8

Objective 2

11. –15 13. $-\frac{1}{2}$ 15. False

Objective 3

17. 15 19. $2\frac{5}{8}$ 21. –95 23. –10 25. $\frac{3}{4}$

Objective 4

27. Allen 29. 1753

1.5 Adding and Subtracting Real Numbers

Key Terms

1. additive inverses
2. different signs
3. difference
4. same sign

Objective 1

1. 15
3. -10

Objective 2

5. 2
7. $-\frac{1}{6}$
9. $1\frac{3}{8}$

Objective 3

11. -6
13. -13
15. 16
17. $-\frac{1}{30}$

Objective 4

19. 1
21. -20
23. -13.8

Objective 5

25. $[-2+(-3)]+10$; 5
27. $[2+(-3)]-(-6)$; 5

Objective 6

29. 41°F

1.6 Multiplying and Dividing Real Numbers

Key Terms

1. product 2. negative 3. multiplicative inverse 4. positive

Objective 1

1. –28 3. –17.5

Objective 2

5. 182

Objective 3

7. –40, –20, –10, –8, –5, –4, –2, –1, 1, 2, 4, 5, 8, 10, 20, 40

Objective 4

9. 6 11. $\frac{3}{7}$

Objective 5

13. 70 15. 0 17. $\frac{5}{4}$

Objective 6

19. –10 21. $\frac{3}{4}$

Objective 7

23. $-2-(10)(-2)$; 18

25. $\frac{2}{3}[16+(-10)]-(-34)$; 38

27. $\frac{(-4)(7)}{-3+14}$; $-\frac{28}{11}$

Objective 8

29. $x(-1)=7$

1.7 Properties of Real Numbers

Key Terms

1. distributive property
2. identity property
3. associative property
4. inverse property
5. commutative property

Objective 1

1. $y+4=4+y$
3. $ab(2)=2(ab)$
5. $10(\frac{1}{4}\cdot 2)=(\frac{1}{4}\cdot 2)(10)$
7. $2+[10+(-9)]=[10+(-9)]+2$

Objective 2

9. $(4\cdot 5)(-7)=4[5(-7)]$
11. $(2m)(-7)=(2)[m(-7)]$
13. $[x+(-4)]+3y=x+[(-4)+3y]$

Objective 3

15. -7
17. 7

Objective 4

19. $-\frac{1}{7}+\frac{1}{7}=0$; inverse
21. $\frac{2}{7}\cdot\frac{7}{2}=1$; inverse
23. $-14+14=0$; inverse

Objective 5

25. $r(10-4)$; $6r$
27. $4(c-d)$
29. $-10y+18z$

1.8 Simplifying Expressions

Key Terms

1. numerical coefficient
2. combining like terms
3. term
4. like terms
5. unlike terms

Objective 1

1. $6+3y$ 3. $-17+4b$ 5. $10x+3$

Objective 2

7. 4 9. .3

Objective 3

11. Like 13. Like 15. Unlike

Objective 4

17. $-11x+10$ 19. $\frac{3}{10}r-\frac{1}{2}s$ 21. $5r-4$ 23. $3x-4$

Objective 5

25. $7x+2x$; $9x$ 27. $(5x-3)+4(x+2)$; $9x+5$

29. $10-12(4-2x)$; $-38+24x$

Chapter 2 LINEAR EQUATIONS AND INEQUALITIES IN ONE VARIABLE

2.1 The Addition Property of Equality

Key Terms

1. addition property of equality 2. solution set 3. equivalent equations

Objective 1

1. Yes 3. No 5. No

Objective 2

7. 20 9. –5 11. $\frac{3}{2}$ 13. –5 15. –12.8

Objective 3

17. –4 19. 7 21. –8 23. 2 25. 0

27. $\frac{1}{3}$ 29. 7.2

2.2 The Multiplication Property of Equality

Key Terms

1. multiplication property of equality 2. coefficient 3. reciprocal

Objective 1

1. 3 3. 3 5. –14 7. –36

9. $\frac{7}{9}$ 11. 4.3 13. 6.4

Objective 2

15. 9 17. 9 19. 7 21. –3.9

23. $-\frac{7}{4}$ 25. 18 27. 8 29. –24

2.3 More on Solving Linear Equations

Key Terms

1. conditional equation
2. LCD
3. contradiction
4. simplify each side separately
5. empty set
6. identity

Objective 1

1. $\frac{5}{2}$
3. 10
5. –1
7. 2
9. –7

Objective 2

11. 2
13. $\frac{53}{11}$
15. 30
17. –6

Objective 3

19. No solution
21. All real numbers
23. No solution
25. No solution
27. All real numbers

Objective 4

29. $\frac{17}{p}$

2.4 An Introduction to Applications of Linear Equations

Key Terms

1. complementary
2. degree
3. supplementary
4. solving an applied problem
5. straight angle
6. consecutive integers

Objective 2

1. $4+3x=7$; 1
3. $-2(4-x)=24$; 16
5. $-3(x-4)=-5x+2$; -5
7. $83-x=19+37$; 27

Objective 3

9. 27 inches
11. 52
13. $4.25
15. Mark 3 laps; Pablo 14 laps; Faustino 12 laps

Objective 4

17. 133°
19. 49°
21. 76°
23. 43°

Objective 5

25. 76, 78
27. 27, 28
29. 13, 15

2.5 Formulas and Applications from Geometry

Key Terms

1. formula
2. area
3. vertical angles
4. perimeter
5. solving a literal equation

Objective 1

1. 24
3. 7
5. 2400
7. 12

Objective 2

9. 36 ft
11. 12 m
13. 3052.08 cu cm
15. 4 cm

Objective 3

17. $(3x+5)^{\circ}=35^{\circ}$; $(6x-25)^{\circ}=35^{\circ}$

19. $(6x)^{\circ}=72^{\circ}$; $(10x-48)^{\circ}=72^{\circ}$

21. $(2x+16)^{\circ}=48^{\circ}$; $(7x+20)^{\circ}=132^{\circ}$

Objective 4

23. $H=\dfrac{V}{LW}$

25. $r=\dfrac{S-a}{S}$

27. $n=\dfrac{a_n-a_1+d}{d}$

29. $F=\frac{9}{5}C+32$

2.6 Ratios and Proportions

Key Terms

1. extremes
2. means
3. ratio
4. proportion
5. cross products
6. terms

Objective 1

1. $\frac{7}{9}$
3. $\frac{3}{4}$
5. $\frac{5}{24}$
7. 5-lb box
9. 20-count box
11. 48-oz jar

Objective 2

13. 15
15. 4
17. $\frac{10}{3}$
19. $-\frac{41}{3}$
21. –13

Objective 3

23. 75 min
25. 15 inches
27. 4.5 oz
29. $5.75

2.7 Further Applications of Linear Equations

Key Terms

1. simple interest
2. mixture problem
3. distance problems
4. denomination

Objective 1

1. 22.5 liters
3. \$800
5. \$3.15

Objective 2

7. $\frac{1}{2}$ liter
9. 625 lb
11. 18 liters

Objective 3

13. \$28,500 at 8%; \$18,500 at 6%
15. \$51,000 at 8%; \$17,000 at $9\frac{1}{2}$%
17. \$7000 at 9%; \$6000 at 11%

Objective 4

19. 24 nickels; 46 dimes
21. 185
23. 15 nickels; 21 dimes; 30 quarters

Objective 5

25. 288 mi
27. 9.52 m/sec
29. 8 mph

2.8 Solving Linear Inequalities

Key Terms

1. linear inequality in one variable
2. negative infinity
3. addition property of inequality
4. interval notation
5. three part inequality
6. multiplication property of inequality
7. interval

Objective 1

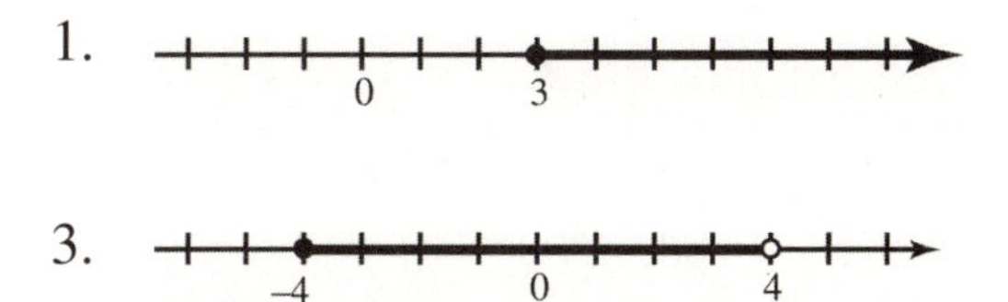

Objective 2

5. $j \le 5$

7. $a \ge 3$

9. $b < -2$

Objective 3

11. $s < -2$

13. $k \geq -4$

Objective 4

15. $y > 4$

17. $m \geq 4$

19. $x < -14$

Objective 5

21. $315 or more

23. 89 or more

25. 55 or more

Objective 6

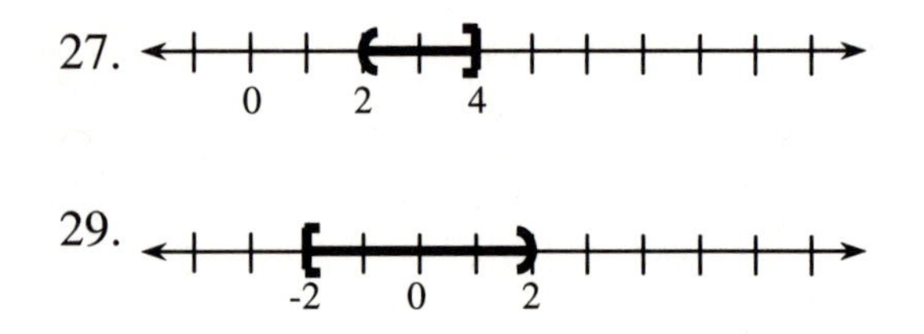

Chapter 3 LINEAR EQUATIONS IN TWO VARIABLES

3.1 Reading Graphs; Linear Equations in Two Variables

Key Terms

1. ordered pair
2. quadrant
3. *y*-axis
4. line graph
5. coordinates
6. *x*-axis
7. origin
8. bar graph
9. rectangular coordinate system
10. linear equation in two variables

Objective 1

1. 1991 – 1992
3. 1993 – 1994
5. 2655 students

Objective 2

7. $(4,7)$
9. $(0,\frac{1}{3})$

Objective 3

11. No
13. No

Objective 4

15. (a) $(-4,-5)$ (b) $(2,7)$ (c) $(-\frac{3}{2},0)$ (d) $(-2,-1)$ (e) $(-5,-7)$

17. (a) $(-2,-2)$ (b) $(-2,0)$ (c) $(-2,19)$ (d) $(-2,3)$ (e) $(-2,-\frac{2}{3})$

Objective 5

19.

x	0	$\frac{4}{3}$	4
y	2	0	–4

21.

x	y
0	$\frac{3}{2}$
–2	0
2	3

23.

x	y
–4	4
0	4
6	4

Objective 6

24. – 30.

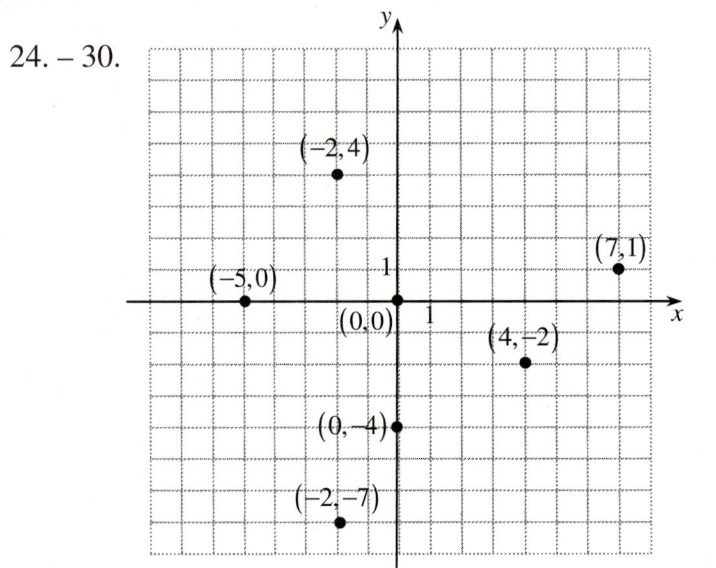

3.2 Graphing Linear Equations in Two Variables

Key Terms

1. horizontal line
2. graph of a linear equation
3. *x*-intercept
4. vertical line
5. *y*-intercept
6. line through the origin

Objective 1

1. $(0,3)$, $(3,0)$, $(2,1)$

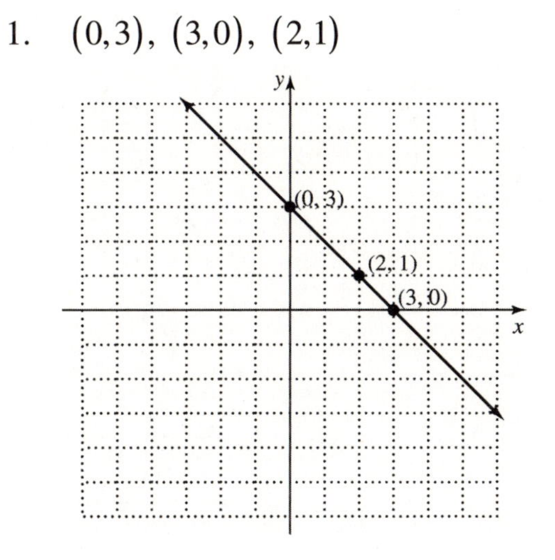

3. $(0,2)$, $(-4,0)$, $(-2,1)$

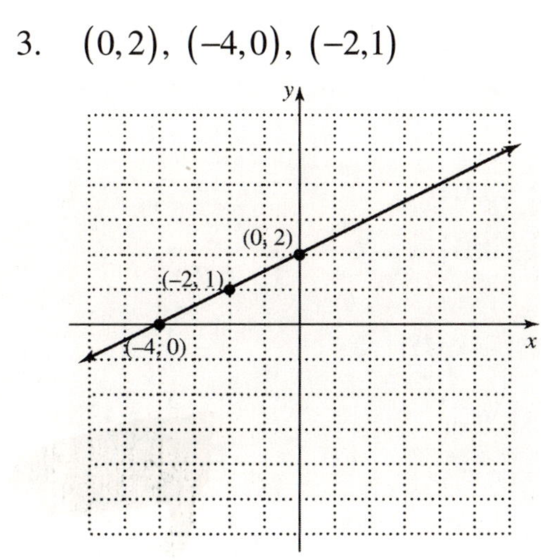

5. $(0,-4)$, $(4,0)$, $(-2,-6)$

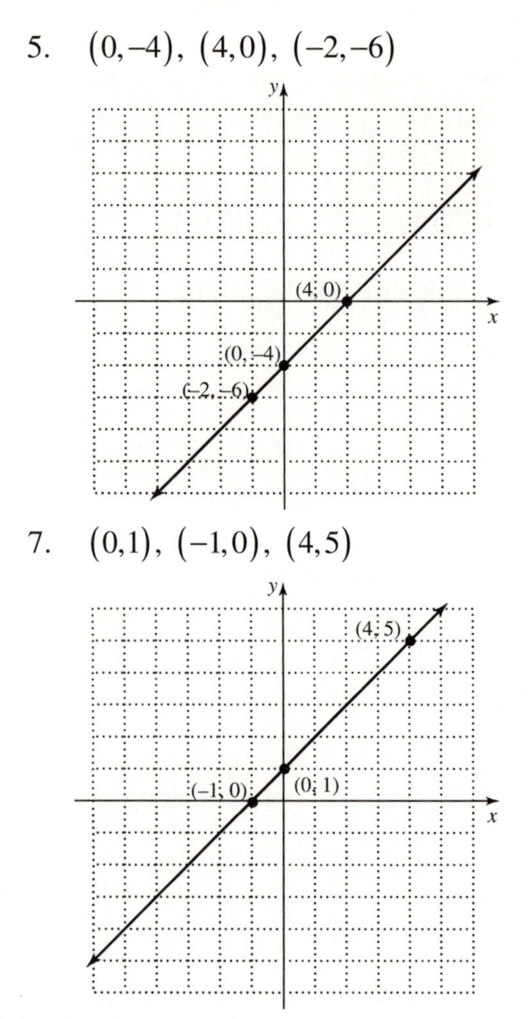

7. $(0,1)$, $(-1,0)$, $(4,5)$

Objective 2

9. x-intercept: $(-2,0)$;
   y-intercept: $(0,5)$

11. x-intercept: $(0,0)$;
    y-intercept: $(0,0)$

13. x-intercept: $(2,0)$;
    y-intercept: $(0,-5)$

15. x-intercept: $\left(-\frac{2}{3},0\right)$;
    y-intercept: $(0,-1)$

17. x-intercept: $\left(-\frac{9}{2},0\right)$;
    y-intercept: $(0,-1)$

Objective 3

19. x-intercept: $(2,0)$

    y-intercept: $(0,6)$

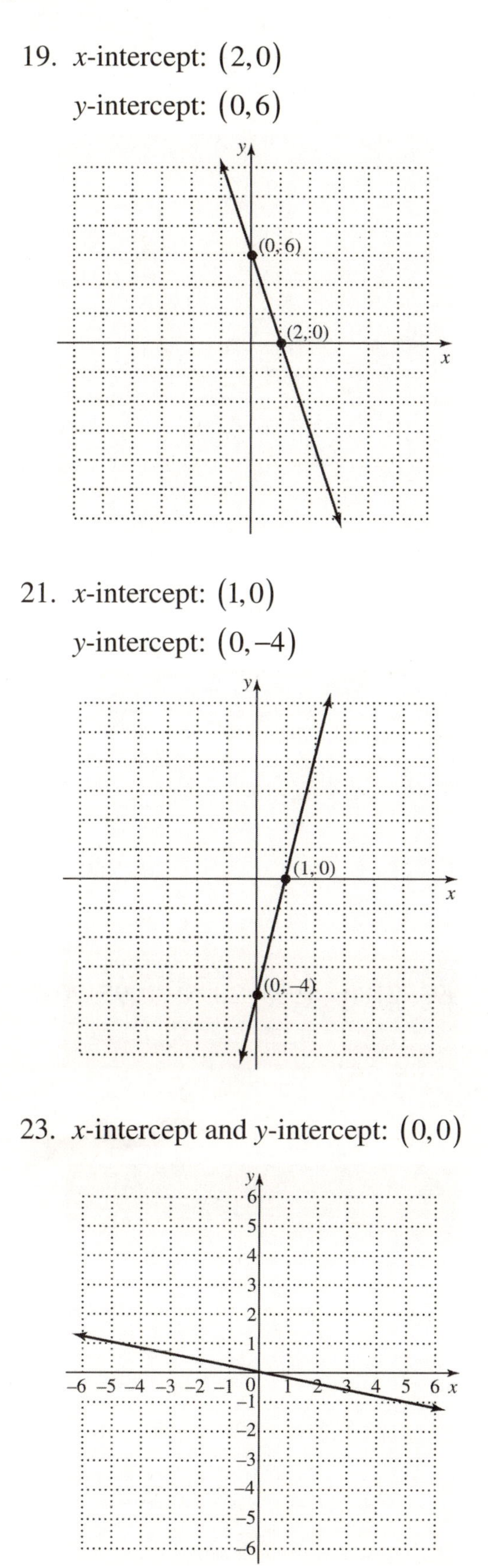

21. x-intercept: $(1,0)$

    y-intercept: $(0,-4)$

23. x-intercept and y-intercept: $(0,0)$

Objective 4

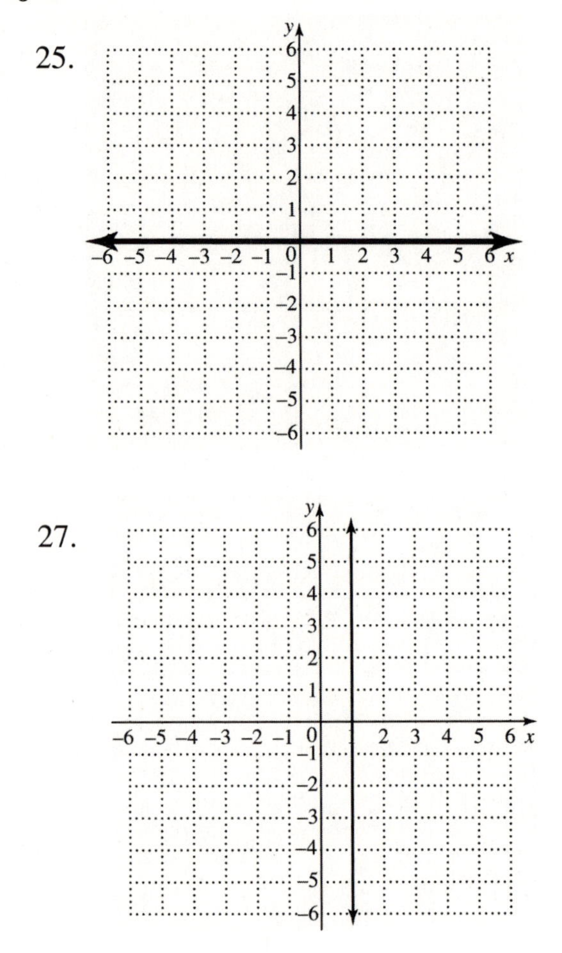

Objective 5

29. 1990: 2435; 1991: 2350; 1992: 2265; 1993: 2180; 1994: 2095; 1995: 2010

3.3 The Slope of a Line

Key Terms

1. negative slope
2. slope
3. run
4. vertical line
5. subscript notation
6. positive slope
7. parallel
8. rise
9. horizontal line
10. perpendicular

Objective 1

1. -2 3. $\frac{1}{5}$ 5. $-\frac{3}{5}$ 7. 0 9. Undefined slope

Objective 2

11. -5 13. $-\frac{2}{5}$ 15. $\frac{3}{4}$ 17. $\frac{4}{3}$ 19. Undefined slope

Objective 3

21. -5; 5; neither
23. 1; 1; parallel
25. -2; $-\frac{1}{4}$; neither
27. -2; $-\frac{5}{3}$; neither
29. 0; 0; parallel

3.4 Equations of a Line

Key Terms

1. slope-intercept form
2. point-slope form
3. standard form

Objective 1

1. $y = \frac{2}{3}x - 4$

3. $y = -7x - 2$

5. $y = -4x$

Objective 2

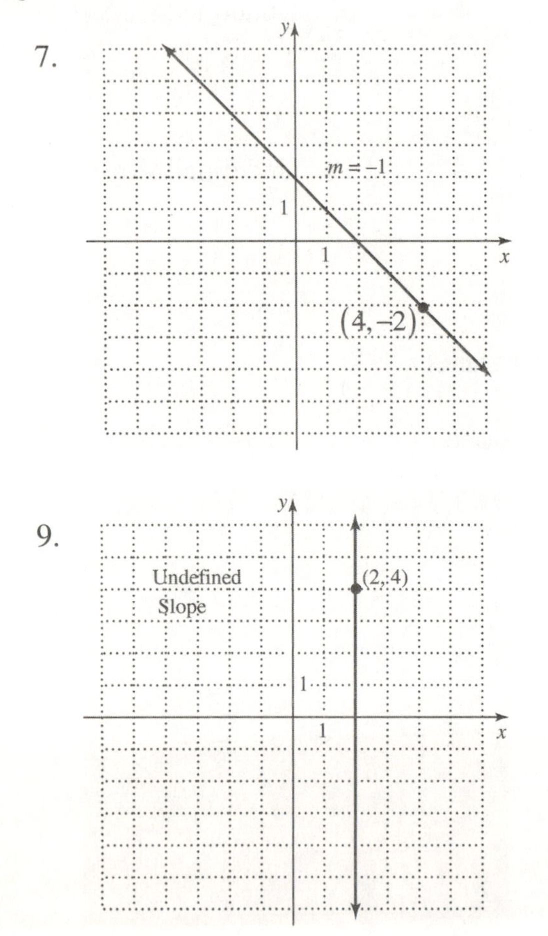

11. 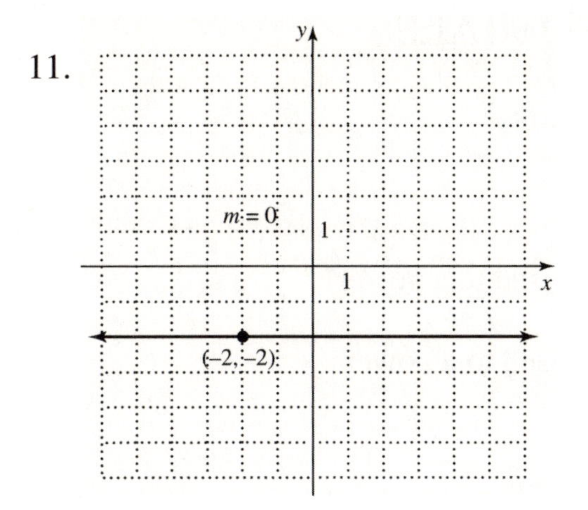

Objective 3

13. $x-3y=-7$ 15. $2x+3y=-9$ 17. $y=0$ 19. $x+y=0$

Objective 4

21. $11x+y=29$ 23. $x=-3$ 25. $y=6$

Objective 5

27. $m=19$; $y=19x+82$

29. $m=\frac{58}{3}$; $y=\frac{58}{3}x+\frac{253}{3}$

Chapter 4 EXPONENTS AND POLYNOMIALS

4.1 The Product Rule and Power Rules for Exponents

Key Terms

1. product rule for exponents
2. quotient raised to a power
3. power raised to a power
4. product raised to a power

Objective 1

1. $2^6 = 64$
3. $(-2y)^5$
5. 256; base, –4; exponent, 4

Objective 2

7. 4^{11}
9. $100n^{12}$
11. $-7m^6$; $12m^{12}$

Objective 3

13. 3^{12}
15. $(-2)^{35}$

Objective 4

17. y^3z^{12}
19. $(-2)^6 r^{18} s^6$

Objective 5

21. $\dfrac{(-2)^7 a^7}{b^{14}}$
23. $\dfrac{x^4}{2^4 y^4}$

Objective 6

25. $\dfrac{3^4 b^8}{11^4}$
27. $\dfrac{7^7 a^{14} b^{21}}{2^7}$

Objective 7

29. $10x^7$

4.2 Integer Exponents and the Quotient Rule

Key Terms

1. negative exponent
2. zero exponent
3. quotient rule for exponents

Objective 1

1. 1
3. –1
5. 0
7. –1

Objective 2

9. $\frac{7}{2}$
11. $\frac{1}{20}$
13. $2r^7$

Objective 3

15. 16
17. $\frac{1}{2x^6}$
19. a^6b^6
21. $8^5b^4c^7$

Objective 4

23. x^9
25. $\frac{4^2}{5^2w^2y^{10}}$
27. c^{25}
29. $\frac{1}{k^4t^{20}}$

4.3 An Applications of Exponents: Scientific Notation

Key Terms

1. scientific notation
2. right
3. left

Objective 1

1. 3.25×10^{2}
3. 2.3651×10^{4}
5. 4.296×10^{11}
7. 2.46×10^{-1}
9. 4.26×10^{-3}

Objective 2

11. 25,000
13. $-2,450,000$
15. 45,000,000
17. .000724
19. .4752

Objective 3

21. 210,000,000
23. 253
25. .02
27. .0313
29. 2.1

4.4 Adding and Subtracting Polynomials; Graphing Simple Polynomials

Key Terms

1. monomial
2. degree of a term
3. degree of a polynomial
4. coefficient
5. binomial
6. term
7. trinomial
8. descending powers

Objective 1

1. 1; –7
3. 4; 9, 3, –4, 2

Objective 2

5. $6s^3$
7. $13y^4 - 7y^3 - 9y^2 - 3y + 4$
9. $-\frac{1}{4}r^3$

Objective 3

11. (c)
13. $n^8 - n^2$; degree 8; binomial

Objective 4

15. (a) 16 (b) 21
17. (a) 72 (b) 17

Objective 5

19. $2m^4 + 7m^3 - 7$
21. $9p^4 - 3p^3 - 3p^2 - 5p + 22$
23. $-4a^5 - 2a^3 - 2a - 4$
25. $4m^2 - 2$
27. $-4x^3 - 4x^2 + 7x + 8$

Objective 6

29.

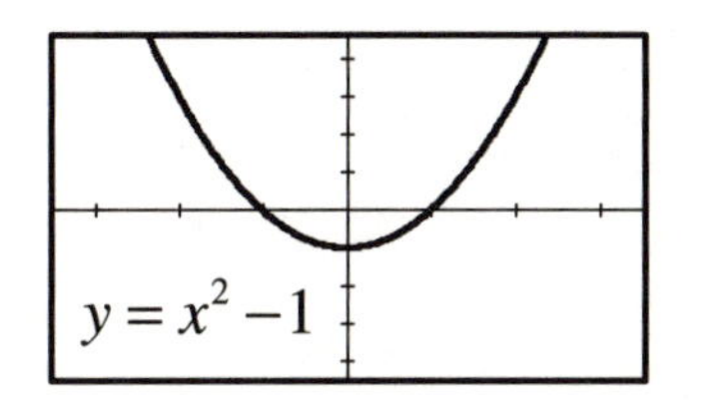

4.5 Multiplying Polynomials

Key Terms

1. multiply two polynomials
2. FOIL method

Objective 1

1. $12y^5$
3. $-6x^4 - 12x^5 - 4x^6$
5. $-6y^5 - 9y^4 + 12y^3 - 33y^2$
7. $-35b^4 + 7b^2 - 28b^3$
9. $-24r^4s^5 + 12r^3s^4 - 6r^3s^5$

Objective 2

11. $3p^2 + 10p + 8$
13. $y^3 + 64$
15. $6m^5 + 4m^4 - 5m^3 + 2m^2 - 4m$
17. $6y^4 - 5y^3 - 8y^2 + 7y - 6$
19. $4r^2 + 4rs + s^2$
21. $4x^4 - 4x^3 + 9x^2 - 4x + 4$

Objective 3

23. $6x^2 - 5xy - 6y^2$
25. $121k^2 - 16$
27. $2v^4 - 5v^2w^2 - 3w^4$
29. $x^2 - \frac{5}{3}x + \frac{4}{9}$

4.6 Special Products

Key Terms

1. product of the sum and difference of two terms
2. square of a binomial

Objective 1

1. $z^2 + 6z + 9$
3. $25y^2 - 30y + 9$
5. $25m^2 + 30mn + 9n^2$
7. $4p^2 + 12pq + 9q^2$
9. $16y^2 - 5.6y + .49$
11. $9a^2 + 3ab + \frac{1}{4}b^2$

Objective 2

13. $144 - x^2$
15. $16p^2 - 49q^2$
17. $81 - 16y^2$
19. $49m^2 - \frac{9}{16}$
21. $y^4 - 4$

Objective 3

23. $x^3 + 6x^2 + 12x + 8$
25. $8x^3 - 36x^2 + 54x - 27$
27. $k^4 + 8k^3 + 24k^2 + 32k + 16$
29. $81b^4 - 216b^3 + 216b^2 - 96b + 16$

4.7 Dividing Polynomials

Key Terms

1. dividing a polynomial by a monomial
2. long division
3. dividing a polynomial by a polynomial

Objective 1

1. $2m+3$
3. $5m^3-3m$
5. $-18m+10+\frac{2}{m}$
7. $2p+6p^4$
9. $2y^6-\frac{9}{4}y$
11. $\frac{1}{2}m+\frac{3}{2}-\frac{6}{m}$
13. $4z^2+\frac{8}{3}z-2+\frac{5}{3z^2}$
15. $-2+\frac{2}{y}-\frac{10}{y^2}$

Objective 2

17. $6a-5$
19. $a-7+\frac{37}{2a+3}$
21. $5w-2+\frac{-4}{w-4}$
23. $2x+5$
25. $9p^3-2p+6$
27. $3x^2-6x+2+\frac{13x-7}{2x^2+3}$
29. y^2-y+1

Chapter 5 FACTORING AND APPLICATIONS

5.1 The Greatest Common Factor; Factoring by Grouping

Key Terms

1. greatest common factor
2. factored form
3. factoring
4. common factor
5. factoring by grouping

Objective 1

1. 6
3. 1
5. 28
7. m^4
9. $k^2m^4n^4$
11. $9xy^2$

Objective 2

13. $6y^3$
15. $13(2r+3t)$
17. $8a(3b-a+5c)$
19. No common factor (except 1)
21. $(a+b)(3-x)$

Objective 3

23. $(x+2)(x+5)$
25. $3(x-3)(x+4)$
27. $(2a-3b)(a^2+b^2)$
29. $(2x^2+3y)(x^2+2y^2)$

5.2 Factoring Trinomials

Key Terms

1. prime polynomial
2. coefficient

Objective 1

1. 1 and 42, –1 and –42, 2 and 21, -2 and –21, 7 and 6, –7 and –6, 14 and 3, –14 and –3; the pair with sum of 17 is 3 and 14.

3. –8 and 8, 1 and –64, –1 and 64, 2 and –32, –2 and 32, 16 and –4, –16 and 4; the pair with a sum of 12 is –4 and 16.

5. $x+4$

7. $x+2$

9. $(x+2)(x+9)$

11. $(x-2)(x+1)$

13. $(x-7)(x+5)$

15. $(x+5)(x+1)$

17. $(x-7y)(x+3y)$

Objective 2

19. $2(x-2)(x+7)$

21. Prime

23. $2ab(a-3b)(a-2b)$

25. $5(r+4)(r+3)$

27. $10k^4(k+2)(k+5)$

29. $2y^2(x+2y)(x-3y)$

5.3 More on Factoring Trinomials

Key Terms

1. binomial factor
2. squared term of a trinomial

Objective 1

1. $(4b+3)(2b+3)$
3. $(5a+2)(3a+2)$
5. $(3b+2)(b+2)$
7. $p(3p+2)(p+2)$
9. $b(7a+4)(a+2)$
11. $3(c+2d)(3c+2d)$
13. $(3r-1)(3r+5)$

Objective 2

15. $x+3$
17. $4x-2$
19. $(5x+2)(2x+3)$
21. $(a+6)(2a+1)$
23. $(4m+1)(2m-3)$
25. $(5q+6)(3q-4)$
27. $(5c+2d)(2c-d)$
29. $(6x-y)(3x-4y)$

5.4 Special Factoring Techniques

Key Terms

1. difference of squares
2. sum of cubes
3. perfect square trinomial
4. difference of cubes

Objective 1

1. $(x+7)(x-7)$
3. $(3j+\frac{4}{7})(3j-\frac{4}{7})$
5. $(3m^2+1)(3m^2-1)$

Objective 2

7. $(y+3)^2$
9. $(z-\frac{2}{3})^2$
11. $-4(2x+3)^2$

Objective 3

13. $(a-1)(a^2+a+1)$
15. $(c-6)(c^2+6c+36)$
17. $(c^3-d^2)(c^6+c^3d^2+d^4)$
19. $(4x-3y)(16x^2+12xy+9y^2)$
21. $(10a-3b)(100a^2+30ab+9b^2)$

Objective 4

23. $(m+4)(m^2-4m+16)$
25. $(2b+1)(4b^2-2b+1)$
27. $(t^2+1)(t^4-t^2+1)$
29. $(6m+5p)(36m^2-30mp+25p^2)$

5.5 Solving Quadratic Equations by Factoring

Key Terms

1. zero-factor property
2. standard form
3. quadratic equation

Objective 1

1. $-9, \frac{3}{2}$
3. $-7, 7$
5. $-7, 9$
7. $3, -\frac{2}{3}$
9. $-\frac{2}{3}$
11. $\frac{4}{3}, -\frac{4}{3}$
13. $-\frac{2}{7}, \frac{3}{2}$
15. $-3, \frac{1}{2}$

Objective 2

17. $0, -\frac{3}{2}, 5$
19. $-7, 0, 7$
21. $-4, 0, 2$
23. $-\frac{3}{2}, \frac{3}{2}, 2$
25. $-\frac{4}{3}, 0, 1$
27. $0, 7, 8$
29. $-\frac{5}{2}, \frac{3}{2}, -3$

5.6 Applications of Quadratic Functions

Key Terms

1. consecutive odd integers
2. consecutive integers
3. hypotenuse
4. consecutive even integers

Objective 1

1. Length, 17 cm; width, 9 cm
3. Length, 18 ft; width, 14 ft
5. First square, 5 m; second square, 3 m
7. Length, 9 cm; width, 3 cm
9. 6 ft

Objective 2

11. 0, 1 or 7, 8
13. –2, –1
15. 7, 9 or –7, –5

Objective 3

17. 6 inches
19. 7 m, 24 m, 25 m
21. 15 ft
23. 18 ft
25. 20 mi

Objective 4

27. 33.4 mpg
29. $2\frac{1}{2}$ sec

Chapter 6 RATIONAL EXPRESSIONS AND APPLICATIONS

6.1 The Fundamental Property of Rational Expressions

Key Terms

1. rational expression
2. fundamental property of rational expressions
3. lowest terms

Objective 1

1. (a) 5
   (b) $-\frac{11}{2}$
3. (a) $\frac{1}{3}$
   (b) $-\frac{1}{11}$
5. (a) Undefined
   (b) Undefined
7. (a) $-\frac{3}{10}$
   (b) $\frac{11}{10}$

Objective 2

9. 0
11. 0, 4
13. –5, 5
15. –3, 0, –6

Objective 3

17. $3a^4b^6$
19. $2r-s$
21. $\dfrac{x+5}{x+2}$
23. $\dfrac{3y+2}{2y+1}$
25. $\dfrac{5x+3y}{x-3y}$

Objective 4

27. $\dfrac{-(2x-3)}{x+2}; \dfrac{-2x+3}{x+2}; \dfrac{2x-3}{-(x+2)}; \dfrac{2x-3}{-x-2}$

29. $\dfrac{-(2x-1)}{3x+5}; \dfrac{-2x+1}{3x+5}; \dfrac{2x-1}{-(3x+5)}; \dfrac{2x-1}{-3x-5}$

6.2 Multiplying and Dividing Rational Expressions

Key Terms

1. multiply
2. divide

Objective 1

1. $\dfrac{10m^3n}{3}$
3. $-\frac{3}{2}$
5. $\dfrac{a-1}{a+1}$
7. $\dfrac{3-x}{2x+3}$
9. $\dfrac{-1}{8+2x}$
11. $\dfrac{x+4}{2x-8}$
13. $-\frac{3}{5}$

Objective 2

15. $\dfrac{12m^3}{5}$
17. $-\frac{1}{2}$
19. $\dfrac{2(m-5)}{m+3}$
21. $\dfrac{(m-1)(m+n)}{m(m-n)}$
23. $\dfrac{3k+1}{3k-1}$
25. $\dfrac{y^2-9yz+18z^2}{y^2+yz-20z^2}$
27. $\dfrac{-3(6k-1)}{3k-1}$
29. $\dfrac{4(y-1)}{3(y-3)}$

6.3 Least Common Denominators

Key Terms

1. greatest
2. least common denominator

Objective 1

1. 48
3. 180
5. $24a^2b^2$
7. $15r^3(r-5)$
9. $x-y$ or $y-x$
11. $m(2m-1)(m+5)$
13. $z^2(z-2)(z+4)^2$
15. $(2q-5)(q+2)(q-2)$
17. $m^2(m-2)(m+7)$

Objective 2

19. 28
21. $44a+4$
23. $8z(4z+1)$ or $32z^2+8z$
25. $9(y+2)=9y+18$
27. $30r$
29. $2(3x+1)$ or $6x+2$

6.4 Adding and Subtracting Rational Expressions

Key Terms

1. different denominators
2. parentheses
3. same denominator

Objective 1

1. $\frac{11}{3w^2}$
3. $\frac{1}{b-2}$
5. $\frac{4}{m+1}$
7. $\frac{1}{2y+1}$

Objective 2

9. $\frac{5x-6}{15}$
11. $\frac{3a-4}{a^2-4}$
13. $\frac{3y^2+10y-12}{(y+4)(y+2)(y-2)}$
15. $\frac{16s^2+19s-1}{(3s-2)(s-4)(2s+3)}$
17. $\frac{3x^2-18x+7}{(2x+1)(2x-1)(x+2)}$

Objective 3

19. 5
21. $\frac{1}{3x+4y}$
23. $\frac{5-20s}{18}$
25. $\frac{q-22}{(2q+1)(q+2)(q-2)}$
27. $\frac{y^2-18y-3}{(y+3)(y+1)(y-2)}$
29. $\frac{11x^2-x-11}{(2x-1)(x+3)(3x+2)}$

6.5 Complex Fractions

Key Terms

1. method 1
2. method 2
3. complex fraction

Objective 1

1. $-\frac{2}{3}$
3. $\dfrac{7m^2}{2n^3}$
5. $\dfrac{2y-5}{3y-8}$
7. $\dfrac{3p-2}{2p+1}$
9. $\dfrac{9s+12}{6s^2+2s}$
11. $(a+2)^2$
13. $\dfrac{24}{w-3}$
15. $\dfrac{5(3a+4)}{2a+5}$

Objective 2

17. $-\frac{7}{3}$
19. $\dfrac{4}{r^2s^2}$
21. $\dfrac{r^2+3}{5+r^2t}$
23. $\dfrac{2s^2+3}{1-3s^2}$
25. $\dfrac{2(1-4h)}{h(1+4h)}$
27. $\dfrac{4m-3}{6-4m}$
29. $\dfrac{(k-23)(k+2)}{5k(k+1)}$

6.6 Solving Equations with Rational Expressions

Key Terms

1. simplified 2. adding or subtracting 3. solved 4. LCD

Objective 1

1. equation; 4 3. operation; $\frac{3x}{10}$

Objective 2

5. All real numbers 7. 5 9. 4, 6 11. $-\frac{13}{6}$

13. 4 15. No solution 17. –2, 1

Objective 3

19. $t = \frac{I}{Pr}$ 21. $r = \frac{S - a_1}{S}$ or $r = 1 - \frac{a_1}{S}$

23. $G = \frac{Fd^2}{mM}$ 25. $T_1 = \frac{V_1 P_1 T_2}{V_2 P_2}$

27. $G = \frac{Fd^2}{m_1 m_2}$ 29. $b_2 = \frac{2A}{h} - b_1$ or $b_2 = \frac{2A - b_1 h}{h}$

6.7 Applications of Rational Expressions

Key Terms

1. rate of work
2. check
3. read
4. smaller

Objective 1

1. 15
3. $\frac{3}{4}$ or $\frac{1}{3}$
5. $\frac{9}{13}$
7. $\frac{1}{2}$ or 4
9. Sharon, $57,000; Elaine, $95,000

Objective 2

11. 3 mph
13. 350 mi
15. 50 mph
17. 24 mph
19. 8 mph

Objective 3

21. $\frac{12}{5}$ or $2\frac{2}{5}$ hr
23. $\frac{24}{11}$ or $2\frac{2}{11}$ hr
25. $\frac{2}{5}$ hr
27. 3 hr
29. 6 hr

Chapter 7 EQUATIONS OF LINES; FUNCTIONS

7.1 Review of Graphs and Slopes of Lines

Key Terms

1. linear equation in two variables
2. first degree equation
3. slope
4. ordered pair
5. perpendicular lines
6. rectangular coordinate system
7. y-intercept
8. origin
9. parallel lines
10. coordinates
11. quadrant
12. x-intercept

Objective 1

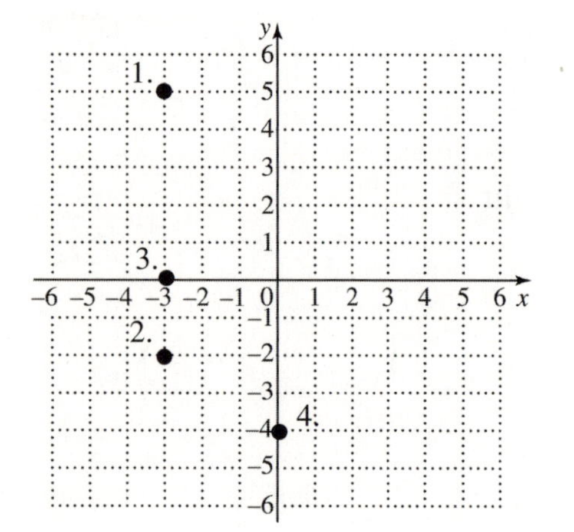

Objective 2

5. x-intercept: $(4, 0)$; y-intercept: $(0, -5)$

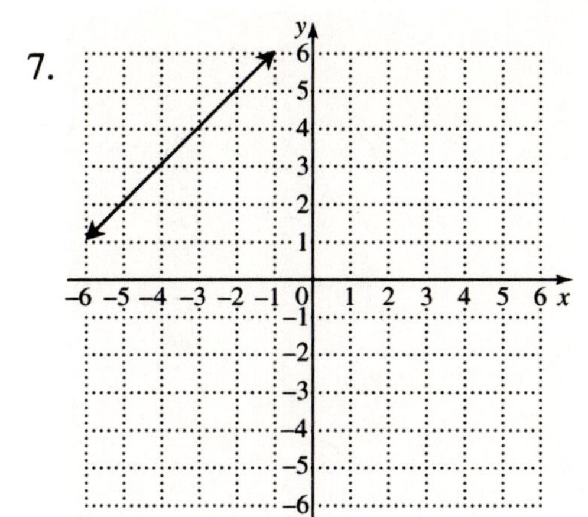

Objective 3

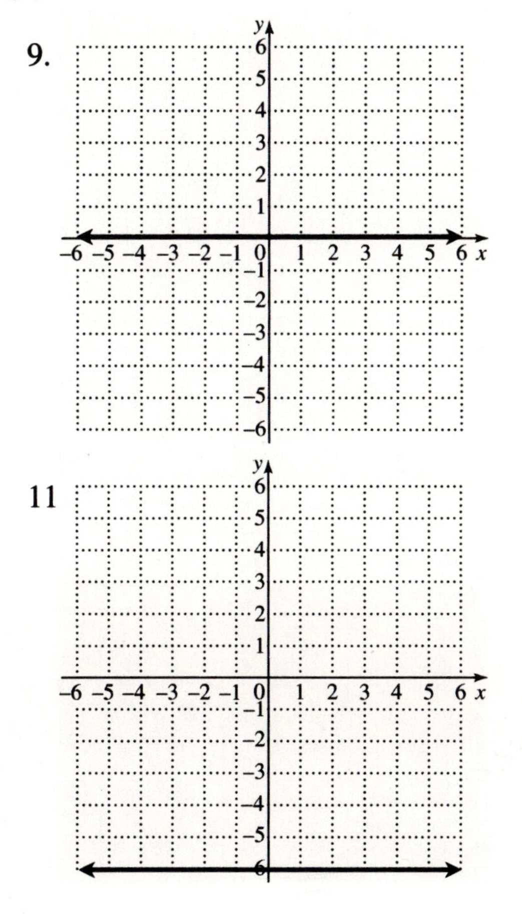

Objective 4

13. $(0, -6)$ 15. $(-4, 7)$

Objective 5

17. $\frac{3}{2}$ 19. 0

Objective 6

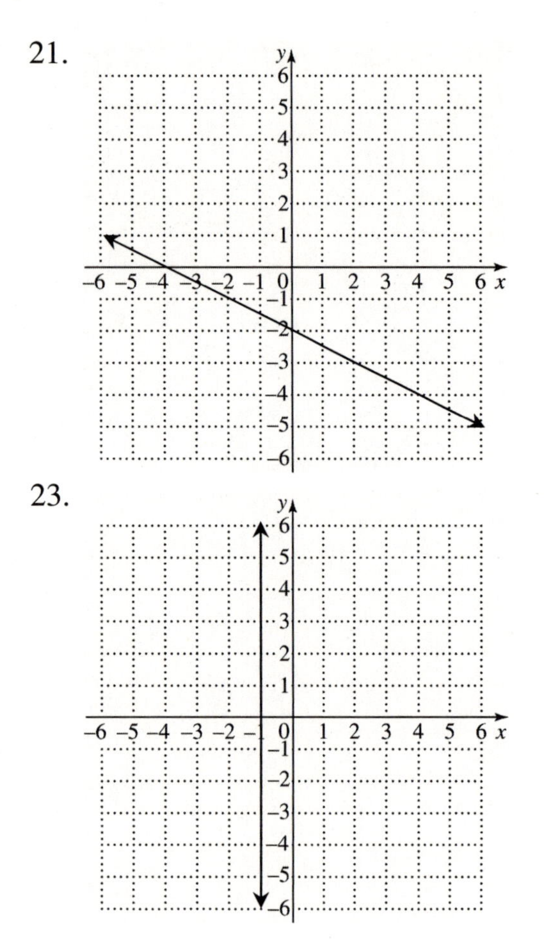

Objective 7

25. neither   27. perpendicular

Objective 8

29. 113.5 feet per minute

7.2 Review Equations of Lines; Linear Models

Key Terms

1. standard form 2. slope-intercept form 3. vertical line

4. horizontal line 5. point-slope form

Objective 1

1. $2x - y = 5$ 3. $3x - 5y = -2$

Objective 2

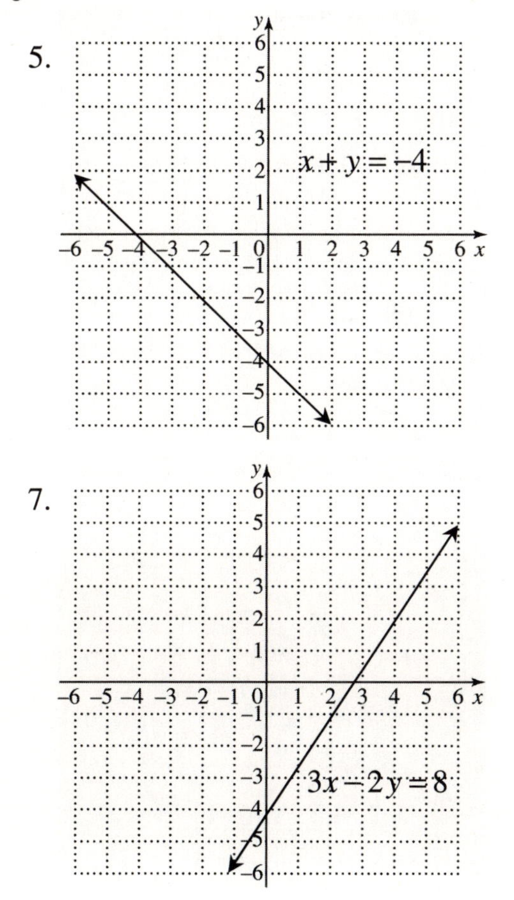

9.

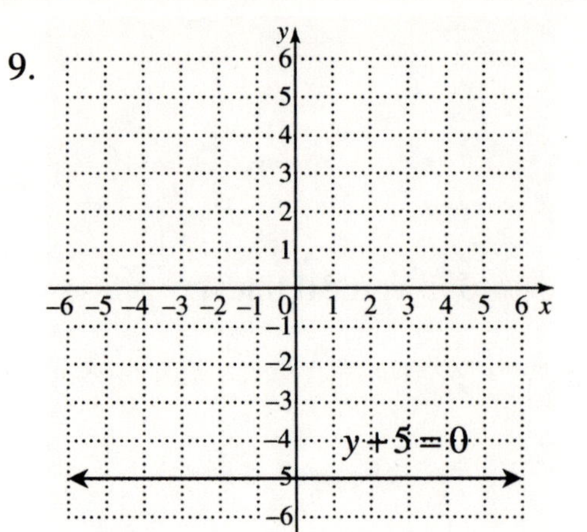

Objective 3

11. $5x - y = 21$

13. $2x + 3y = -13$

15. $y = -4$

Objective 4

17. $4x + y = 29$

19. $x + 2y = -2$

21. $y = -5$

Objective 5

23. $x + 3y = -5$

25. $x + 2y = -1$

27. $y = 5$

Objective 6

29. $351.45

7.3 Functions

Key Terms

1. relation 2. vertical line test 3. function 4. dependent variable

5. linear function 6. domain 7. constant function 8. function notation

9. independent variable 10. range

Objective 1

1. independent variable: outside temperature; dependent variable: cost of heating your home

Objective 2

3. not a function 5. function 7. function 9. function

Objective 3

11. Function; domain: $\{3, 2, 1, -1\}$; range: $\{0, 4, 6, 3\}$

13. Not a function; domain: $\{1, 2, -1\}$; range: $\{3, -1, 4\}$

15. Function; domain: $\{4, 3, 2, 1, 0\}$; range: $\{2\}$

Objective 4

17. function 19. function 21. not a function

Objective 5

23. 1; 13; $2x+7$ 25. $\frac{4}{5}$; $\frac{4}{17}$; $\frac{4}{x^2+2x+2}$

27. $f(x)=-\frac{3}{2}x-3$

Objective 6

29. Domain: $(-\infty, \infty)$; range: $\{-2\}$

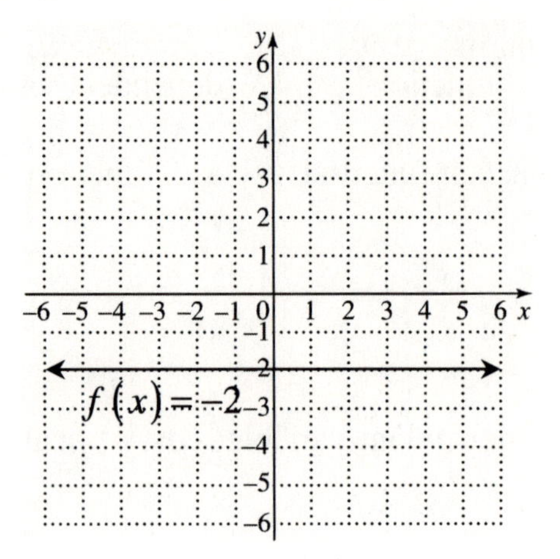

7.4 Operations on Functions and Composition

Key Terms

1. polynomial function 2. operations on functions 3. composite function

Objective 1

1. 10; 1 3. 2; 41 4. −4; −28

Objective 2

5. $(f+h)(x)=2x^2+4x-28$ 7. $(g-h)(x)=-x-7$

9. −36 11. 0 13. $(f-f)(x)=0$

15. $(fh)(x)=x^3+4x^2-16x-64$

17. $\left(\frac{f}{h}\right)(x)=x-4, x\neq 4$ 19. 96 21. −16

Objective 3

23. 258 25. 7 27. $(f\circ g)(x)=9x^2+6x+3$

29. $(f\circ h)(x)=x^2+8x+18$

7.5 Variation

Key Terms

1. direct variation
2. inverse variation as the nth power of x
3. inverse variation
4. joint variation
5. direct variation as a power

Objectives 1 and 2

1. $y = 3x$ 3. $y = 50x$ 5. 45 7. 147

Objective 3

9. $y = \frac{20}{x}$ 11. $y = \frac{5}{6x}$ 13. $\frac{4}{9}$ 15. 100 amperes

Objective 4

17. $y = \frac{1}{2}xz$ 19. 24 21. 128

Objective 5

23. $y = \frac{3x}{z}$ 25. $y = \frac{2x}{3z}$ 27. $\frac{1}{2}$ hour 29. 6,000,000 dynes

Chapter 8 SYSTEMS OF LINEAR EQUATIONS

8.1 Solving Systems of Linear Equations by Graphing

Key Terms

1. consistent system
2. solution set
3. independent equations
4. inconsistent system
5. dependent equations
6. system of linear equations
7. set-builder notation

Objective 1

1. solution
3. not a solution
5. not a solution

Objective 2

7. $(1, 2)$
9. $(1, 0)$
11. $(-4, 2)$

Objective 3

13. $\{(x, y)|4x-3y=12\}$
15. $\{(x, y)|5x-2y=10\}$
17. $\varnothing$

Objective 4

19. (a) inconsistent (b) parallel lines (c) no solution
21. (a) inconsistent (b) parallel lines (c) no solution
23. (a) dependent (b) one line (c) infinite number of solutions

Objective 5

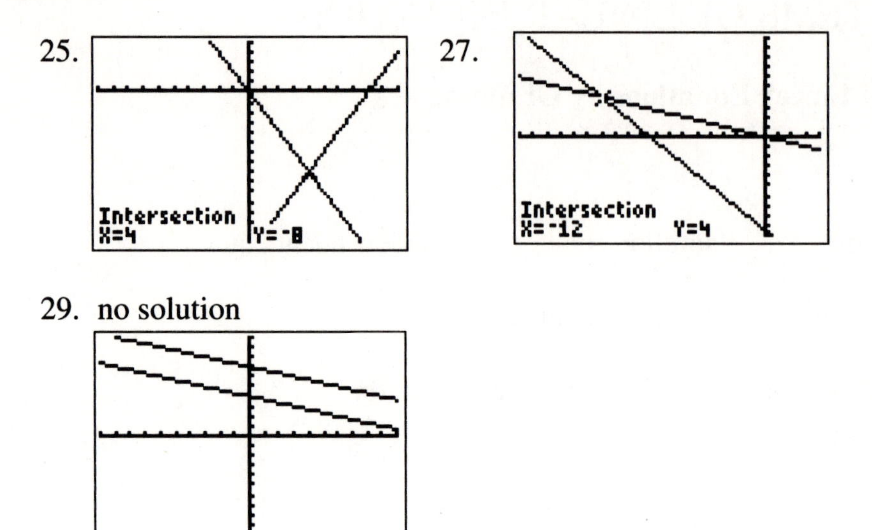

8.2 Solving Systems of Linear Equations by Substitution

Key Terms

1. substitution method
2. substitute
3. check

Objective 1

1. $\{(-4,-8)\}$ 3. $\{(-1, 3)\}$ 5. $\{(3, 1)\}$ 7. $\{(3, 4)\}$ 9. $\{(2, 1)\}$

Objective 2

11. $\{(x, y)|2x+y=3\}$ 13. $\varnothing$ 15. $\{(x, y)|36x-20y=12\}$

17. $\{(x, y)|48x-56y=32\}$ 19. $\{(x, y)|72x-60y=-12\}$

Objective 3

21. No solution 23. $\{(-44,-60)\}$

25. $\left\{\left(\frac{22}{9},\frac{8}{9}\right)\right\}$ 27. Infinite number of solutions 29. $\{(-12,0)\}$

8.3 Solving Systems of Linear Equations by Elimination

Key Terms

1. opposites 2. false 3. elimination method 4. true

Objective 1

1. $\{(3,1)\}$ 3. $\{(8,2)\}$ 5. $\left\{\left(-\frac{1}{2},5\right)\right\}$ 7. $\{(0,3)\}$

Objective 2

9. $\{(1,-8)\}$ 11. $\{(-3,1)\}$ 13. $\{(2,-5)\}$

Objective 3

15. $\{(4,2)\}$ 17. $\{(-8,-2)\}$ 19. $\left\{\left(\frac{19}{8},\frac{9}{8}\right)\right\}$ 21. $\left\{\left(-\frac{2}{13},\frac{23}{13}\right)\right\}$

Objective 4

23. No solution 25. Infinite number of solutions

27. No solution 29. No solution

8.4 Systems of Linear Equations in Three Variables

Key Terms

1. common point 2. coincide 3. ordered triple

4. line in common 5. no points common

Objective 2

1. $\{(-1, 2, 1)\}$ 3. $\{(3, 1, -2)\}$ 5. $\{(0, -2, 5)\}$ 7. $\{(4, -4, 1)\}$

9. $\{(-12, 18, 0)\}$

Objective 3

11. $\{(2, -1, 5)\}$ 13. $\{(2, -5, 3)\}$ 15. $\{(3, 5, 0)\}$ 17. $\{(-3, 5, -6)\}$

19. $\{(2, -3, 1)\}$

Objective 4

21. $\varnothing$ 23. $\{(x, y, z) | 3x-2y+4z=5\}$

25. $\{(0, 0, 0)\}$ 27. $\varnothing$ 29. $\{(x, y, z) | x-5y+2z=0\}$

8.5 Applications of Systems of Linear Equations

Key Terms

1. state the answer 2. read 3. assign variables

Objective 1

1. 12 ft 3. 15 in 5. 21 in

Objective 2

7. 12 fives; 20 tens 9. Small: \$.60; large: \$1.50

11. Marigold: \$12.29; carnation: \$17.60

Objective 3

13. 80 oz of 20%; 40 oz of 50% 15. $\frac{10}{9}$ L

17. 80 kg of \$12; 40 kg of \$15

Objective 4

19. 65 mph: 195 mi; 60 mph: 180 mi 21. 5.5 hr

23. 20 mph

Objective 5

25. 9, 10, 12 27. 50°, 70°, 60°

29. 52 cm, 71 cm, 74 cm

8.6 Solving Systems of Linear Equations by Matrix Methods

Key Terms

1. row operations 2. elements 3. row echelon form

4. augmented matrix 5. square matrix 6. dimensions

Objective 1

1. Examples are $\begin{bmatrix} 2 & 1 \\ -3 & 0 \\ 5 & 2 \end{bmatrix}$ and $\begin{bmatrix} 5 & -2 & 0 & 7 \end{bmatrix}$.

In Exercises 3 and 5, the number of rows is given first.

3. 2, 2 5. 2, 1

Objective 2

7. $\left[\begin{array}{cc|c} 5 & -1 & 6 \\ 2 & 3 & 4 \end{array}\right]$ 9. $\left[\begin{array}{cc|c} 1 & 0 & -6 \\ 0 & 1 & 2 \end{array}\right]$ 11. $\left[\begin{array}{ccc|c} 4 & -7 & 1 & 1 \\ 2 & 3 & -5 & -2 \\ 6 & -1 & 8 & 5 \end{array}\right]$

Objective 3

13. $\{(4, 1)\}$ 15. $\{(3, 1)\}$ 17. $\left\{\left(\frac{2}{3}, \frac{3}{4}\right)\right\}$

Objective 4

19. $\{(1, -1, 2)\}$ 21. $\{(1, 1, 3)\}$ 23. $\{(2, 1, 2)\}$

Objective 5

25. $\varnothing$ 27. $\left\{(x, y) \mid \frac{2}{3}x - \frac{4}{5}y = 7\right\}$ 29. $\varnothing$

Chapter 9 INEQUALITIES AND ABSOLUTE VALUE

9.1 Set Operations and Compound Inequalities

Key Terms

1. intersection of two sets
2. compound inequality
3. union of two sets

Objective 1

1. $\{2, 3\}$
3. $\varnothing$
5. $\{1, 3, 5\}$
7. $\varnothing$

Objective 2

9. $(-\infty, -4]$

11. $(-2, -1)$

13. $(-\infty, 4]$

15. $\varnothing$

Objective 3

17. $\{-6, -5, -4, -3, -2, -1\}$

19. $\{0, 1, 2, 3, 4, 5, 6, 8, 10\}$

21. $\{0, 1, 2, 3, 4, 5, 6, 7, 8, 9, 10\}$

Objective 4

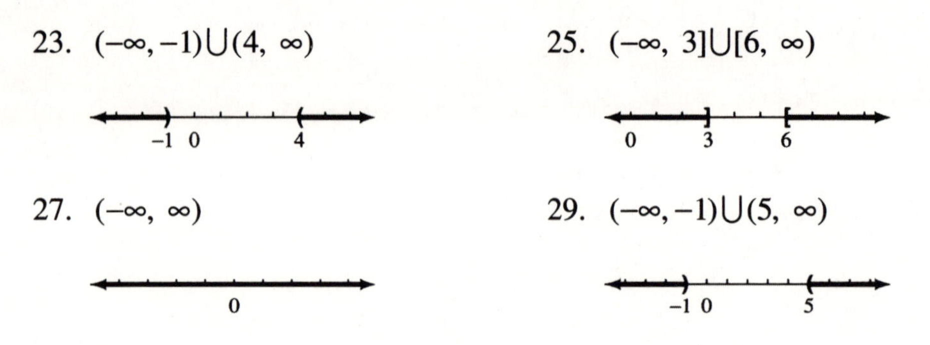

23. $(-\infty, -1) \cup (4, \infty)$

25. $(-\infty, 3] \cup [6, \infty)$

27. $(-\infty, \infty)$

29. $(-\infty, -1) \cup (5, \infty)$

9.2 Solving Systems of Linear Equations by Substitution

Key Terms

1. negative 2. absolute value equation 3. absolute value inequality

Objective 1

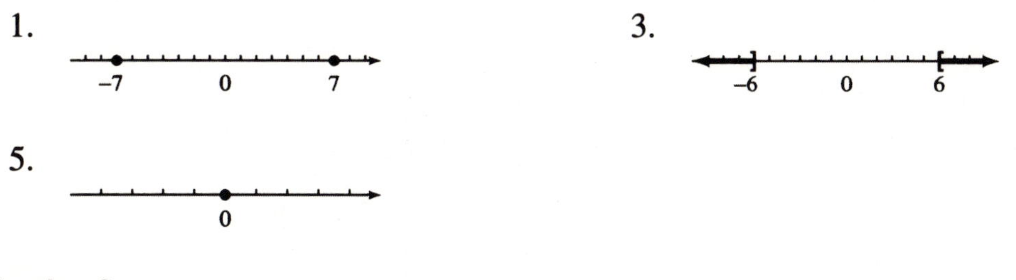

Objective 2

7. $\left\{\frac{7}{3}, -\frac{5}{3}\right\}$ 9. $\left\{-\frac{3}{2}\right\}$

Objective 3

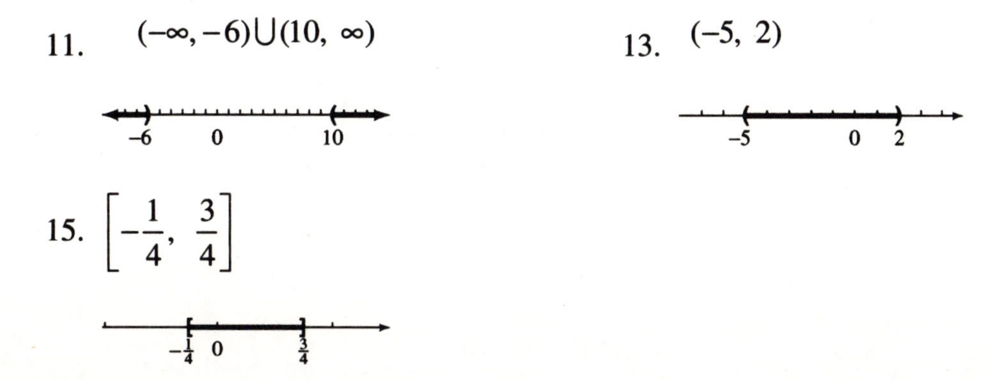

Objective 4

17. $\varnothing$ 19. $\{3, -2\}$

Objective 5

21. $\left\{\frac{7}{2}\right\}$ 23. $\left\{2, -\frac{3}{2}\right\}$ 25. $\left\{\frac{5}{2}, -\frac{1}{2}\right\}$

Objective 6

27. $\{0\}$ 29. $\varnothing$

9.3 Solving Systems of Linear Equations by Elimination

Key Terms

1. test point 2. boundary line 3. linear inequality in two variables

Objective 1

1.

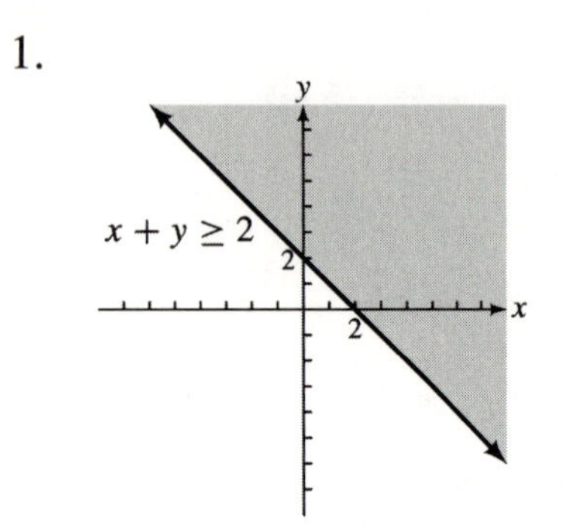

3.

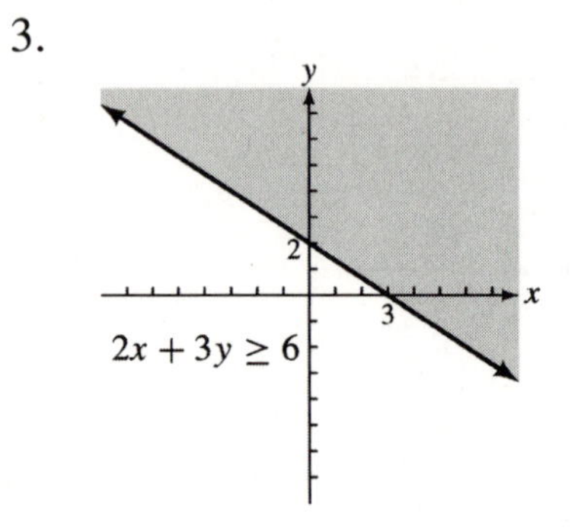

5.

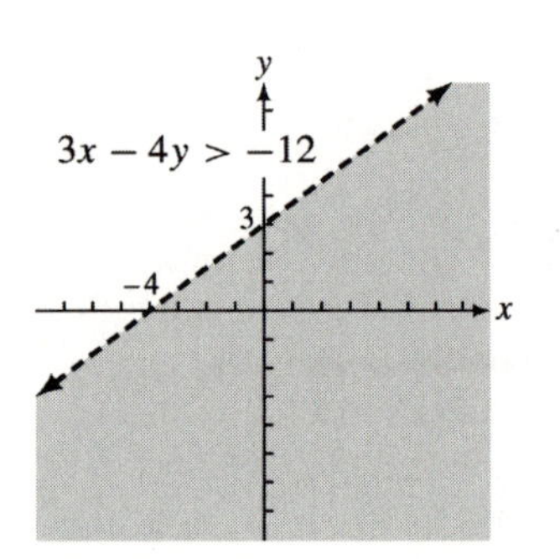

7.

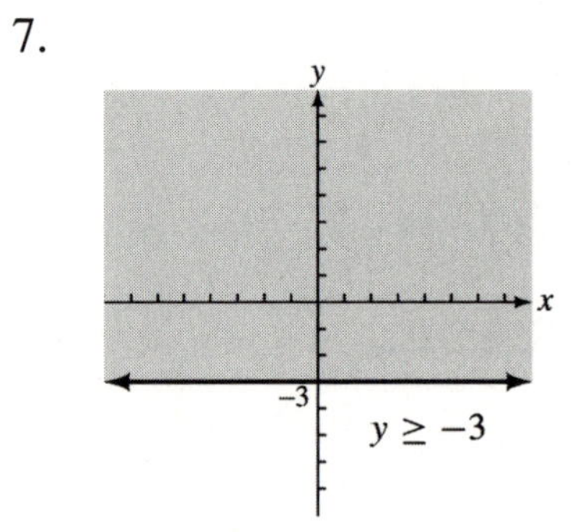

Objective 2

9.

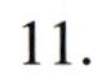
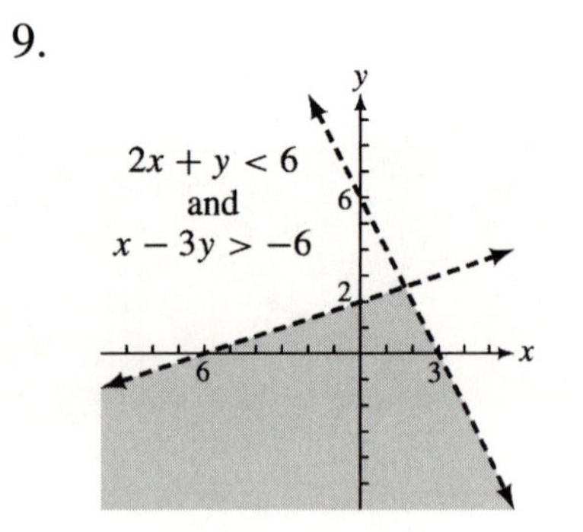

11.

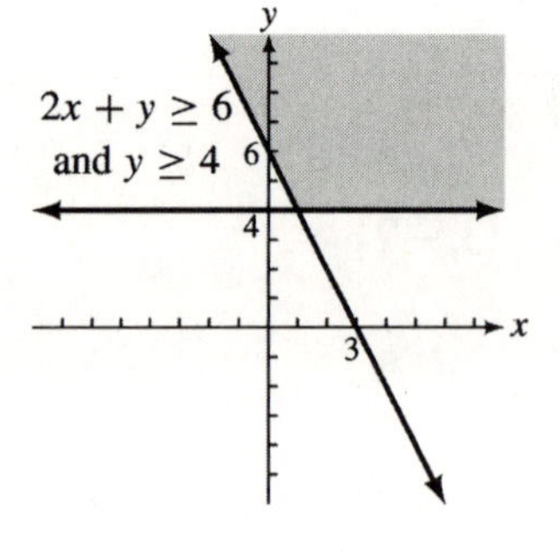

13. 15.

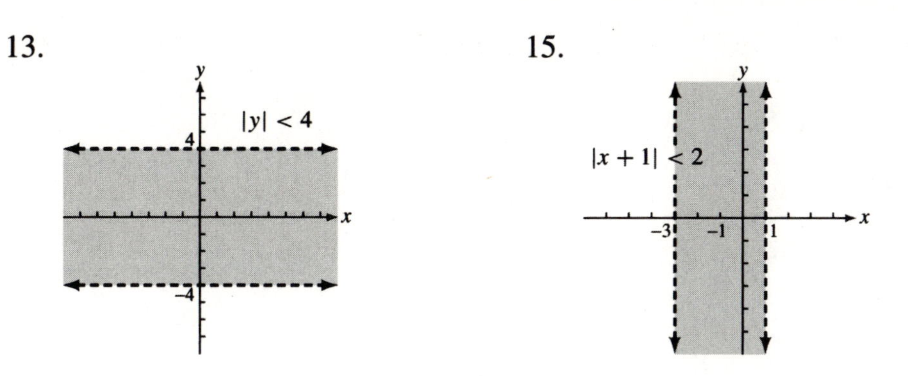

Objective 3

17.

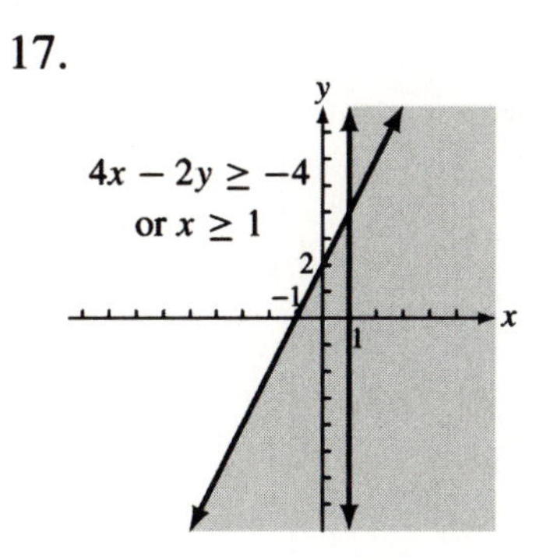

19.

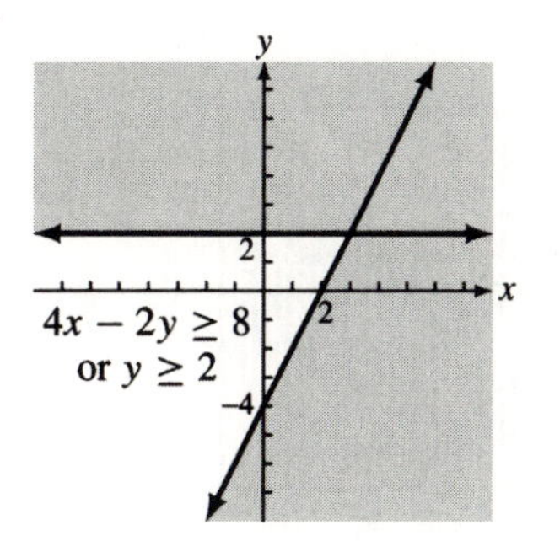

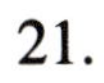

21.

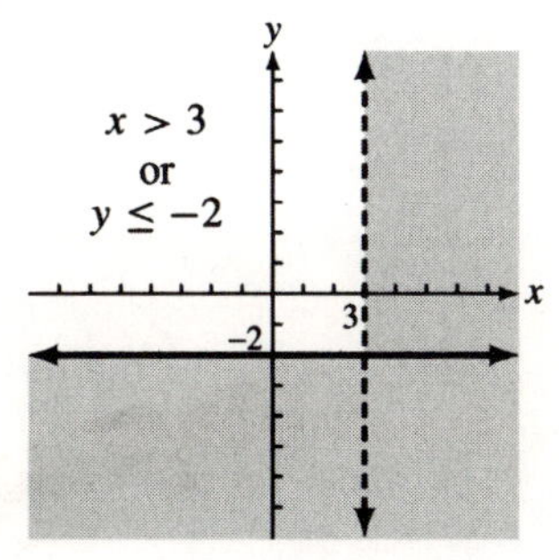

23.

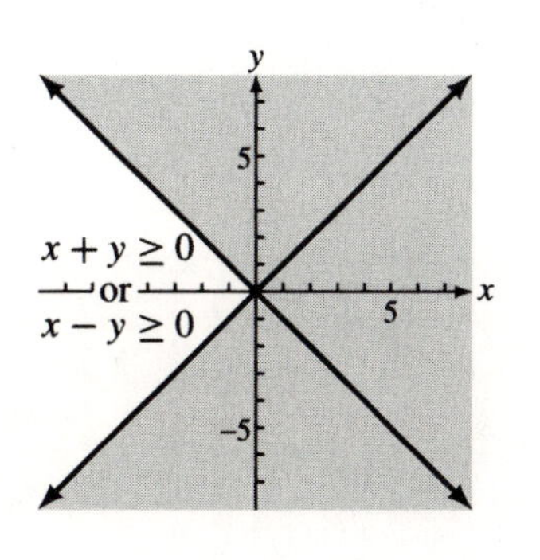

25.

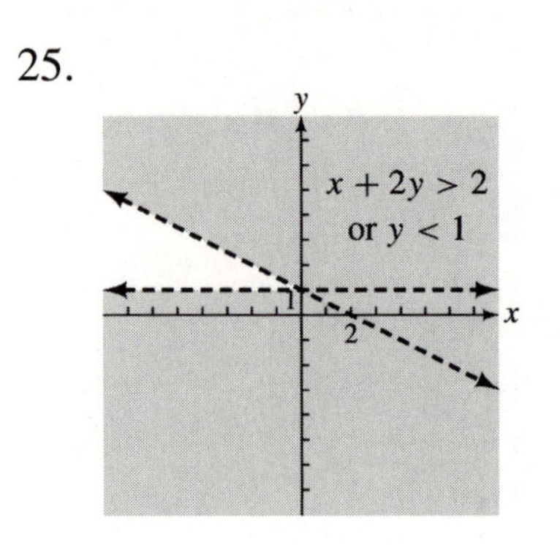

Objective 4

27. $(-\infty,\ 2.57)$

29.

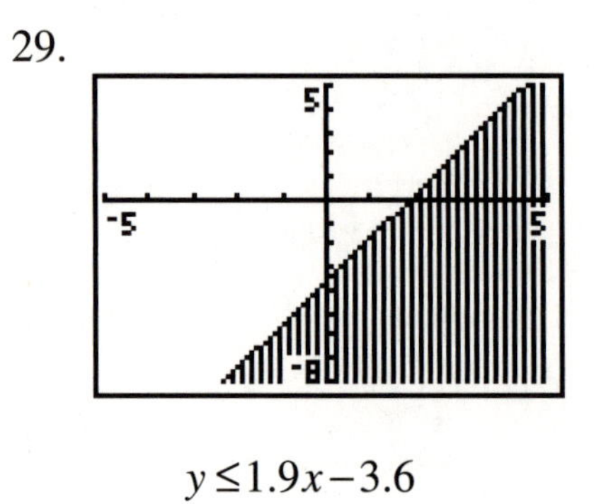

$y \le 1.9x - 3.6$

Chapter 10 ROOTS, RADICALS, AND ROOT FUNCTIONS

10.1 Radical Expressions and Graphs

Key Terms

1. radical sign 2. perfect square 3. index 4. principal roots

5. radical expression 6. *n*th root 7. radicand

Objective 1

1. ±13 3. ±35 5. 10 7. −23

Objective 2

9. rational 11. rational

Objective 3

13. 3 15. −6 17. 2

Objective 4

19. Domain: $x \geq -6$; Range: $f(x) \geq 0$ 21. Domain: $(-\infty, \infty)$; Range: $(-\infty, \infty)$

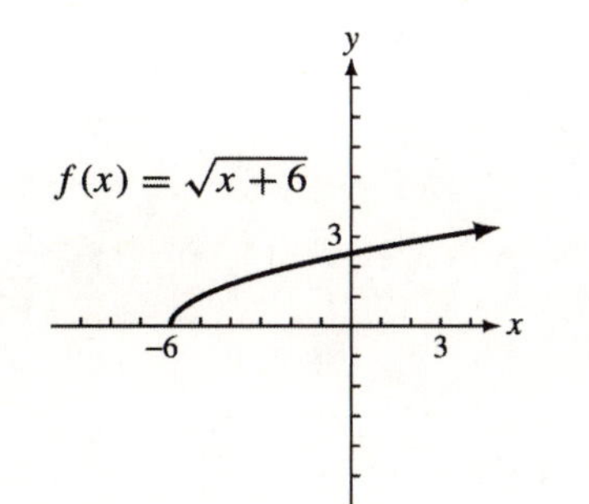

Objective 5

23. $-y^3$

Objective 6

25. 8.718 27. −31.464 29. 3.368

10.2 Rational exponents

Key Terms

1. $a^{1/n}$ 2. $a^{-m/n}$ 3. radical form of $a^{m/n}$ 4. $a^{m/n}$

Objective 1

1. 2 3. 5 5. –12

Objectives 2 and 3

7. –243 9. –125 11. 4 13. $-\frac{11}{5}$ 15. $\left(\sqrt[3]{7x^2y}\right)^2$

17. $12x^{\frac{5}{4}}$ 19. $n^{-\frac{4}{3}}$

Objective 4

21. 8 23. a^4 25. $y-3y^3$

27. $d^{\frac{6}{7}}+4d^{\frac{13}{7}}$ 29. $x^{\frac{3}{2}}\left(x^{\frac{3}{2}}-2\right)$

10.3 Simplifying radical expressions

Key Terms

1. Pythagorean formula
2. product rule for radicals
3. quotient rule for radicals
4. distance formula

Objective 1

1. $\sqrt{70}$ 3. $\sqrt{\frac{33}{(rp)}}$ 5. Cannot be simplified

Objective 2

7. $\frac{5}{4}$ 9. $\frac{a^2}{5}$ 11. $\frac{\sqrt[4]{p}}{2}$

Objective 3

13. $3\sqrt{3}$ 15. $2\sqrt[3]{3}$ 17. $2x^3y^2\sqrt[4]{y^2}$

Objective 4

19. $x\sqrt[3]{x}$ 21. $\sqrt[4]{x^3y^2}$

Objective 5

23. 26 25. $\sqrt{39}$

Objective 6

27. 5 29. $7\sqrt{2}$

10.4 Adding and subtracting radical expressions

Key Terms

1. same radicand
2. sum of the roots

Objective 1

1. $5\sqrt{7}$
3. $27\sqrt{3}$
5. Cannot be simplified
7. $22\sqrt[3]{3}$
9. $3\sqrt{13}$
11. $9\sqrt{3y}$
13. $9\sqrt[3]{3}$
15. $12\sqrt{x}$
17. $60\sqrt{2}$
19. Cannot be simplified
21. $\frac{\left(4-\sqrt{3}\right)}{3}$
23. 2
25. $\frac{7\sqrt{7}}{4}$
27. $\frac{-\sqrt{3}}{35}$
29. $\sqrt{5}$

10.5 Multiplying and dividing radical expressions

Key Terms

1. conjugates
2. FOIL method
3. rationalizing the denominator

Objective 1

1. $6+3\sqrt{7}+2\sqrt{2}+\sqrt{14}$

3. $9-\sqrt[3]{25}$

Objective 2

5. $\frac{(6\sqrt{5})}{5}$

7. $\frac{7\sqrt{3}}{15}$

9. $\frac{\sqrt{5}}{2}$

11. $\frac{3}{4}$

13. $\frac{(2\sqrt{2m})}{m}$

15. $\frac{(3\sqrt{14})}{7}$

17. $\frac{\sqrt[3]{3}}{3}$

19. $\frac{\sqrt[3]{3}}{2}$

21. $\frac{(t^2\sqrt[3]{x^2})}{x^3}$

Objective 3

23. $3(\sqrt{7}-3)$

25. $\frac{(4-\sqrt{5})}{11}$

Objective 4

27. $4-\sqrt{2}$

29. $\frac{(1-\sqrt{2})}{2}$

10.6 Solving equations with radicals

Key Terms

1. power rule for solving an equation with radicals

2. extraneous solutions 3. isolate the radical

Objective 1

1. {25} 3. {15} 5. {41} 7. {7} 9. {15}

Objective 2

11. {2} 13. {6} 15. {2} 17. {−2} 19. {−1}

Objective 3

21. {3} 23. {33}

Objective 4

25. {5} 27. {−3}

Objective 5

29. $C = \frac{1}{4\pi^2 f^2 L}$

10.7 Complex numbers

Key Terms

1. real part 2. complex number 3. pure imaginary number

4. imaginary unit 5. imaginary part 6. $i\sqrt{b}$ 7. standard form

Objective 1

1. $7i$ 3. $-3i\sqrt{7}$ 5. $-6i\sqrt{2}$ 7. $i\sqrt{42}$ 9. 5

Objective 2

11. Nonreal Complex 13. Pure Imaginary

Objective 3

15. $7+6i$ 17. $3\sqrt{3}-4i\sqrt{2}$

Objective 4

19. $11+13i$ 21. $-1-2i\sqrt{6}$

Objective 5

23. $\frac{37}{97}+\left(\frac{38}{97}\right)i$ 25. $\frac{1}{5}+\left(\frac{1}{5}\right)i$

Objective 6

27. 1 29. 1

Chapter 11 QUADRATIC EQUATIONS, INEQUALITIES, AND FUNCTIONS

11.1 Solving Quadratic Equations by the Square Root Property

Key Terms

1. standard form
2. quadratic equation
3. square root property
4. zero-factor property

Objective 1

1. $\{-9, 8\}$ 3. $\left\{-\frac{5}{2}, -\frac{2}{3}\right\}$ 5. $\left\{-\frac{5}{4}, \frac{1}{3}\right\}$ 7. $\left\{-\frac{5}{4}, \frac{5}{4}\right\}$

Objective 2

9. $\{4, -4\}$ 11. $\{2, -2\}$ 13. $\{3, -3\}$

Objective 3

15. $\{1, 9\}$ 17. $\left\{\frac{(-3+\sqrt{5})}{4}, \frac{(-3-\sqrt{5})}{4}\right\}$

19. $\left\{-\frac{11}{3}, 1\right\}$ 21. $\left\{\frac{(-5+2\sqrt{3})}{4}, \frac{-5-2\sqrt{3}}{4}\right\}$

Objective 4

23. $\{3i, -3i\}$ 25. $\{5i, -5i\}$ 27. $\left\{\frac{3i}{2}, -\frac{3i}{2}\right\}$ 29. $\left\{\frac{(1+i\sqrt{5})}{7}, \frac{(1-i\sqrt{5})}{7}\right\}$

11.2 Solving Quadratic Equations by Completing the Square

Key Terms

1. completing the square 2. *a* 3. square root property

Objective 1

1. $\{-3, -1\}$ 3. $\{-5, 2\}$ 5. $\{9, 3\}$ 7. $\{3, -1\}$ 9. $\{5, 9\}$

Objective 2

11. $\left\{0, \frac{5}{3}\right\}$ 13. $\left\{\frac{2+\sqrt{6}}{2}, \frac{2-\sqrt{6}}{2}\right\}$ 15. $\left\{-\frac{2}{3}, -\frac{3}{2}\right\}$

17. $\left\{4, -\frac{3}{4}\right\}$ 19. $\left\{-2, -\frac{4}{3}\right\}$

Objective 3

21. $\{-3, -4\}$ 23. $\left\{\frac{1+3\sqrt{37}}{2}, \frac{1-3\sqrt{37}}{2}\right\}$ 25. $\{-11, 5\}$

27. $\left\{\frac{3}{4}, -4\right\}$ 29. $\{5, -3\}$

11.3 Solving Quadratic Equations by the Quadratic Formula

Key Terms

1. quadratic formula
2. discriminant
3. two irrational solutions
4. one rational solution

Objective 2

1. $\{-5, -7\}$

3. $\left\{\frac{3}{5}, 2\right\}$

5. $\left\{\frac{3}{4}, -\frac{3}{4}\right\}$

7. $\left\{\frac{\left(-5+3\sqrt{5}\right)}{10}, \frac{\left(-5-3\sqrt{5}\right)}{10}\right\}$

9. $\{3+i,\ 3-i\}$

11. $\left\{\frac{\left(5+i\sqrt{7}\right)}{4}, \frac{\left(5-i\sqrt{7}\right)}{4}\right\}$

13. $\left\{\frac{\left(-9+i\sqrt{39}\right)}{10}, \frac{\left(-9-i\sqrt{39}\right)}{10}\right\}$

Objective 3

15. (b)
17. (d)
19. (d)
21. (c)
23. Cannot be factored
25. $(3k-1)(4k+9)$
27. Cannot be factored
29. $(5y+3)(6y-5)$

11.4 Equations in Quadratic Form

Key Terms

1. quadratic in form 2. quadratic formula 3. assign a variable

Objective 1

1. $\{-3, 2\}$ 3. $\left\{-\frac{7}{2}, -\frac{1}{3}\right\}$ 5. $\left\{-\frac{7}{2}, 4\right\}$ 7. $\left\{-3, -\frac{1}{5}\right\}$

Objective 2

9. 9 hr 11. 15 hr for one, 30 hr for the other

Objective 3

13. $\{3\}$ 15. $\left\{\frac{3}{2}\right\}$ 17. $\left\{\frac{1}{2}, 1\right\}$ 19. $\left\{\frac{7}{3}, 7\right\}$

Objective 4

21. $\{3, -3, 4, -4\}$ 23. $\{-6, 14\}$ 25. $\{-9, -7\}$ 27. $\{1\}$

29. $\{2, -2, i\sqrt{5}, -i\sqrt{5}\}$

11.5 Formulas and Further Applications

Key Terms

1. quadratic function
2. check
3. square both sides

Objective 1

1. $\sqrt{2p^2+k^2}, -\sqrt{2p^2+k^2}$

3. $\dfrac{\left(\sqrt{2k^2+18}\right)}{2}, \dfrac{\left(-\sqrt{2k^2+18}\right)}{2}$

5. $\dfrac{\left(\sqrt{2y^2-6y}\right)}{2}, \dfrac{\left(-\sqrt{2y^2-6y}\right)}{2}$

7. $c=\dfrac{A^2}{b}$

9. $D=\pm\dfrac{\sqrt{ABC}}{A}$

Objective 2

11. South: 72 mi; east: 54 mi

15. 48 cm

Objective 3

17. 3 m by 5 m

19. 12 in by 15 in

21. 9 in by 14 in

23. 2 ft

Objective 4

25. 2.5 sec

27. 3.1 sec

29. 9.1 sec

11.6 Graphs of Quadratic Functions

Key Terms

1. vertical shift
2. horizontal shift
3. axis
4. parabola
5. vertex

Objectives 1 and 2

1. $(0, -1)$ 

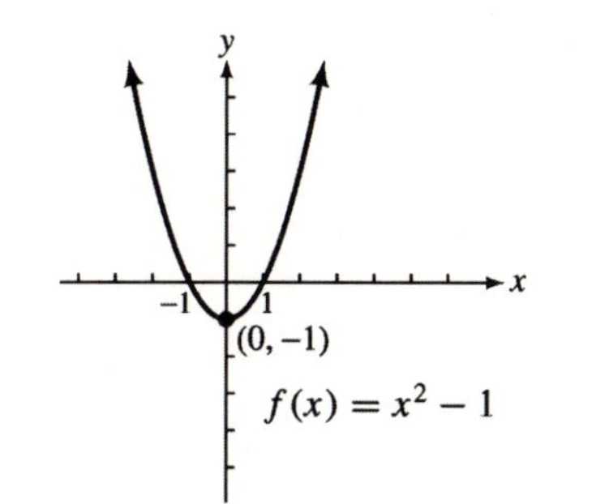

3. $(0, 3)$

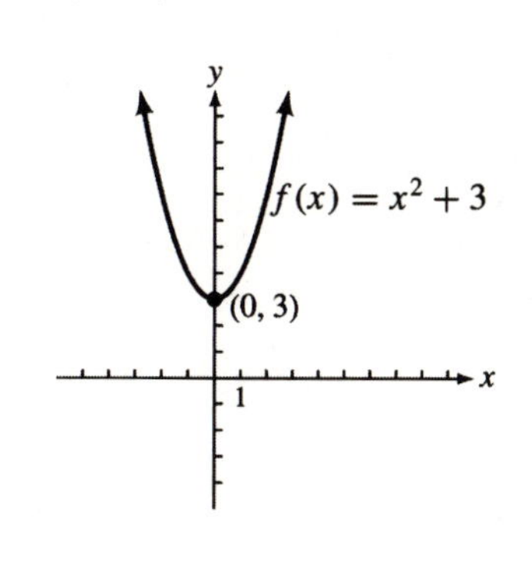

5. $(0, 2)$

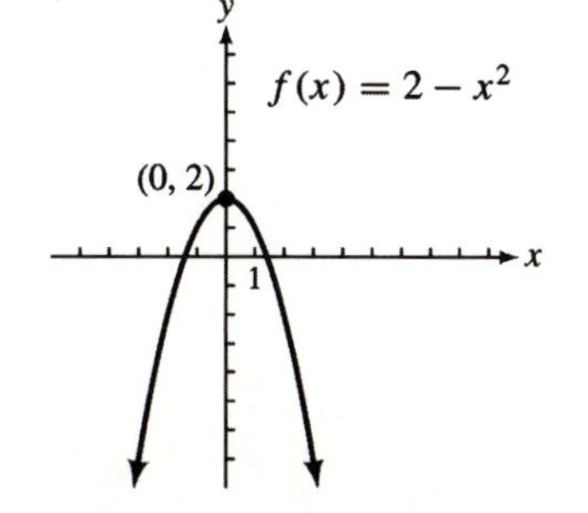

7. $(-2, 0)$ 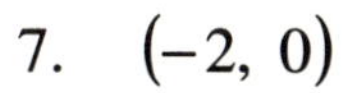

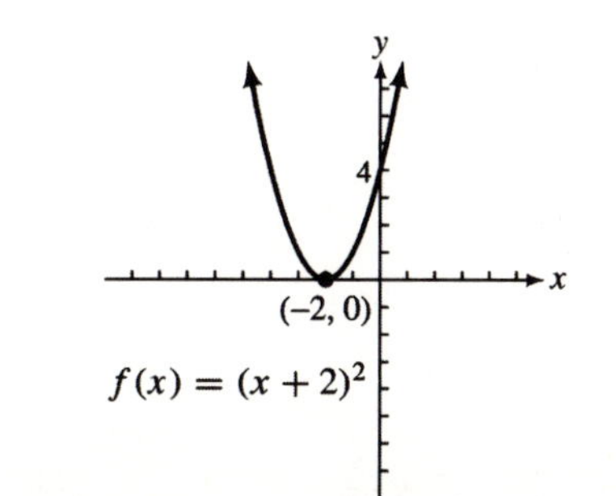

9. $(-3, -1)$

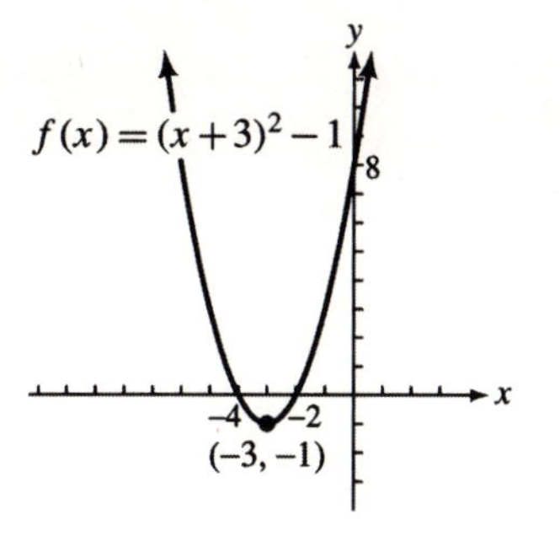

11. $(-2, 3)$

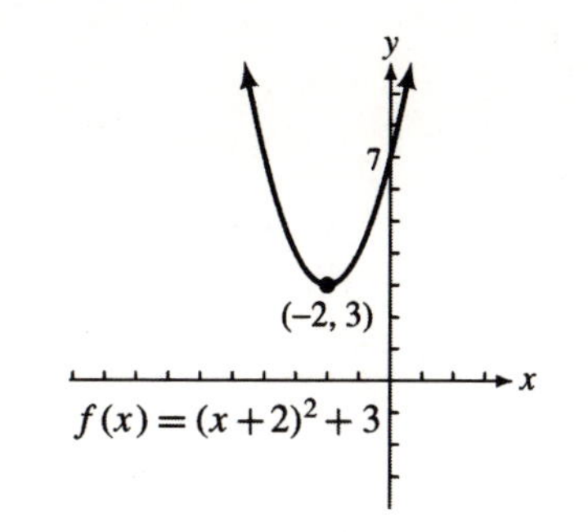

Objective 3

13. Upward; wider
15. Downward; narrower
17. Upward; wider
19. Downward; narrower
21. Upward; narrower
23. Upward; narrower
25. Downward; wider

Objective 4

27.

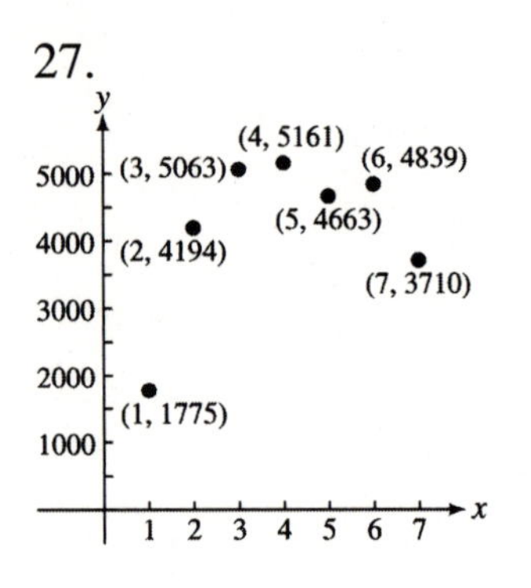

29. $y = -268.72x^2 + 2472.28x - 428.56$

11.7 More About Parabolas and Their Applications

Key Terms

1. finding any intercepts
2. horizontal axis
3. vertex

Objectives 1 and 2

1. $(-2, 1)$

y
5
(−2, 1)
x
$f(x) = x^2 + 4x + 5$

3. $(4, 6)$

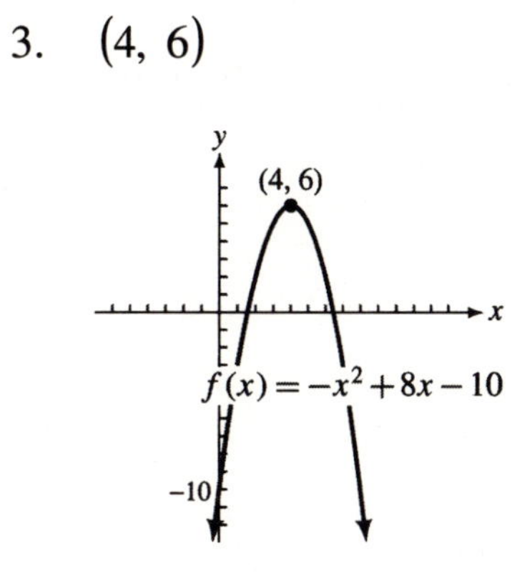

5. $(-1, -1)$

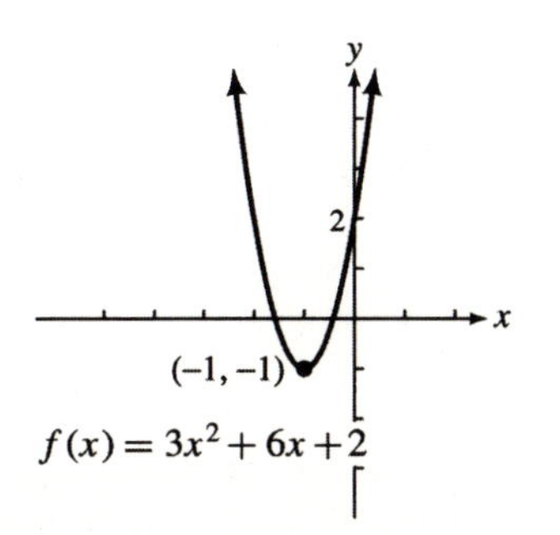

7. $(-2, -2)$

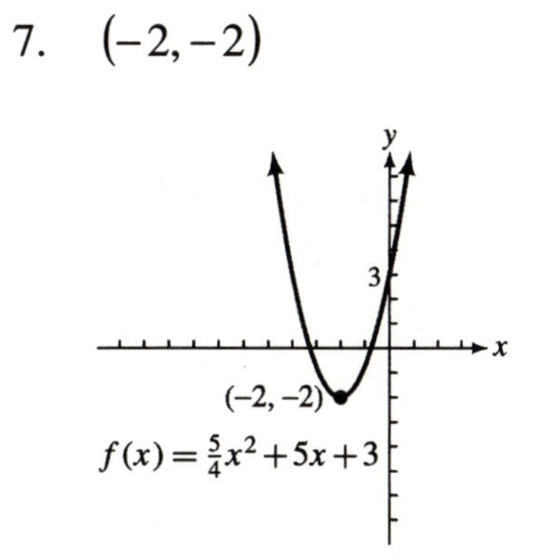

Objective 3

9. 0 11. 0 13. 0

Objective 4

15. 30 units; $6600 17. 64 ft; 2 sec

19. 256 ft; $\frac{5}{2}$ sec 21. 9 (a square)

Objective 5

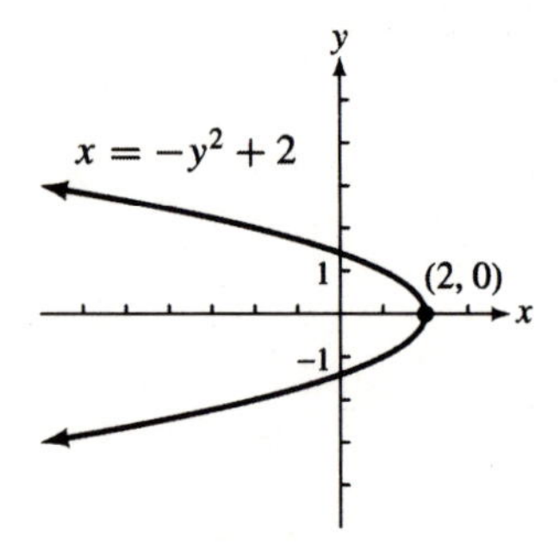

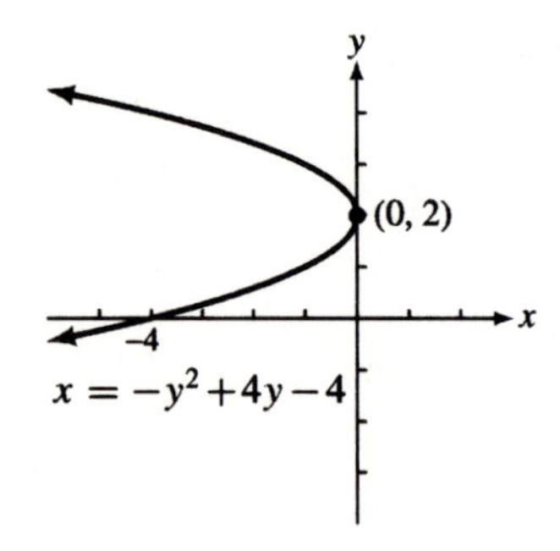

27.

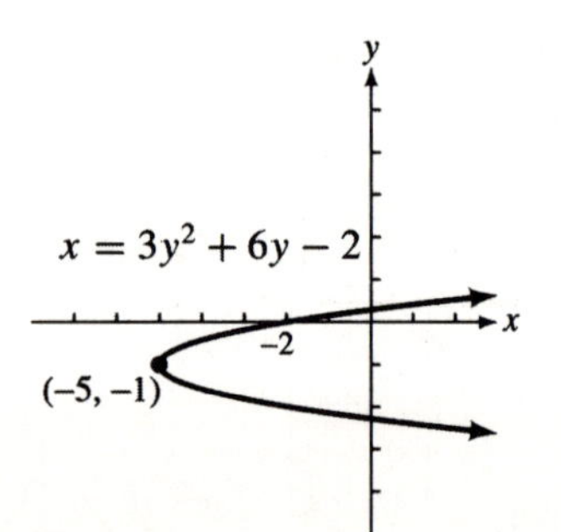

29. $(1, -1)$

11.8 Quadratic and Rational Inequalities

Key Terms

1. quadratic inequality 2. test number 3. rational inequality

Objective 1

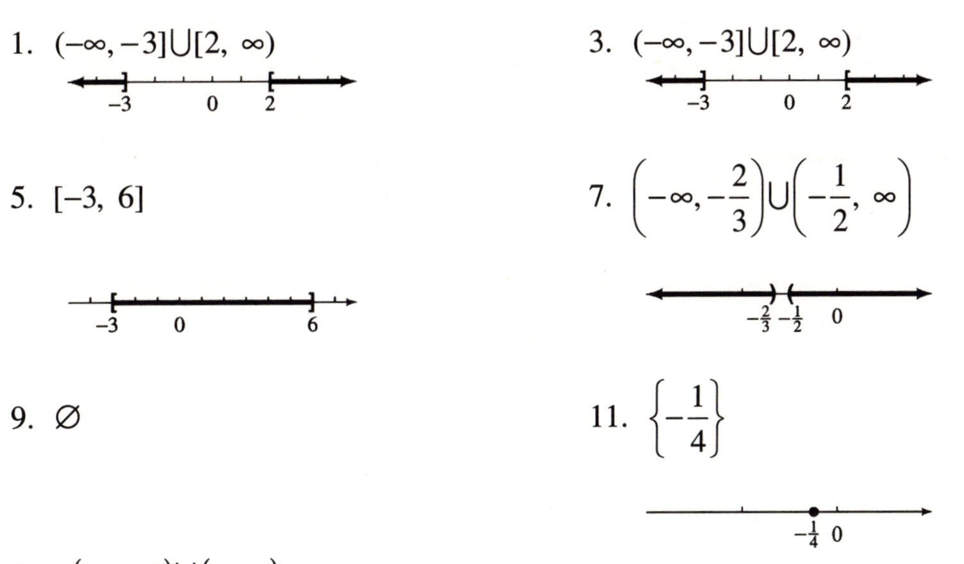

1. $(-\infty, -3] \cup [2, \infty)$

3. $(-\infty, -3] \cup [2, \infty)$

5. $[-3, 6]$

7. $\left(-\infty, -\frac{2}{3}\right) \cup \left(-\frac{1}{2}, \infty\right)$

9. $\varnothing$

11. $\left\{-\frac{1}{4}\right\}$

13. $(-\infty, 3) \cup (3, \infty)$

0 3

Objective 2

15. $(-\infty, -4] \cup [-1, 2]$

−4 −1 0 2

17. $(-\infty, -3] \cup [-1, 4]$

−3 −1 0 4

19. $\left(-\infty, -\frac{3}{2}\right] \cup \left[-\frac{1}{3}, \frac{1}{2}\right]$

$-\frac{3}{2}$ $-\frac{1}{3}$ 0 $\frac{1}{2}$

21. $\left[\frac{3}{4}, \frac{10}{3}\right] \cup \left[\frac{7}{2}, \infty\right)$

0 $\frac{3}{4}$ $\frac{10}{3}$ $\frac{7}{2}$

Objective 3

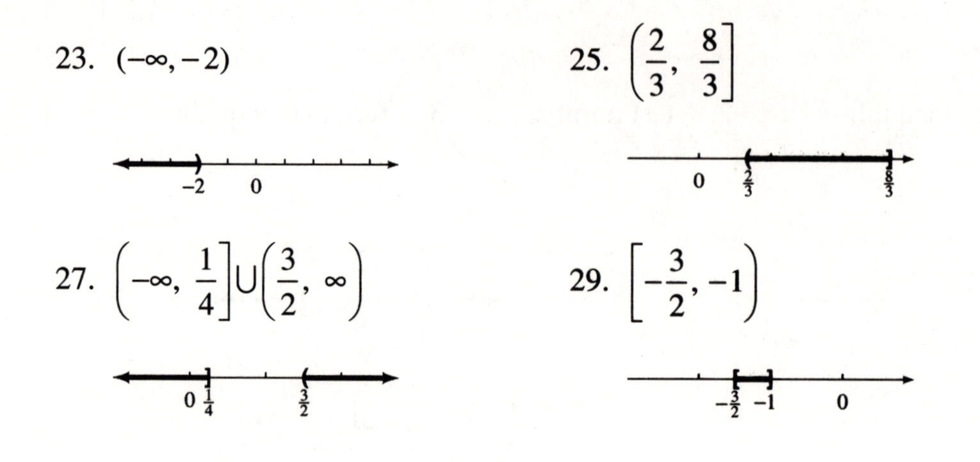

23. $(-\infty, -2)$

25. $\left(\frac{2}{3}, \frac{8}{3}\right]$

27. $\left(-\infty, \frac{1}{4}\right] \cup \left(\frac{3}{2}, \infty\right)$

29. $\left[-\frac{3}{2}, -1\right)$

Chapter 12 INVERSE, EXPONENTIAL, AND LOGARITHMIC FUNCTIONS

12.1 Inverse Functions

Key Terms

1. interchange x and y
2. inverse
3. one-to-one function
4. horizontal line test

Objective 1

1. Not one-to-one
3. Not one-to-one
5. $\{(1,-3),\ (2,-2),\ (3,-1),\ (4,\ 0)\}$

Objective 2

7. One-to-one
9. Not a Function
11. Not one-to-one

Objective 3

13. $f^{-1}(x)=\dfrac{(x+5)}{2}$
15. Not one-to-one
17. $f^{-1}(x)=\sqrt[3]{x+1}$
19. $f^{-1}(x)=\dfrac{(3+x)}{x}$

Objective 4

21.

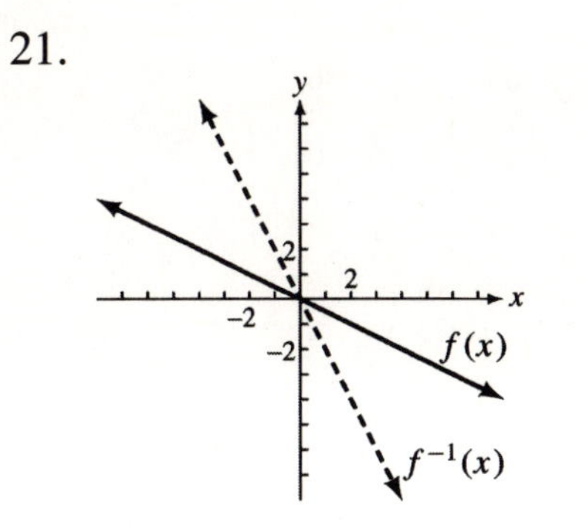

23. Not one-to-one

25. Not one-to one

Objective 5

27. $y_1 = 2x - 5$; INV y_1

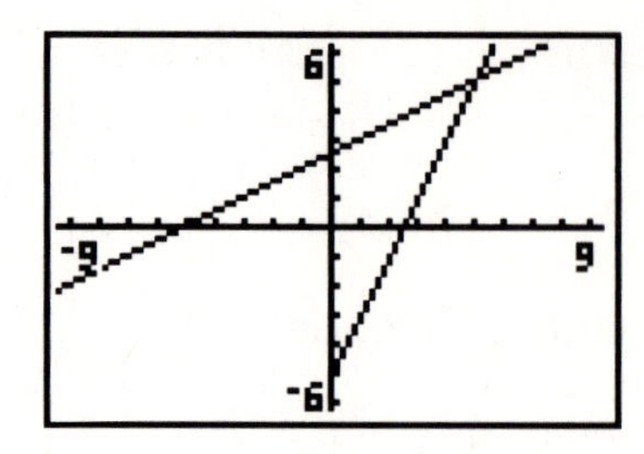

29. $y_1 = x^3 - 2x^2 + 3$; INV y_1

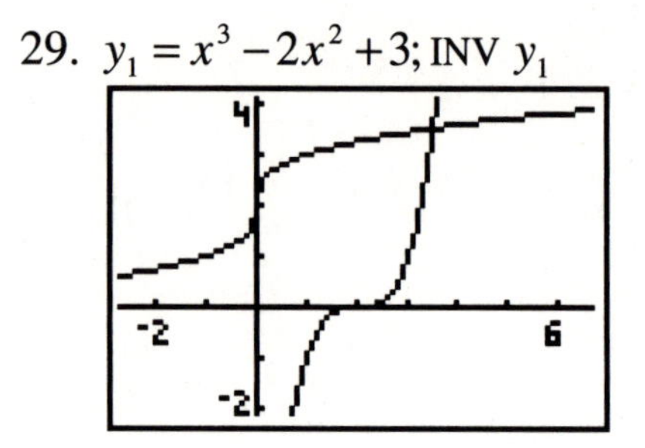

12.2 Exponential Functions

Key Terms

1. exponential function
2. exponential equation
3. asymptote
4. same base

Objective 1

1. Exponential function
3. Not an exponential function
5. Exponential function
7. Not an exponential function

Objective 2

9.

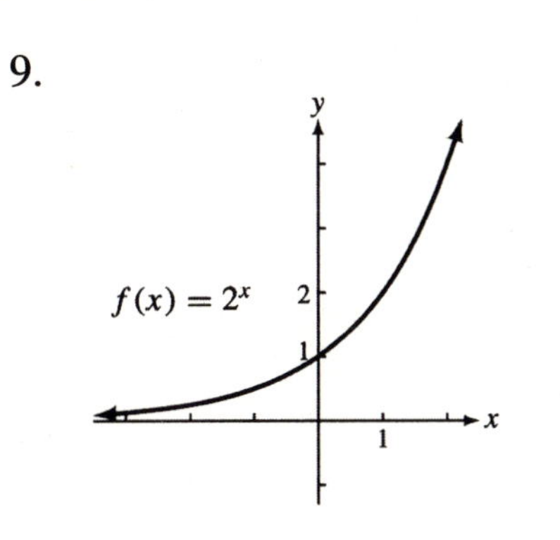

11.

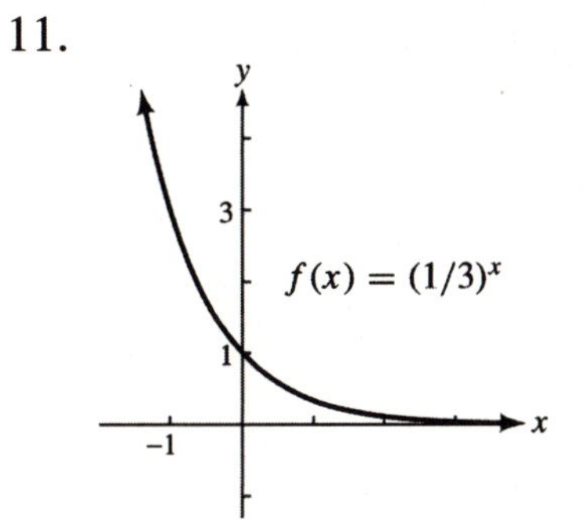

13.

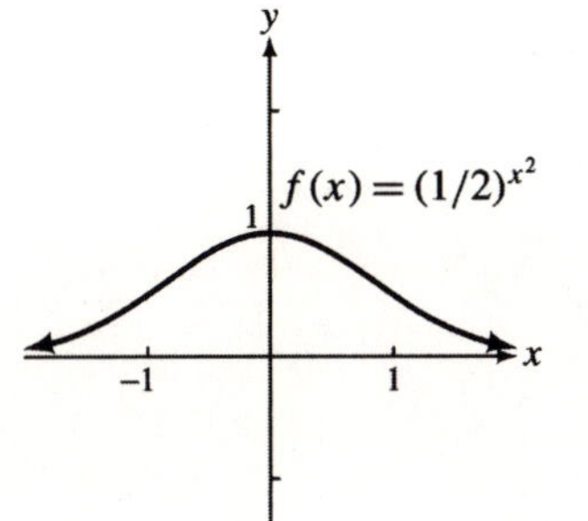

15.

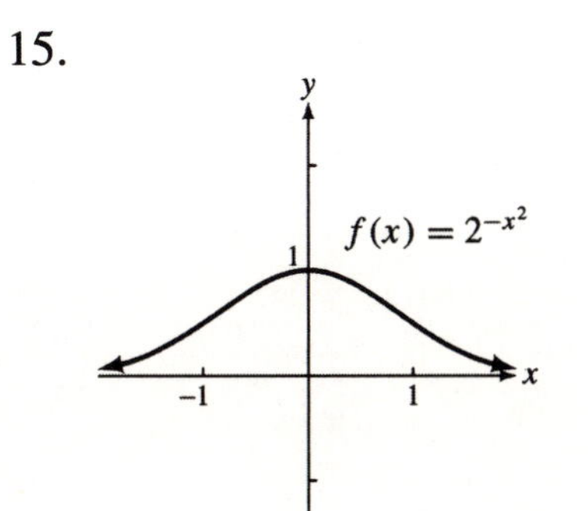

Objective 3

17. $\left\{\frac{2}{3}\right\}$ 19. $\{1\}$ 21. $\left\{\frac{1}{2}\right\}$

Objective 4

23. 515 geese 25. 7750 bacteria 27. 1 g 29. 2500 bacteria

12.3 Logarithmic Functions

Key Terms

1. base 2. logarithm 3. rise 4. fall 5. logarithmic function

Objective 1

1. 3 3. –1 5. –2

Objective 2

7. $\log_3 9 = 2$ 9. $16^{\frac{1}{4}} = 2$ 11. $10^{-3} = .001$

Objective 3

13. $\{6\}$ 15. $\{5\}$ 17. $\{2\}$

Objective 4

19.

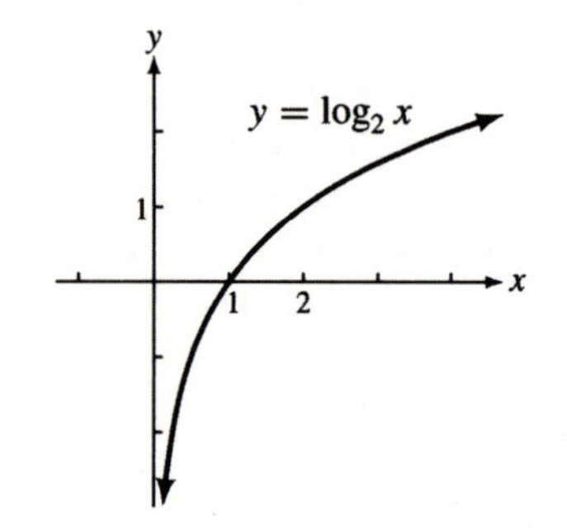

21.

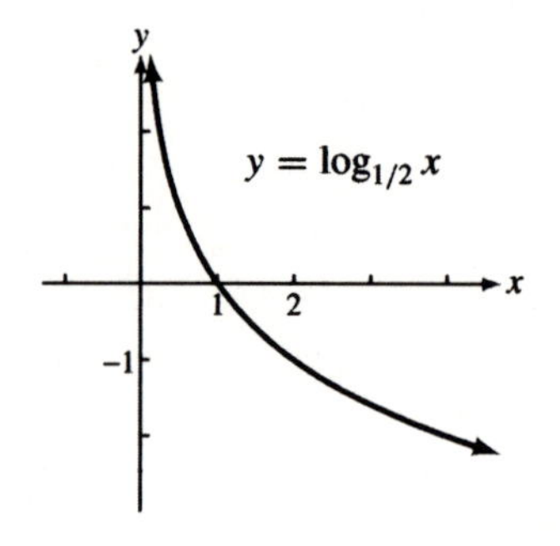

23.

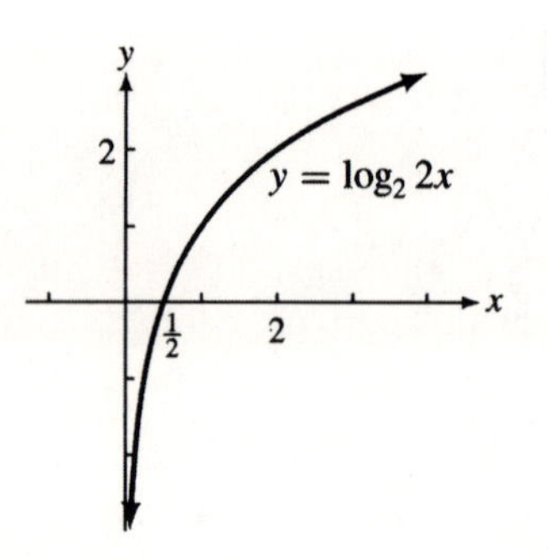

Objective 5

25. 28 squirrels 27. 1200 cm/sec 29. 8 students

12.4 Properties of Logarithms

Key Terms

1. special properties
2. quotient rule for logarithms
3. product rule for logarithms
4. power rule for logarithms

Objective 1

1. $\log_3 6 + \log_3 5$ 3. $1 + \log_6 r$ 5. $\log_4 21$ 7. $\log_7 66y^3$

Objective 2

9. $\log_3 m - \log_3 n$ 11. $\log_3 10 - \log_3 x$ 13. $\log_{10} \frac{12m}{7}$

Objective 3

15. $2 \cdot \log_5 3$ 17. $7 \cdot \log_m 2$ 19. $\frac{1}{3} \log_3 7$

Objective 4

23. $\frac{1}{3}(\log_a 2 + \log_a k)$ 25. $\log_5 7 + 3 \cdot \log_5 m - \log_5 8 - \log_5 y$

27. $\log_2 15mk$ 29. $\log_4 \frac{4}{y}$ or $1 - \log_4 y$

12.5 Common and Natural Logarithms

Key Terms

1. always negative   2. universal constant   3. change-of-base rule

4. common logarithm   5. natural logarithm

Objective 1

1. 1.7576   3. 2.9262   5. 5.4472

Objective 2

7. 7.3   9. 8.8   11. 4.0×10^{-4} 13. 5.0×10^{-2}

Objective 3

15. 4.9628   17. 6.0591

Objective 4

19. 600   21. 74.1

Objective 5

23. 1.189   25. 2.322   27. 3.227   29. −1.730

12.6 Exponential and Logarithmic Equations; Further Applications

Key Terms

1. $x = y$ 2. single logarithm 3. continuous compound interest formula

4. half-life 5. power rule 6. compound interest formula

Objective 1

1. $\{.49\}$ 3. $\{1.23\}$ 5. $\{3.47\}$

Objective 2

7. $\{5\}$ 9. $\left\{\frac{1}{4}\right\}$ 11. $\{\sqrt[4]{10}\}$ or $\{1.778\}$

Objective 3

13. \$1259.71 15. \$12,905.41 17. \$11,229.05

Objective 4

19. 400 g 21. 13.9 wk 23. 13.7 days

Objective 5

25. $\{12.36\}$ 27. $\{18.52\}$ 29. $\{3.49\}$

Chapter 13 NONLINEAR FUNCTIONS, CONIC SECTIONS, AND NONLINEAR SYSTEMS

13.1 Additional Graphs of Functions

Key Terms

1. absolute value function
2. asymptotes
3. greatest integer function
4. reciprocal function
5. square root function

Objective 1

1.

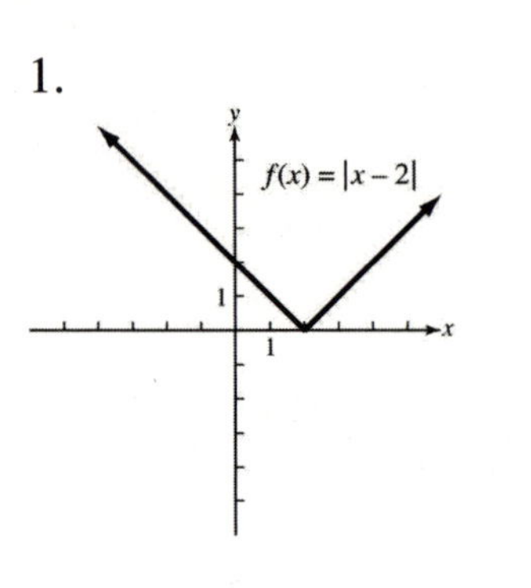

3.

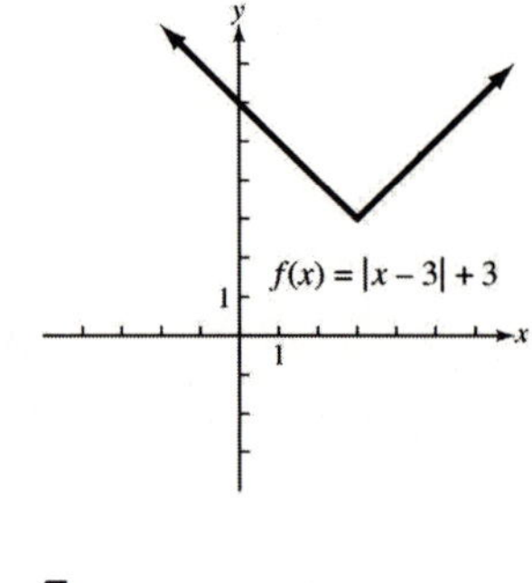

5.

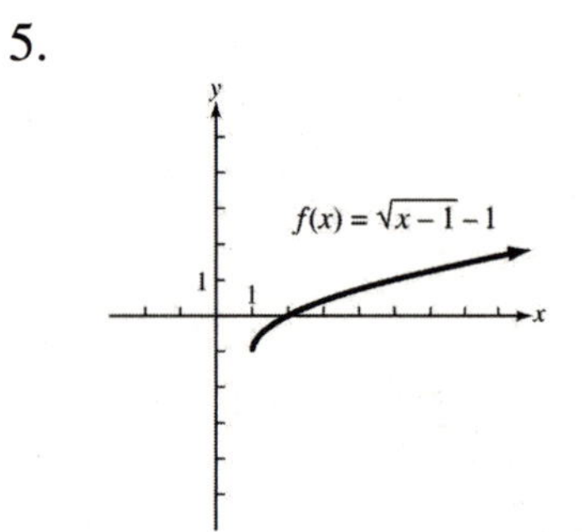

7.

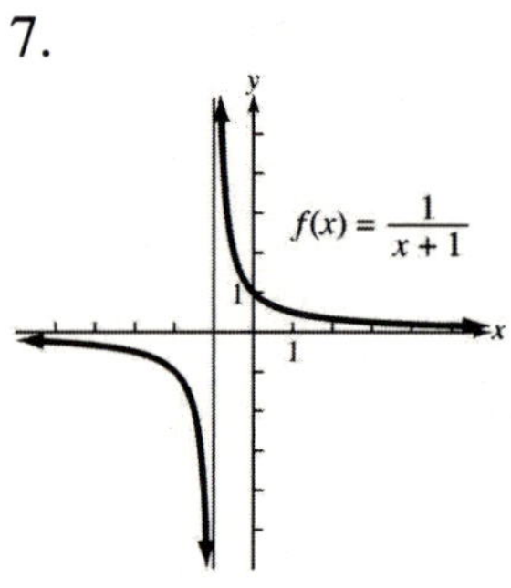

9.

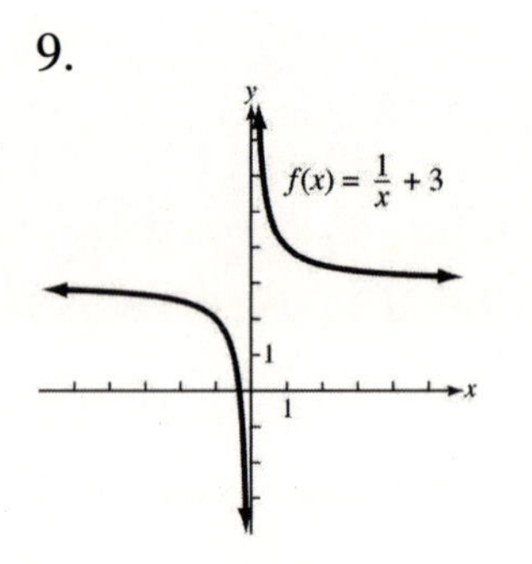

11.

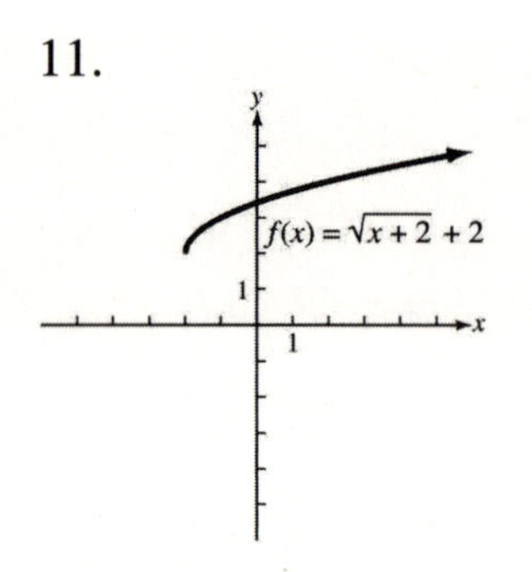

13.

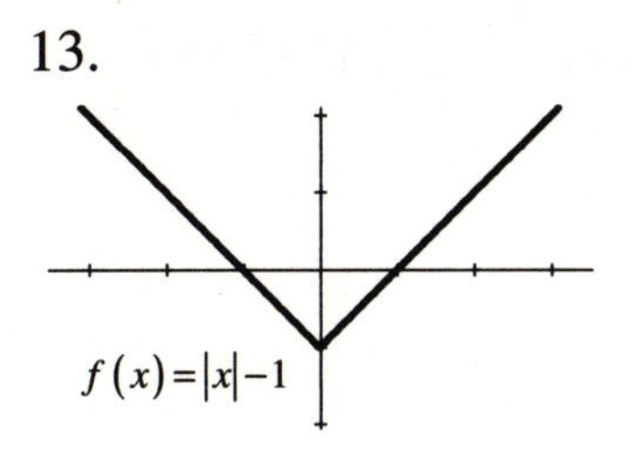

15.

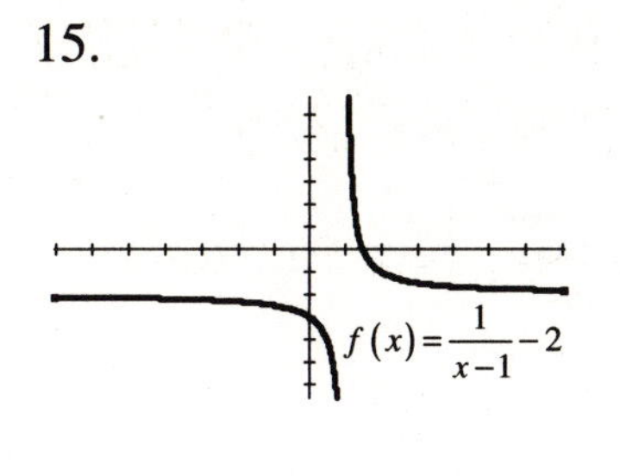

17.

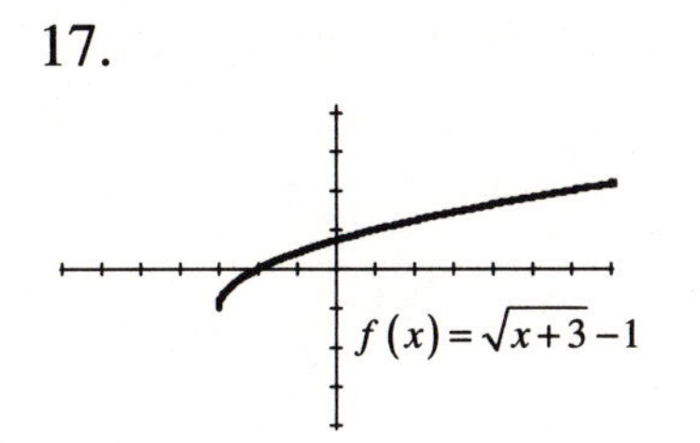

19.

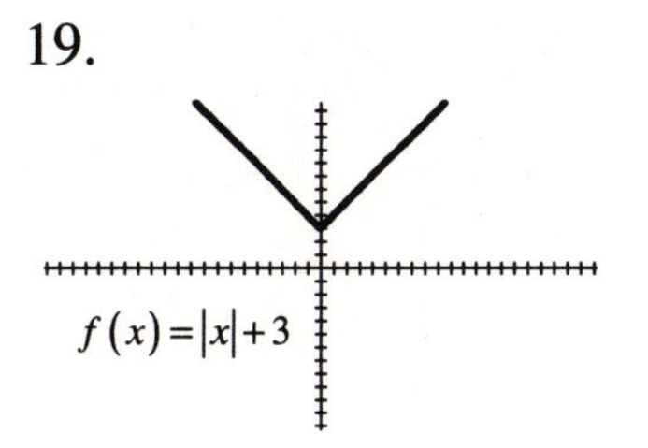

Objective 2

21.

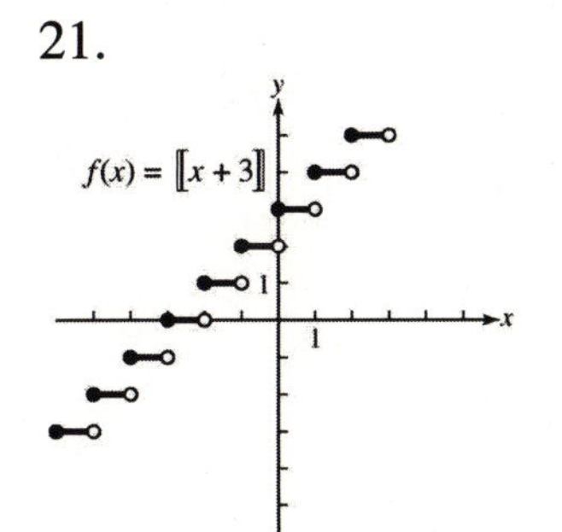

23.

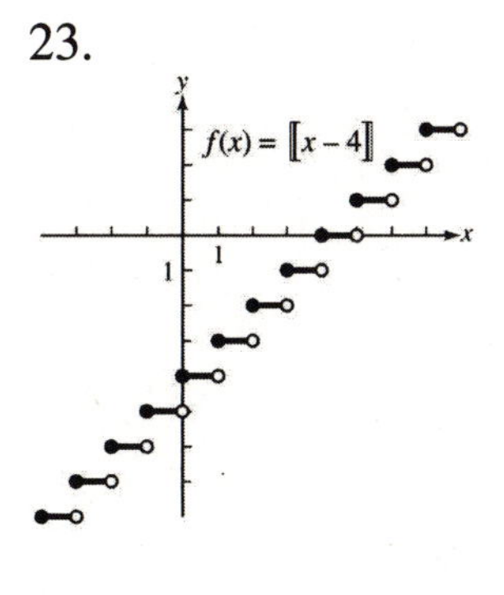

25.

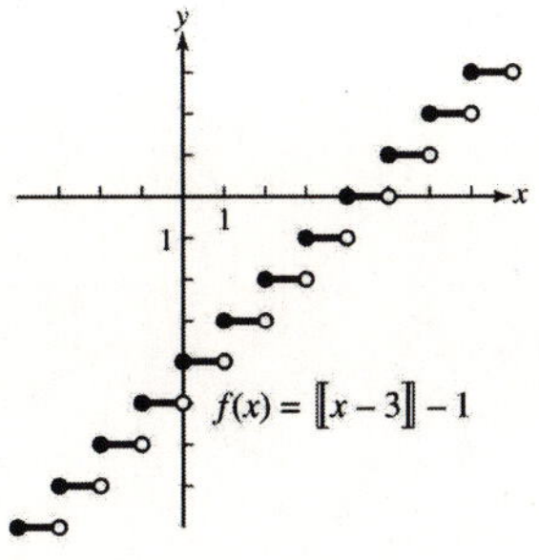

27.

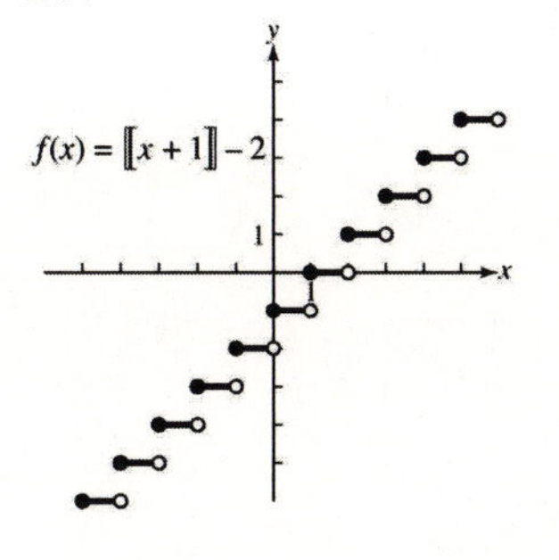

29.

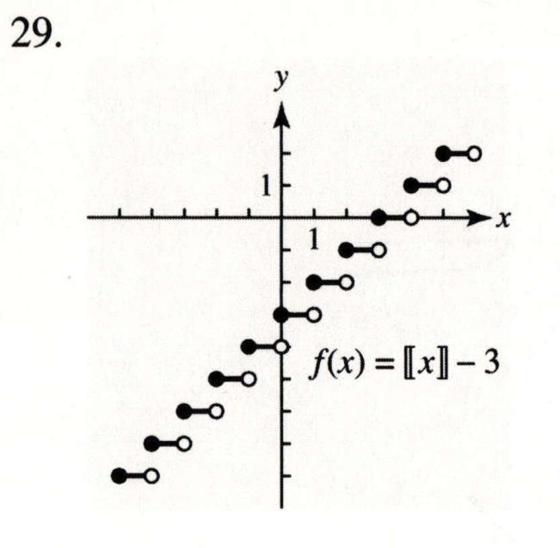

13.2 The Circle and the Ellipse

Key Terms

1. center-radius form
2. circle
3. conic sections
4. foci
5. ellipse

Objective 1

1. $(x-2)^2+(y+3)^2=25$
3. $x^2+(y-5)^2=9$
5. $(x+5)^2+(y-4)^2=16$
7. $(x-3)^2+(y+4)^2=25$

Objective 2

9. $(1,\ 3);\ 5$
11. $(2,\ 1);\ 6$
13. $(5,-6);\ 3$
15. $\left(-\frac{3}{2},\ -1\right);\frac{\sqrt{21}}{2}$

Objective 3

17.

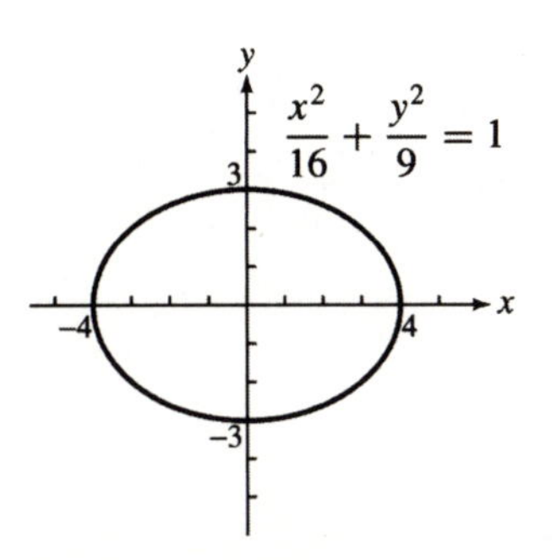

19.

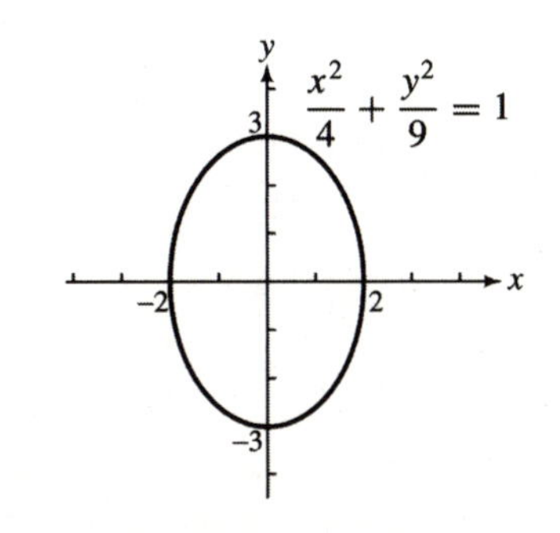

21.

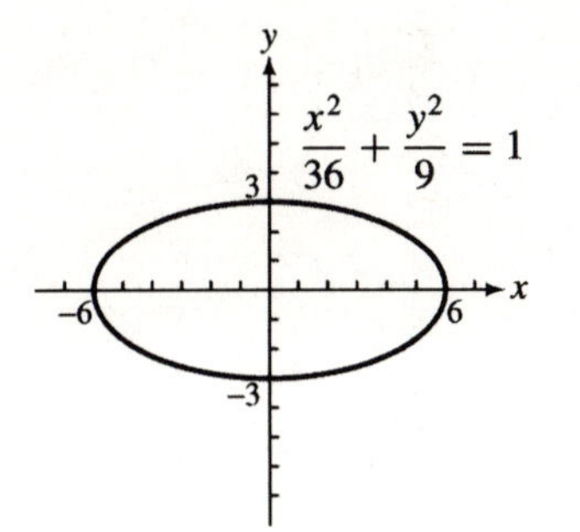

Objective 4

23. $y_1 = \sqrt{17 - x^2}$; $y_2 = -\sqrt{17 - x^2}$

25. $y_1 = \sqrt{92 - \left(\frac{92}{79}\right)x^2}$; $y_2 = -\sqrt{92 - \left(\frac{92}{79}\right)x^2}$

Objective 5

27.

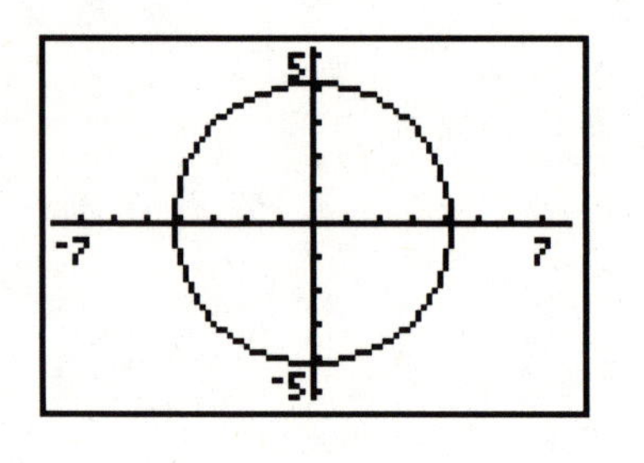

$x^2 + y^2 = 17$

29.

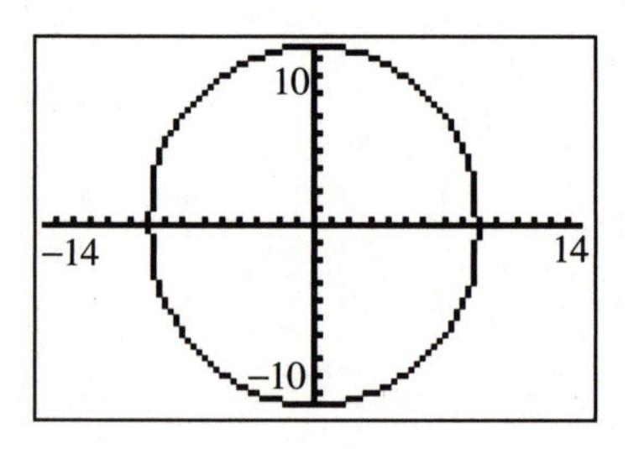

13.3 The Hyperbola and Functions Defined by Radicals

Key Terms

1. fundamental rectangle 2. hyperbola 3. asymptotes (of a hyperbola)

Objectives 1 and 2

1.

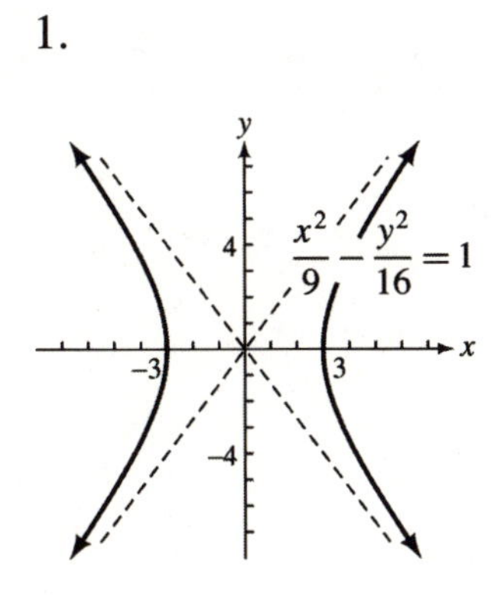

3.

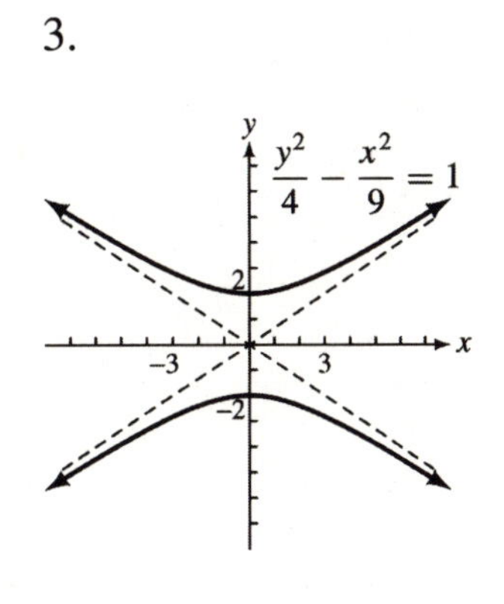

5.

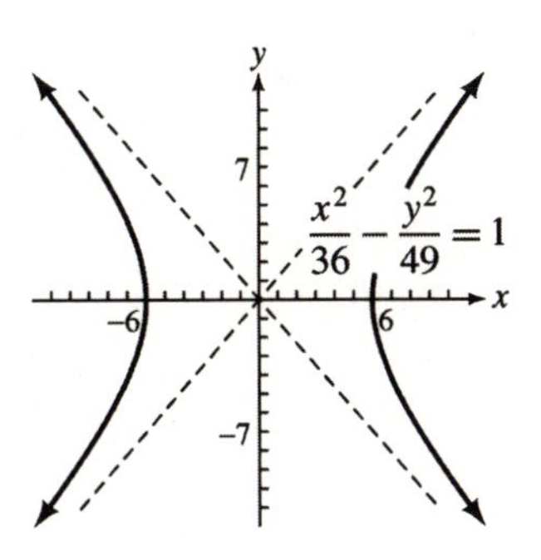

7.

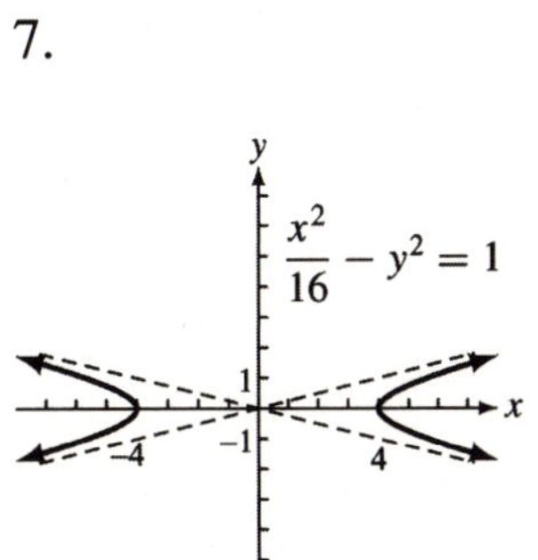

9.

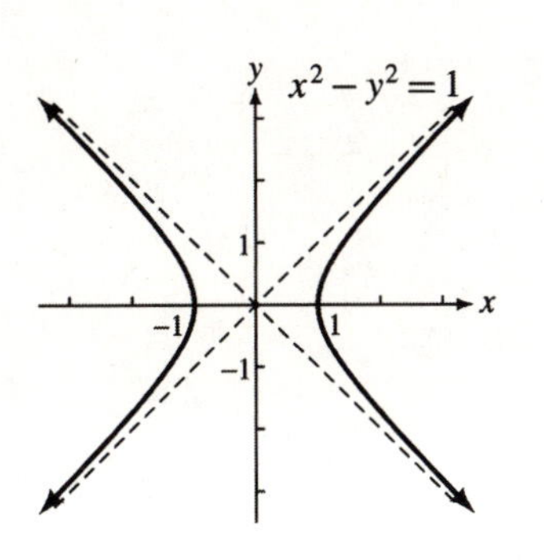

Objective 3

11. Hyperbola   13. Parabola   15. Hyperbola   17. Circle

19. Circle

Objective 4

21.

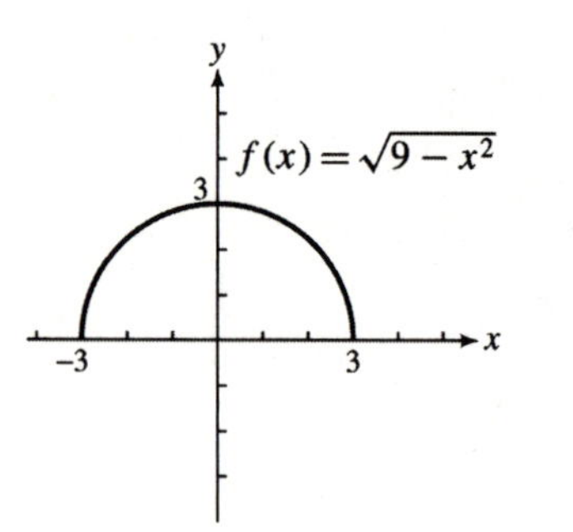

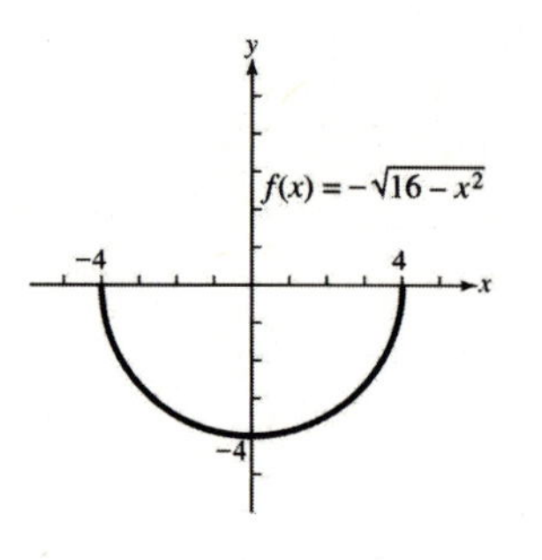

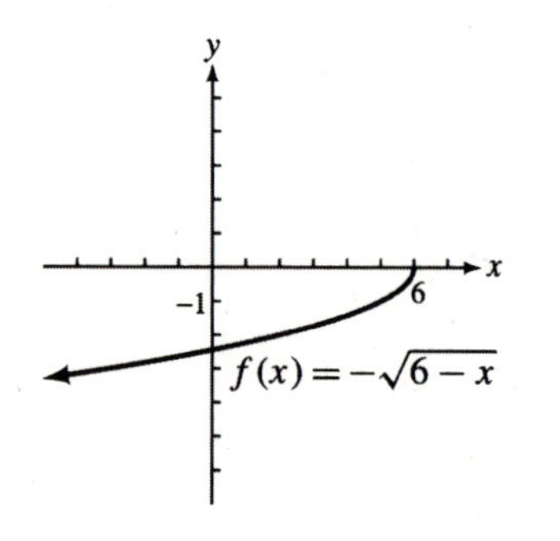

27.

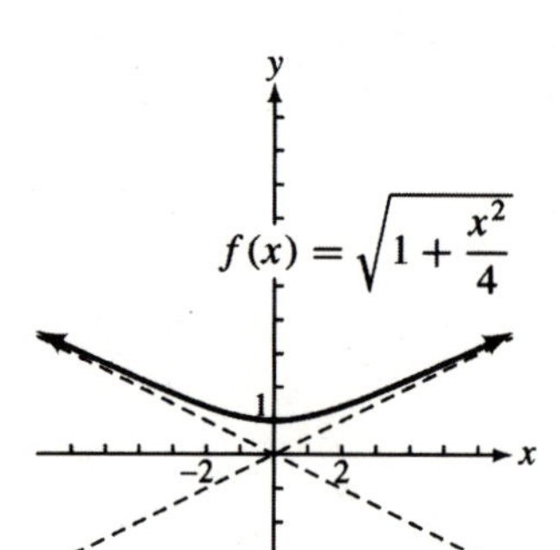

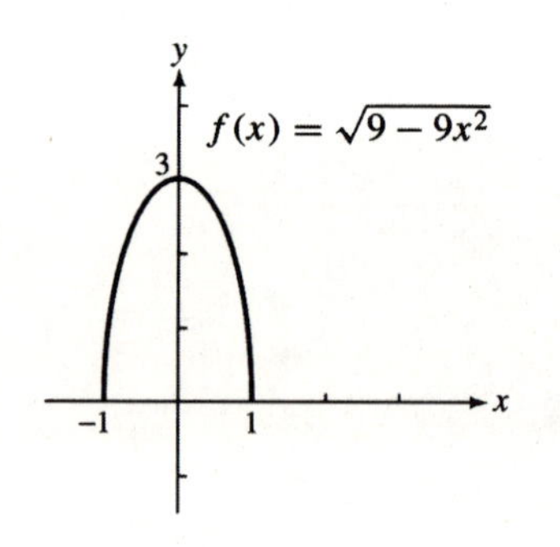

13.4 Nonlinear Systems of Equations

Key Terms

1. check all solutions
2. substitution method
3. elimination method
4. nonlinear system of equations

Objective 1

1. $\left\{(4,-1), \left(\frac{16}{5}, -\frac{13}{5}\right)\right\}$

3. $\left\{\left(\frac{1}{4}, \frac{3}{2}\right), (-1, 1)\right\}$

5. $\{(5, 2), (-1, -4)\}$

7. $\{(3, -2), (-2, 3)\}$

Objective 2

9. $\{(1, 3), (1, -3), (-1, 3), (-1, -3)\}$

11. $\{(2, 3), (2, -3), (-2, 3)\ (-2, -3)\}$

13. $\{(3, 1), (3, -1)\ (-3, 1), (-3, -1)\}$

15. $\{(1, 1), (1, -1), (-1, 1), (-1, -1)\}$

Objective 3

17. $\{(6, 1), (1, 6), (-6, -1), (-1, -6)\}$

19. $\left\{(1, 1), (-1, -1), \left(\sqrt{3}, \frac{\sqrt{3}}{3}\right), \left(-\sqrt{3}, -\frac{\sqrt{3}}{3}\right)\right\}$

21. $\left\{(3, -2)\ (-3, 2), \left(\frac{2\sqrt{6}}{3}, -\frac{3\sqrt{6}}{2}\right), \left(-\frac{2\sqrt{6}}{3}, \frac{3\sqrt{6}}{2}\right)\right\}$

23. $\{(2, 2), (-2, -2), (2i, -2i), (-2i, 2i)\}$

Objective 4

25. $\{(2.88, 2.08), (2.88, -2.08), (-2.88, 2.08), (-2.88, -2.08)\}$

27. $\{(1.02, 2.10), (-2.25, 6.01)\}$

29. $\{(1.00, 1.73), (-1.00, -1.73)\}$

13.5 Second-Degree Inequalities and Systems of Inequalities

Key Terms

1. second-degree inequality
2. system of inequalities

Objective 1

1.

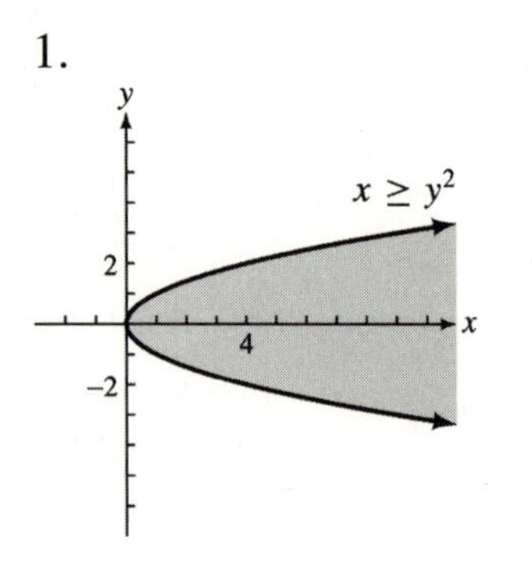

3.

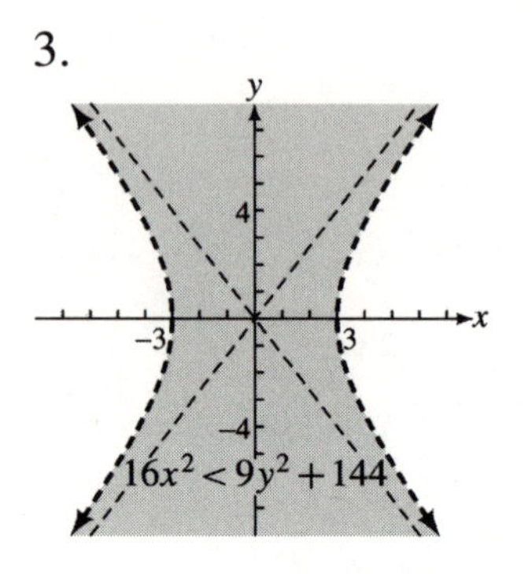

5.

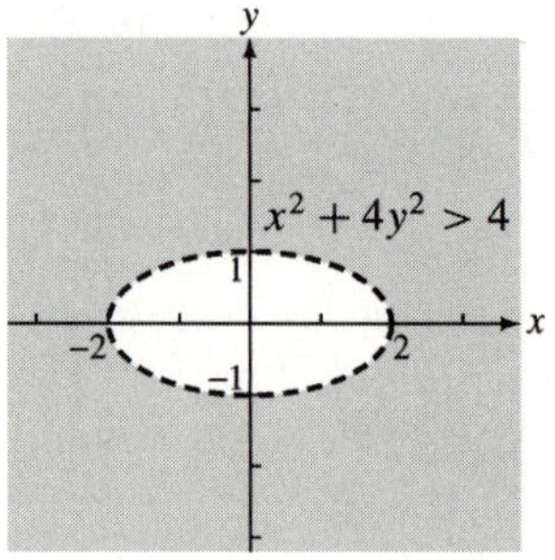

7.

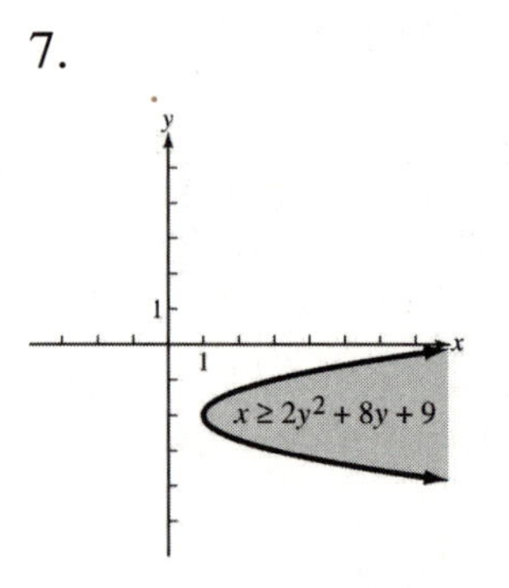

9.

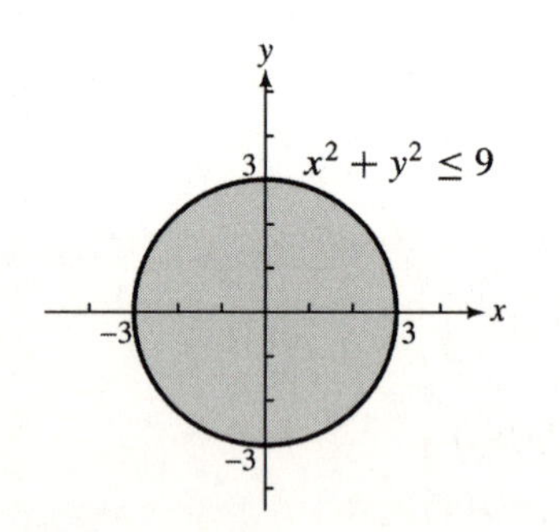

11.

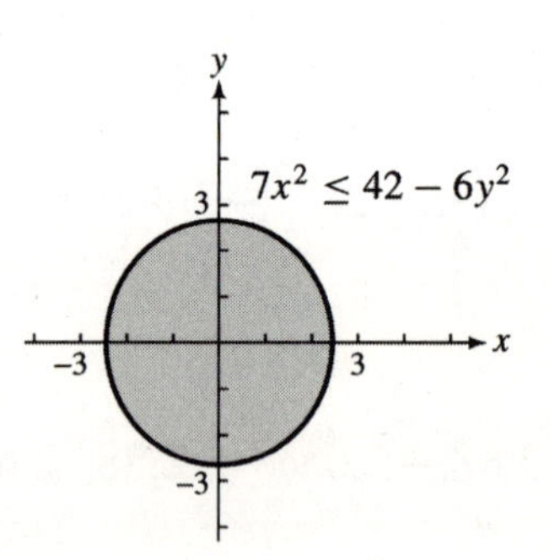

13.

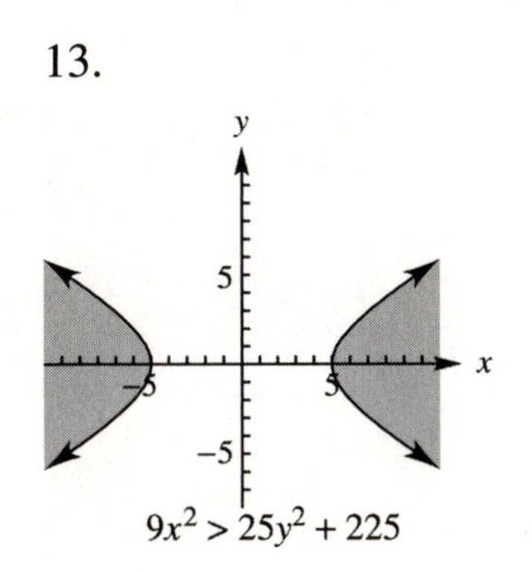

15.

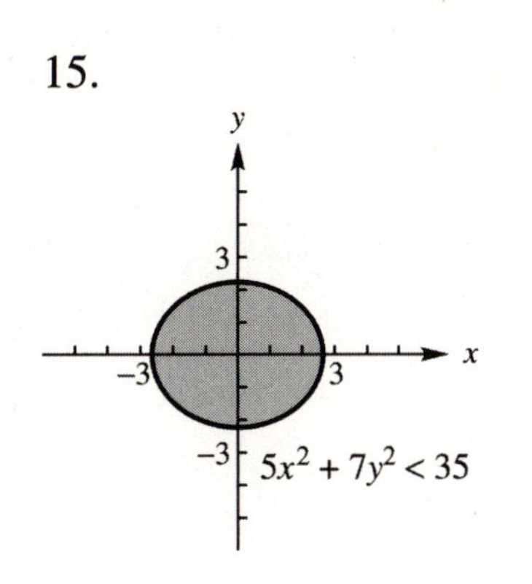

Objective 2

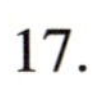

17.

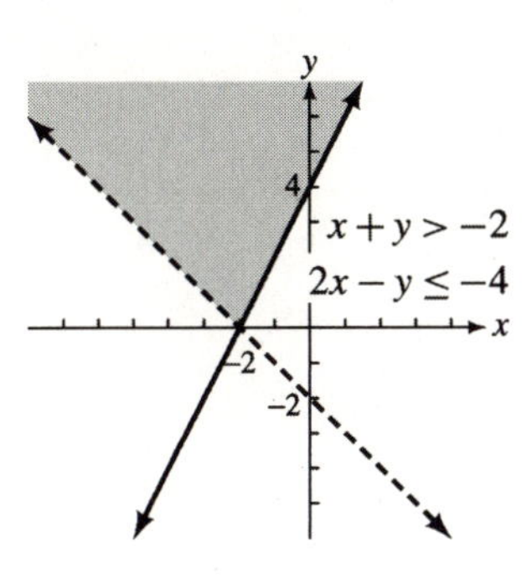

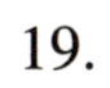

19.

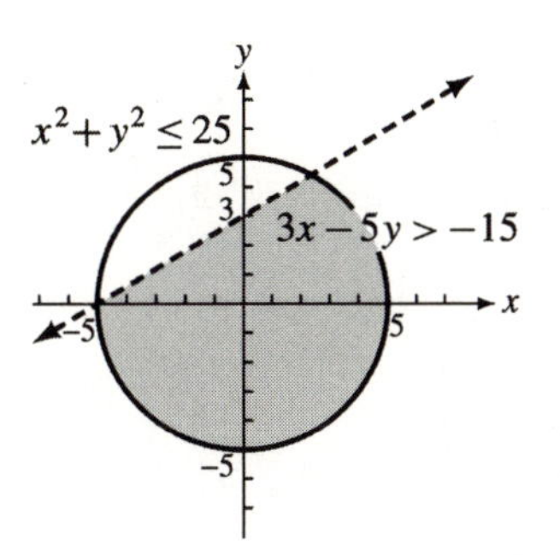

21.

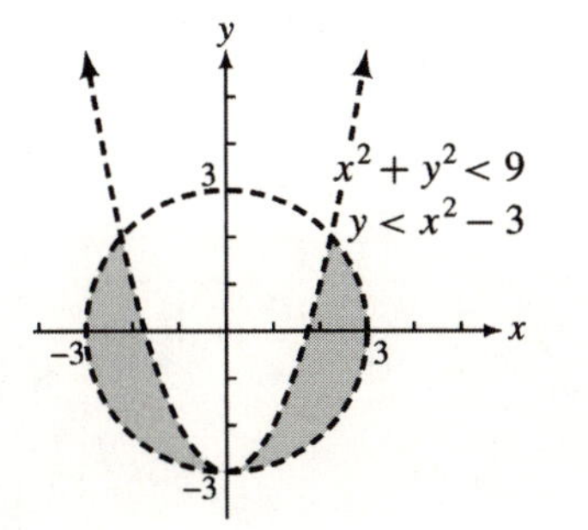

23.

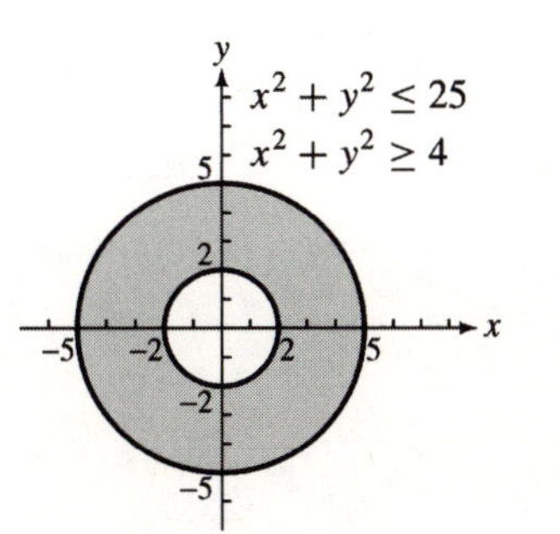

25.

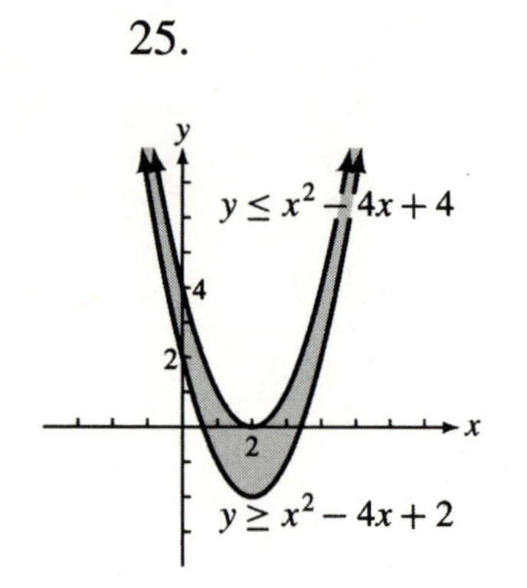
y
$y \leq x^2 - 4x + 4$
4
2
2
x
$y \geq x^2 - 4x + 2$

27.

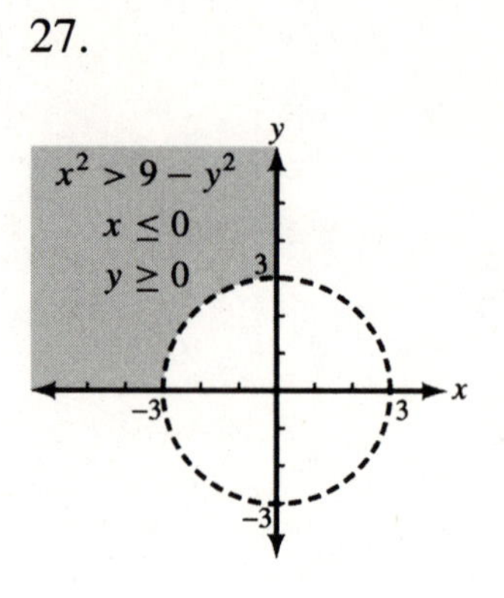
y
$x^2 > 9 - y^2$
$x \leq 0$
$y \geq 0$
3
−3
3
x
−3

29.

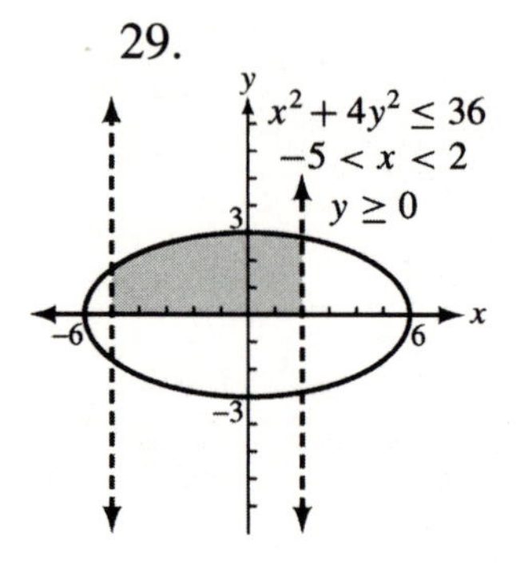
y
$x^2 + 4y^2 \leq 36$
$-5 < x < 2$
$y \geq 0$
3
−6
6
x
−3

Chapter 14 SEQUENCES AND SERIES

14.1 Sequences and Series

Key Terms

1. summation notation 2. general term 3. series

4. finite sequence 5. arithmetic mean 6. index of summation

Objective 1

1. 4, 7, 10, 13, 16 3. 2, 4, 8, 16, 32 5. –12

Objective 2

More than one answer is possible in some of these.

7. $a_n = n$ 9. $a_n = 2n+1$ 11. $a_n = \left(\sqrt{3}\right)^n$

Objective 3

13. 16 g 15. $a_n = 12n$

Objective 4

17. $1+5+9+13+17+21+25=91$

19. $-\frac{1}{2}+\left(-\frac{1}{4}\right)+\left(-\frac{1}{6}\right)+\left(-\frac{1}{8}\right)=-\frac{25}{24}$

Objective 5

Several different answers may be possible for Exercises 21-25.

21. $\sum_{i=1}^{5}(i+1)$ 23. $\sum_{i=1}^{5}i^2$ 25. $\sum_{i=1}^{5}i \cdot x^{i-1}$

Objective 6

27. 5.5 29. .3

14.2 Arithmetic Sequences

Key Terms

1. arithmetic sequence (or arithmetic progression) 2. common difference

3. sum of the first *n* terms 4. general term

Objective 1

1. 4 3. 4 5. $-\frac{1}{2}$

Objective 2

7. $a_n = 3n + 1$ 9. $a_n = 3n - 11$ 11. $a_n = -.3n + 1.1$

Objective 3

13. Week Eight

Objective 4

15. –5, –7, –9, –11, –13 17. –1, 5, 11, 17, 23

19. 62 21. –23

Objective 5

23. 115 25. 114 27. 250,249.5 29. 80

14.3 Geometric Sequences

Key Terms

1. sum of the first n terms 2. common ratio 3. general term

4. geometric sequence (or geometric progression) 5. annuity

6. sum of an infinite geometric sequence 7. future value of an annuity

Objective 1

1. 2 3. $\sqrt{2}$

Objective 2

More than one answer may be possible in some of these.

5. $a_n = 6 \cdot 2^{n-1}$ 7. $a_n = \left(\sqrt{3}\right)^n$

Objective 3

9. 486 11. –162 13. 3, 6, 12, 24, 48

15. $\frac{1}{2}, \frac{5}{2}, \frac{25}{2}, \frac{125}{2}, \frac{625}{2}$

Objective 4

17. $-\frac{315}{32}$ 19. 8190 21. 682

Objective 5

23. \$193,430.27 25. \$3,464.34

Objective 6

27. $\frac{32}{3}$ 29. $-\frac{5}{4}$

14.4 The Binominal Theorem

Key Terms

1. *r*th term of the binomial expansion
2. binomial theorem
3. *n* factorial

Objective 1

1. $x^3 - 9x^2 + 27x - 27$
3. $r^3 + 6r^2s + 12rs^2 + 8s^3$
5. $64m^3 - 240m^2p + 300mp^2 - 125p^3$
7. $\frac{z^3}{8} - \frac{z^2x}{4} + \frac{zx^2}{6} - \frac{x^3}{27}$
9. $81r^4 - 216r^3s + 216r^2s^2 - 96rs^3 + 16s^4$
11. $81m^{12} - 540m^9n^2 + 1350m^6n^4 - 1500m^3n^6 + 625n^8$
13. $243x^5 + 405x^4y + 270x^3y^2 + 90x^2y^3 + 15xy^4 + y^5$
15. $y^{12} - 6y^{10}x^2 + 15y^8x^4 - 20y^6x^6 + 15y^4x^8 - 6y^2x^{10} + x^{12}$

Objective 2

17. $r^7 + 14r^6 + 84r^5 + 280r^4$
19. $6561r^8 + 87,480r^7 + 510,300r^6 + 1,701,000r^5$
21. $1024z^{10} - 5120z^9y + 11,520z^8y^2 - 15,360z^7y^3$
23. $-8k$
25. $-4320x^2y^3$
27. $135r^2$
29. $-5103m^6r^5$